DR. WOODWARD'S SHIELD

DR. WOODWARD'S SHIELD

History, Science, and Satire in Augustan England

JOSEPH M. LEVINE

CORNELL UNIVERSITY PRESS
Ithaca and London

Copyright © 1977 by The Regents of the University of California
Preface to Paperback Edition and Appendix copyright
© 1991 by Cornell University

First published, Cornell Paperbacks, 1991 by Cornell University Press.

Library of Congress Cataloging-in-Publication Data

Levine, Joseph M.

　　Dr. Woodward's shield : history, science, and satire in Augustan
England / Joseph M. Levine.
　　　　p.　cm.
　　Reprint, with new pref. Originally published: Berkeley :
University of California Press, 1977.
　　Includes bibliographical references and index.
　　ISBN 0-8014-9935-6 (pbk. : alk. paper)
　　1. Woodward, John, 1665–1728. 2. Antiquarians—Great Britain—Biography.
3. Great Britain—History—Roman period, 55 B.C.–449 A.D.—Historiography.
4. Forgery of antiquities—England—History. 5. Satire, English—History and
criticism. 6. Romans—England—London. 7. Classicism—England. 8. Shields.
　　I. Title. II. Title: Doctor Woodward's shield.
DA93.W66L48　1991
941.07′092—dc20
[B]　　　　　　　　　　　　　　　　　　　　　　　　　　　　91-8938

Contents

Illustrations

I am grateful to the following institutions for the courtesy of allowing me to reproduce illustrations from prints and books in their collections: The British Museum and the British Library, The Sedgwick Museum Cambridge, the Society of Antiquaries, the Bodleian Library, The Ashmolean Museum, and the Ministry of Public Building and Works.

Frontispiece: *Dr. Woodward's Shield*: From a drawing by K. Howard, engraved by P. van Gunst, published in Amsterdam 1705. (Courtesy of the Society of Antiquaries.)

Dr. Woodward: From a mezzotint by W. Humphrey (1774) after an original painting which was in the possession of Col. King. (Courtesy of the Trustees of the British Museum.)

Dr. Woodward: From an anonymous painting in the Sedgwick Museum. (Courtesy of the Sedgwick Museum, Cambridge, photograph by David Bursill.)

Gresham College: From a print by George Vertue (1739) in John Ward, *Lives of the Professors of Gresham College* (London, 1740). (Courtesy of the Society of Antiquaries.)

The Roman Wall at Bishopsgate: From the frontispiece to John Edward Price, *On a Bastion of London Wall* (Westminster, 1880). (Courtesy of the Society of Antiquaries.)

Dr. Woodward's Cabinet of Fossils: The earliest of four cabinets built for Dr. Woodward and transferred at his death to the Woodward Museum Cambridge. (Courtesy of the Sedgwick Geological Museum.)

Fossil Plants: From a plate in Edward Lhwyd's *Lithophylacii Britanici Ichnographia* (London, 1699). (Courtesy of the Bodleian Library.)

Dr. Mead's Shield Depicting Scipio: From Frances Grose, *A Treatise on Ancient Armour* (London, 1786). (Courtesy of the Society of Antiquaries.)

vii

Dr. Woodward's Shield (front): (Courtesy of the Trustees of the British Museum.)

Dr. Woodward's Shield (reverse): (Courtesy of the Trustees of the British Museum.)

The Capitol and *The Pantheon*: From Basil Kennett, *Romae Antiquae or the Antiquities of Rome* (2nd ed., London, 1699). (Courtesy of the Bodleian Library.)

The Dress and Arms of Roman Soldiers: From Basil Kennett, *Romae Antiquae or the Antiquities of Rome* (2nd ed., London, 1699). (Courtesy of the Bodleian Library.)

Gem Depicting Scipio in Dr. Woodward's Collection: From Thomas Hearne's edition of Livy (Oxford, 1708), vol. VI.

The Shield of Dr. Spon: From Jacob Spon, *Recherches Curieuses* (Lyons, 1683). (Courtesy of the Society of Antiquaries.)

Roman Antiquities and a Shield in Ralph Thoresby's Museum: From Thoresby's *Ducatus Leodiensis* (London, 1715). (Courtesy of the Bodleian Library.)

The Shield with the Life of Camillus in the Tower of London: (Courtesy of the Ministry of Public Building and Works.)

The Death of Decebalus on the Trajan Column: From the atlas of plates accompanying Raffaello Fabretti, *Columna Traiani Syntagma* (Rome, 1683). (Courtesy of the Society of Antiquaries.)

The Helmet of John Kemp: (Courtesy of the Trustees of the British Museum.)

Antediluvian Fish in the Museum of J. J. Scheuchzer: From J. J. Scheuchzer, *Piscium Querelae* (Zurich, 1708). (Courtesy of the Bodleian Library.)

Antediluvian Man: From J. J. Scheuchzer, *Homo Diluvii Testis* (Zurich, 1726). (Courtesy of the Trustees of the British Library.)

The Earl of Pembroke's Sculpture of Curtius Leaping into the Gulf: From Cary Creed, *The Marble Statues of the Right Hon. Earl of Pembroke at Wilton* [1730]. (Courtesy of the Society of Antiquaries.)

Preface to the Cornell Paperbacks Edition

More than a dozen years have passed since Dr. Woodward made his reappearance in the world of ideas. No doubt he would have been happy to find his work remembered after so long a time, although his vanity and self-esteem might have been just a little bruised at the relatively small commotion that it made. Now, however, thanks to this paperback edition, the doctor (and his biographer) can hope for a larger audience and some more of the attention that he always craved and perhaps deserved. As I indicated originally, my own interest in Dr. Woodward and his shield first arose from a grand project for which I had some large hopes and made some rather rash promises. Now at last *Humanism and History* has appeared and *The Battle of the Books* is in the offing, both of which take up some of the themes and many of the characters in *Dr. Woodward's Shield*. Even so, I cannot say that I have been able to discover much more about Dr. Woodward and his immediate setting than first I did, and so I am relieved to be able to send my book forth again with only a few typographical corrections. As far as I can tell, only one new manuscript of consequence has come to light lately, and that largely reinforces what I said before. I have, nevertheless, given a description of it in a new appendix at the end of the book, and I thank Donald McIntyre for first calling it to my attention and Thomas Wyllette for having a transcript made for me.

JOSEPH M. LEVINE

Syracuse, New York

Acknowledgments

This book is an offshoot of a larger venture. I had hoped once to write a work called *Humanism and History* which would treat the whole story of modern historiography from a special point of view. "The disadvantage of over-large projects," writes Fernand Braudel, "is that one can sometimes enjoy the journey too much ever to reach the end." I should perhaps be enjoying myself still except for an accidental encounter in the Manuscript Room of the British Museum with my present subject. From that time on, rather like the ancient mariner, I have lived obsessively with my story and forced it, I suspect, upon even more than the one in three of that modest man. It is in part, therefore, to relieve my friends and students, especially my loyal and patient wife, from that fearful reiteration that I have written out this first instalment now. Should I receive any encouragement, I hope that I may follow it soon with another progress report, a work that I shall call *Ancients, Moderns, and History* and that will deal with the background to my present tale.

With so long a gestation, it is hard now to recall all my debts. But I must not fail to acknowledge all those who contributed to my support — and to my enjoyment — while in those wonderfully hospitable archives: the British Museum, the Bodleian Library, the Cambridge University Library, and the Libraries of the Royal Society and The Society of Antiquaries. To my own university, Syracuse, I owe both time and financial support; to the Penrose Fund of the American Philosophical Society, two summer grants; to the Folger Library, a semester in which the writing of this work was begun. I owe a special obligation to Dr. Martin Brett who helped me through some of the more awkward documents. I owe another debt to those who supplied me with special information, above all to Dr. V. A. Eyles who gave me access to his own collection of Woodward letters and entertained me with his geological lore; to Mr. John Cherry of the British Museum who showed me Dr. Woodward's shield and who furnished me with information and illustrations; and to Dr. C. L. Forbes who guided me expertly through the Woodward Museum at Cambridge University and also helped to supply illustrations. Dr. Woodward left tracks everywhere and among the many who helped me to trace them I am particularly grateful to Dr. J. P. Bodmer of the Zentralbibliothek, Zurich; Mrs. G. Dekker-Piket of the Koninklijke Bibliotheek, The Hague; Mr. Tue Gad, Det Kongelige Bibliotek, Copenhagen; Dr. C. Helen Brock of the Glasgow University Library; Miss E. D. Yeo of the National Library, Scotland; Mr. H. J. R. Wing, Christ Church, Oxford; Dr. B. F. Roberts of the University of Wales at Aberystwyth; and the staffs of the Scottish Record Office, the Society of Antiquaries, and the Yorkshire Archaeological Society, Leeds. For permission to use the Penicuik papers I owe thanks to Sir John Clerk.

Finally, and not least, I owe a debt to many of those who in the course of the years stopped to listen to my story and offer their encouragement, especially to my friends Peter Burke, Donald Kelley, and Henry Horwitz, and to my students John Morrison and John Attig. Above all I must thank my wife who has contributed to every page.

I dedicate the book to the memory of my father who, alas, will not be able to read it.

Introduction

Somewhere in the British Museum, almost forgotten and just a trifle rusty, there lies a small round iron shield, unpretentious enough and understandably neglected, yet notorious in its time. Looking it over now, two and a half centuries after its discovery, it is hard to imagine just what the fuss was all about. The shield is only a bit more than a foot in diameter, and its dark color makes a little difficult discerning the relief that has been hammered out upon its face. The subject is unmistakably classical, and the scene that encircles the monstrous head at its center appears to be drawn from ancient history or legend. The Museum is not rich in such things, and one must look elsewhere for a comparable object. But even to a casual view our example looks — if only it were not made of iron — just a little moth-eaten and a trifle mean.

A large book about a small and unimportant shield requires some apology, and I must hasten to say at once that Dr. Woodward's shield is only the occasion for this study and not its real subject. Thus the subtitle: "History, Science, and Satire in Augustan England." It is true that one can find in the following pages the whole story of that surprising work, and it is a tale that possesses some modest interest in itself, but this book is really about other and larger matters. Essentially it is an attempt to reconstruct a whole segment of the intellectual life of the Augustans, not so much directly, or even completely, as by working outward from a single complicated and richly documented event to the general climate of ideas. In this it has been something like trying to peel an onion outward from the core (if that were possible) through layer upon layer of pulp in an effort to grasp its whole shape. To be fair, Dr. Woodward's shield was not at all the center of Augustan life — not even that segment which is to be considered here — but its discovery about 1700 was both a major and a characteristic event in its time, and the controversy that developed around it seemed to offer an exceptional opportunity to get at some important but neglected Augustan ideas and

1

activities. Here for once, in a single incident, many of the greatest wits and profoundest scholars of the age could be discovered engaged upon a common problem. By reconstructing the attitudes and methods that they brought to bear on Dr. Woodward's shield — along with some similar objects — it seemed possible to expose, in a way especially revealing, some of their basic assumptions and characteristic patterns of thought.

In telling my tale I have therefore set myself some larger objectives beyond my immediate story. In particular I have tried to examine the relationship between two intellectual communities, the virtuosi and the wits (that is to say, the scholars and the men of letters), who were then at odds over some fundamental issues. With my focus on Dr. Woodward's shield, I have tried to explore that intricate web of intellectual and personal relationships which bound together the men of learning and the men of literature and brought the two into conflict. From this perspective, the story of Dr. Woodward's shield is significant as an episode in the celebrated quarrel between the ancients and the moderns (Swift's *Battle of the Books*), and it is meant to illuminate the issues that were at stake in that memorable struggle. The setting is broad — Gresham College and the Royal Society; the drawingrooms and coffeehouses of the men of fashion; London, but also the north of England, the west, Europe, even America — for Dr. Woodward seems to have been known almost everywhere. Much of the incident will be new since most of the story is based upon manuscripts that are widely dispersed and little known. If I have fastened my narrative more upon the scholars than the wits, it is because it is their story that has been most neglected. The quarrel has been told often enough from the vantage point of Swift and Pope, but hardly ever from the perspective of their satirical victims. To reverse the viewpoint is to invest the tale with a new and different meaning. But I have written not to justify so much as to rectify the traditional accounts, and also to clarify what seems to me to be still a living issue in the relationship between learning and life, a dispute which was at the bottom of the whole contest.

I must claim another dimension for my book. If I have tried to set Dr. Woodward in his own context and thus to say something about the intellectual climate of his age, I have tried also to place the man and the incident in that long continuum that appears to lead to modern times. I have thus attempted in the course of telling my tale to estimate its significance for the history of modern learning. In general this is a subject that has not yet received its due, despite its obvious significance to the history of culture. Perhaps this is a result of the difficulties inherent in the task, something I discovered in due course for myself. I set out originally to take the measure of Augustan historiography, to try to discover its capability and achievement, and Dr. Woodward's shield seemed an ideal touchstone. Here was an object which was thought by the greatest scholars in Europe to be a Roman weapon illustrating, among other things, one of the greatest events in Roman history. Yet (it appeared) the shield was not ancient, nor the event historical. Here was an erudition which was at first sight overwhelming in its range and sophistication, yet completely and hopelessly in error over a simple and obvious matter of fact. What had gone wrong? To find the answer, however, was more difficult than appeared, for it seemed to require first recovering the general condition of classical scholarship at the time, something which had hardly yet been tried. No doubt this was worth attempting, since classical learning was probably the most important single ingredient in the general culture of the Augustans. But this meant that a book on Dr. Woodward's shield must be more than a book about historical or critical method: it must be a description of historiography in its relation to classical scholarship at a point of special interest in the development of each.

Yet there remained still another task. Dr. Woodward was a many-sided character who combined sundry interests and several careers. In particular, he spent a lifetime in the service of natural philosophy and antiquities (what we should call science and history) in the confidence that they were helpful and complementary activities. In this again he was entirely typical of his time. Dr. Woodward's special enthusiasm was for natural history, particularly geology and paleontology, to which he made important contributions. He believed that by collecting and examining the stones and fossils he was unlocking the history of the natural world in almost exactly the same way that he was doing with antiquities for the history of early man. The stones were "medals of creation," and Dr. Woodward's shield was like a great coin or fossil which could give testimony to past events, if only it could be deciphered and understood. He thought (with most of his contemporaries) of a single historical past, God-created and accessible to both Revelation and Reason, and he took upon himself the task of recovering, with his science and his history, the course that it had taken from the very beginning of time. The Augustans had not yet abandoned universal history and they were ready to bring together all their "modern" interests, including the new science and the new historiography, to attempt a new synthesis. Thus, in my effort to place Dr. Woodward and his shield, I was compelled to add to the history of historiography and the history of classical scholarship a modest contribution to the history of science.

Thus the reader will find in the following pages a book in several parts. In the first he will meet Dr. Woodward and his circle and find the biographical setting for the shield. Our hero was not a great man but he was a man of near-greatness, a flawed genius who left his mark upon his own time and indeed for a long time afterward, but who slipped (like his shield) gradually and finally into oblivion. Dr. Woodward was unfortunate in several ways. He was drawn for posterity by a constellation of wits whose gift for satire has rarely been surpassed, and they left him in vivid caricature. His papers, which alone might have preserved an alternative portrait, were dispersed and scattered to the winds. Finally, his ideas, which were original and influential on many matters were judged invariably wrong by the nineteenth century and so were dismissed contemptuously and have languished ever since. Nor did Dr. Woodward serve his own cause by intruding into all his work an eccentric and irascible personality which not even his biographer can extenuate. In the first part of this book I have tried therefore to rescue Dr. Woodward from the satire of his contemporaries, the contempt of posterity, and the failings of his own personality — without, I hope, pressing his claims too far. I have used his life as I have used his shield, as a touchstone to some general ideas, especially the relationship of science to history, and I have tried to sketch in the world in which he lived and worked and fought, the world of the London virtuosi, the savants, the *erudits*, collectors, and connoisseurs, who took up so much room in the Augustan intellectual landscape.

In Part Two I have dealt specifically with Dr. Woodward's shield, its discovery and contemporary fame, the controversy that it started, and the learning that was lavished upon it in England and abroad. I have tried to portray the English scholars who took up its cause — especially Thomas Hearne and Henry Dodwell, who published a Latin dissertation upon it — and also the famous Europeans who wrote in its favor: Perizonius, Gronovius, Spanheim, Scheuchzer, Cuperus, Le Clerc, and so on. I have tried especially to describe their methods and so to show the state of a whole group of kindred historical disciplines on the eve of their maturity: archaeology, chronology, numismatics, epigraphy, iconography, textual criticism, in general all that the period ascribed to those specifically "modern" subjects, philology and antiquities. And finally I have tried to show how Dr. Woodward and his shield became the victims of satire, how the "Scriblerians" — Swift, Pope, Arbuthnot, and their friends — chose our hero and his weapon for the centerpiece of their collaborative work, and how he came to epitomize for them all that was wrong in the new science and the new history.

In the concluding part, I have extended the story to a later time, to the generation of the Abbé Winckelmann and afterwards, explained why the fake went undetected, and attempted to draw the moral from the tale.

Now, having set out my purpose, I must quickly offer an apology. I have been acutely conscious, in composing this work, of the peril in attempting to describe scholarship with insufficient learning, and satire with insufficient wit. I can only say, as once my subjects argued, that these virtues are hard to obtain and even harder to reconcile, and since I have been forced to choose, I have preferred the footnote and pedantic detail to the grace of a well-turned narrative. I only hope that I have not killed altogether the amusement that usually accompanied my hero. But in choosing pedantry over polish, I have become painfully aware of the other side of the coin. It must seem arrogant and foolhardly indeed to venture on a theme that requires so much specialized knowledge in so many disparate fields; nor can I pretend to either the classical learning or the natural science that would seem requisite for such an undertaking. I can only plead that if my tale had to await the student with appropriate credentials it would probably never be told, and the risk of error therefore seemed worth the taking. I have done what I could to minimize it by avoiding entanglements and sticking to my story, by reporting what I have found in the manuscripts and printed sources in a detail sufficient to inform the reader with expert knowledge (who may judge for himself) as well as the reader with only a general background. Although my method has been discursive − for it is not easy to track a virtuoso through the thickets of eighteenth-century learning − I have tried to hold it together by joining the narrative to what is really a single large theme. In effect, the basic problem that bothered the Augustans and that underlies much of my story was whether, or in what sense, history was a science, or whether it belonged to literature. That they did not solve it will surprise no one, for the dilemma is with us still. It is for that reason that I have meant, in recounting my tale, not only to restore a forgotten episode in the history of English thought, but also to evaluate its meaning for its own time and ours. If therefore I have not met the demands of the wits for elegance I have at least tried to meet it for relevance, and I offer the story of Dr. Woodward's shield for what it tells us about the Augustans, but also incidentally for what it tells us about ourselves.

Part One

THE COMPLETE VIRTUOSO: THE CAREER OF DR. WOODWARD

I.

Doctor of Physick

In 1717 John Woodward was about fifty years old. Behind him lay a career of achievement and controversy remarkable enough to have brought him, in equal parts, both fame and a sizable fortune, contempt and derision. Self-confident and industrious, stubborn and very contentious, Woodward was destined for success — and for notoriety. He was a man who demanded the center of the stage and had occupied his share of it for more than twenty years. Not only was he known widely through his many activities as a physician and scientist, as a member of the Royal Society and the College of Physicians, as the exponent of a remarkable new view of the world, and as a controversialist in a variety of personal and intellectual causes, but now he trod the stage literally, nightly, the chief character in a boisterous new comedy, the *Three Hours After Marriage*. Whether he enjoyed his latest role is hard to say, but it is very doubtful; like other self-made men he had too much self-esteem to enjoy being laughed at.

A little earlier that year, Woodward read a new medical book by another well-known London physician, Dr. John Freind. It irritated him profoundly, as the opinions of others were likely to do. Freind had used the occasion of a learned new edition of Hippocrates to argue his views about the treatment of London's endemic scourge, the smallpox.[1] Woodward had long held independent ideas on the subject; he thought Freind's treatment ill-founded, to say the least, and dangerous. He was provoked also by what he saw as a slight in the work to an old friend and fellow physician. At once he began to jot down some thoughts on the subject.

After two decades of controversy, Woodward might well have ducked this one. But it would have been very much out of character. Freind, it is true, had pronounced his views without reference to Woodward. But Woodward immediately took them to be an indictment of his own medical practice. He could scarcely do otherwise. Not only had he lectured publicly in favor of an alternative cure; earlier that year he had fallen out with Freind on that very

point in a joint consultation over a patient.[2] Even more profoundly, Woodward recognized an argument subversive of his basic philosophical position.
Freind had relied heavily on the authority of Hippocrates, thereby proclaiming his allegiance to the "Ancients."[3] Indeed, it was not long before that he
had enlisted with his friends at Christ Church against the redoubtable Richard
Bentley in the famous quarrel between the Ancients and Moderns, the "Battle
of the Books." But Woodward was a thorough "Modern." What seemed to
him at stake was nothing less than the whole new philosophy of the last
century, the science of observation and experiment now once again retarded
by a mindless appeal to authority. The fact that lives were at stake, not to say
a lucrative medical practice, only made the case more urgent.

Medicine in Dr. Woodward's day was a very prestigious and often a very
lucrative profession — but it was, all the same, neither very rigorous nor very
scientific. At the universities it was still taught — to the extent that it was
taught at all — as commentary on the classical authors; the training of the
doctor was entirely literary. It is not surprising therefore that the greatest
clinician of the age, Dr. Thomas Sydenham, believed that "one had as good
send a man to Oxford to learn shoemaking as practice physick." He preferred
to take apprentices.[4] And Dr. Radcliffe, who was probably the most successful physician of the day (financially, anyway), was probably also the least
intellectual. When asked by a personal friend "Where was his Study?" he
answered, pointing to a few vials, a skeleton, and an herbal, "Sir, this is
Radcliffe's Library."[5] No doubt it was just as well, since the medical
knowledge of neither ancients nor moderns was sufficient yet to furnish
much understanding about human ailments or much succor to the suffering
patient. Of course, medical science had made some important discoveries: not
long before, Harvey had demonstrated the circulation of the blood, and Willis
and his associates some remarkable things about the brain and nervous
system. Anatomy particularly was advancing, although it was barely known at
the universities and Freind doubted that anyone had yet shown its relevance
to medical practice. Here Dr. Woodward had all the advantage. He had never
known the inside of a university before he became a doctor. He had learned
how to practice instead as apprentice to a successful physician and how to
investigate his subject from the more sophisticated members of the Royal
Society. He was widely read but, like the great doctors of his day and all the
scientists, he was in revolt against the authority of books. Unlike most of his
fellow physicians, he understood perfectly the importance of accurate observations for the clinician and of dissection and experiment for the scientist,
and he even anticipated the day when the two might be successfully combined.

Egged on by his friends, Dr. Woodward soon turned his notes into a book, and the battle was joined. He was determined to use the occasion for a larger purpose. He meant nothing less (so he wrote a friend) than to develop his ideas into a discourse that would treat the whole subject, "Of the Nature of Man, of Diseases, and of Remedies."[6] It appeared in 1718 as *The State of Physick and of Diseases... Particularly of the Small-pox*. Dr. Woodward was very proud of his work, pleased to have been able "to bring Physick into some form of Art and to settle a Standard of Health and of Diseases with the Remedies of them." He was confident that he had at long last supplied a scientific foundation for clinical practice. "I have endeavored to reduce our Study into Form of Art and to fix it upon a Mechanical Basis. This has been hitherto wanting which is the Cause that our Business has been involved in so much Darkness and Uncertainty." It had cost him, he complained, an unimaginable labor. It was not easy to reduce a comprehensive theory of medicine into a single treatise. "But in my usual Manner, I have brought things into so narrow a Compass that the Reader must use great Application to make himself Master of what I have delivered." Whatever else might be said of his work, the author insisted that it was founded entirely "upon Nature, Observations, and Fact, and not spun as usual out of Fancy, or drawn out of the Writings of others."[7]

The State of Physick was the beginning of a veritable "smallpox war." Unfortunately, it generated more heat than light and remained more personal than philosophical. The initial skirmish was succeeded by a barrage of tracts and pamphlets, satires and lampoons, in which the medical world divided while the rest of London laughed.[8] Dr. Woodward alone was satisfied; his work he believed had met with general approval, "excepting those two or three gentlemen whose Method of Purging is opposed and they are very angry but have left off their purging." His critics had been moved by pride and vanity alone; his defenders had effectively silenced them. He did notice that they had extended their malice even into foreign parts, to the *Journal des Scavans* and the *Bibliotheque Angloises*, but he was confident that they too would be answered.[9] Still, the battle was not easily to be won; at its climax it even turned into an armed duel in which Woodward snatched victory from defeat with the one weapon that had so long served him, the quickness of his tongue.

It was, as always, the tone as well as the argument that provoked reply. Woodward could not help making the contrast between truth and error stark and personal. It was no surprise, he wrote in *The State of Physick*, that some should "give themselves up wholly to Fiction and Invention" instead of to real philosophy. Inevitably, ordinary men preferred their pleasures to the labor and time, the hazards, of the honest pursuit of knowledge. One had only to compare the slothful to "a Mind truly great, fill'd with the Love of

Virtue and of Good." That the one was Freind and the other Woodward was not left in doubt. Freind had asserted a remedy for smallpox by purging the patient at one stage in the disease, but he had affirmed it, Woodward argued, merely on the authority of others and without ever assigning a cause. [10] Woodward promised to conduct himself wholly by observations and to keep close to Nature, framing his conclusions entirely upon the basic constitution of man. The result was an altogether different remedy. Of the various possibilities, "Blistering, Bleeding, Purging or Vomiting," it was Woodward's conviction that only the last could be truly efficacious. Thus while both doctors agreed upon "evacuation," the essential argument, as Woodward remarked succinctly, was whether to dislodge the evil upward or downward. Observation and case history – including the one in which the two physicians had collaborated – left no doubt in Woodward's mind. His treatise was devoted largely to a consideration of how to administer his remedy, along with some new oils that had been recently discovered.

Woodward's remedies (and they are more elaborate than either this summary or his critics allowed) rested on a bold and comprehensive theory of disease which he had worked out at least a half-dozen years before. "The great Wisdom and Happiness of Man," he wrote in 1718, "consists in a due Care of the Stomach and Digestion." [11] Here, Woodward thought, was the root of a whole variety of human disorders, psychological as well as physiological. Everything depended upon keeping the right balance of "biliose salts" in the stomach. Too much, as in the case of smallpox, and there was but one relief – the bilious matter had to be ejected. If this was an "hypothesis," it was founded upon – indeed, he insisted, deduced from – copious observations and experiments. He had begun his investigations, though he does not say so here, with his own troubled digestive system, and learned the hard way about the inefficacy of prevailing remedies. Then he had built a thriving practice, keeping careful track of the cases that came his way. (These medical histories were printed posthumously and are impressive for their length and detail.) [12] His was an hypothesis, therefore, only in the sense that Bacon, Gilbert, Boyle and the great Newton had argued theirs. In the *State of Physick*, he contrasts his own approach to medicine with one who "only practices upon receipts." "The former governs and leads the distemper: the latter is entirely led by it." [13] Where Freind had denied the possibility of knowing the causes of the smallpox, Woodward was ready with a complete explanation of the disease. And that, in turn, was based on a still more comprehensive theory which embraced nearly all the disorders that troubled him and his patients. In 1711 he had already presented his views to the Royal College of Physicians in a series of lectures; [14] it was here that he had first worked out his ideas about the digestive balance which he was sure was the key to all health. If his conclusions were not entirely convincing, they at least had the merit of appealing both to clinical experience and to the evidence of close physiological inspection and experiment – including dissection.[15] Where they were weakest, of course, was in their attempt to explain too much too soon. But the example and success of mechanics in the age of Newton (his colleague and associate in the Royal Society) was overwhelming to this very self-conscious Modern. Although he did not say so, he would have liked to become the Newton of contemporary medicine. It was not accidental that he entitled his addresses *Three Physico-Mechanical Lectures*.

Woodward's assurance was too much for his adversaries. He was even ready with a show of Greek learning to correct Freind's reading of Hippocrates. Worse still, he had hinted plainly that Freind was killing his patients. (To be fair, Woodward thought he was merely returning the compliment.) It was better to have no remedy at all, he argued, than that of a lazy or incompetent doctor. Indeed, he was convinced that on the whole those patients in his day who took the least medicine tended to do the best, while very few of the rest recovered. He hastened to add that he didn't mean this as a "Disclaimer against the Art of Physick," only against its abuses. Nevertheless, to an impartial observer it was not very reassuring to watch the doctors publicly brawling; there were some who not unnaturally concluded that the best remedy for disease was no doctor at all.[16] They were probably right, although Woodward recorded in his manuscript notes an astonishing record of success.

But the contest had merely begun. A reply to Woodward was ready almost at once, probably by Freind himself. It was signed by a Dr. Byfield, a doctor who had recently advertised a "true Sal Volatile Oleosum" and was now willing to credit Woodward's "Sal Biliosum," along with his own discovery, as amongst the most profound secrets of nature. The argument was entirely satirical and personal and immediately drew a reply from Woodward's old friend and defender John Harris: *A Letter to the Fatal Triumvirate, in Answer to that Pretended to be Written by Dr. Byfield: and Shewing Reasons why Dr. Woodward should take no Notice of it* (January 1719). The tract, like most of those in the controversy, was unsigned, but Woodward (who remained unnaturally quiet after his initial stroke) noted the authorship in his own copy.[17] Harris' triumvirate included, besides Freind, two other doctors, the obscure Salusbury Cade and the famous Richard Mead.[18] The latter was already the most celebrated physician of his time, like Dr. Woodward a great bibliophile and collector; but a generous and personable man of whom Dr. Johnson once wrote that he had lived more in the broad sunshine of life than almost any other. But Mead had shared the dedication in Freind's book, as well as contributing some case histories to it, and he was soon to encounter one of the few clouds in his otherwise untroubled existence. Harris met the Byfield satire with a plea to attend to Woodward's arguments rather than to his person – though to be sure, he was not himself above the *ad hominem*, transforming Freind, once a schoolmaster, into "Don Pedantio Amichi." Harris recalled an earlier occasion where a different party, under a false name and sham character, had attacked Woodward's geological opinions, only to be worsted by a "merry fellow" (none other than Harris himself). As Woodward's earlier work "hath stood its Ground ever since, supported by Observations, Reason and the common Sense of Mankind," so too would the present book. No serious argument had yet been raised against it.

Nor did one soon appear, although the battle went merrily on. Next in order (February 1719) was *A Letter from the Facetious Dr. Andrew Tripe at Bath to his Loving Brother the Profound Greshamite*. (Woodward was, among other things, Professor of Physick at Gresham College.) Dr. Tripe intended, as the subtitle makes plain, to show "that the Scribendi Cacoethes is a Distemper arising from a Redundancy of Biliose Salts and not to be Eradicated by a Diurnal Course of Oyles and Vomits." He promised (in a reference to Woodward's early years in the linen trade) "an Appendix concerning the Application of Socrates his Clyster and the Use of Clean Linen in Controversy," although the addition· was nowhere to be found. The author was apparently Dr. William Wagstaffe, or possibly his friend, the better-known physician-wit, Dr. Arbuthnot.[19] The satire now is more extravagant, personal, and naughty than before. Bilious salts in the fingertips are suggested as the cause of Woodward's clutching his patients' fees so notoriously. *Scribendi cacoethes* is defined as "an Involuntary Propensity in the Hand to write down something without any Manner of Regard to the two Circumstances what or wherefore." His comprehensive theory of disease is ironically applauded; all distempers, even fractures and dislocations, are shown to be curable through vomiting and diet. So too Woodward's plea for the new science is commended; it is better "to be vers'd in the Ledger-Book of the Moderns than in the Writings of the Ancients." In the course of his diatribe, Dr. Tripe rehearses much of Woodward's past, his obscure origins and peculiar medical training, his strange cosmological theories, his expulsion from the Council of the Royal Society, and so on. There are some shrewd hits, but the medical argument remained untouched.

The reply for Woodward (March 1719) was a very long pamphlet entitled *The Two Sosias, or the True Dr. Byfield*, which at once identified Byfield as Freind and Tripe as Mead.[20] The author defended Woodward's personal background and medical ideas in some detail, answering the more obvious misrepresentations; he also upheld the author of the *Letter to the Fatal Triumvirate*. But in the absence of serious criticism it was hardly possible for the medical argument to advance. Nor was it more seriously considered in the pamphlets that followed: *A Serious Conference between Scaramouche and Harlequin* by Momophilus Carthusiensis; *An Account of the Sickness and Death of Dr. Woodward: as also of What Appeared upon Opening his Body*; and *The Life and Adventures of Don Bilioso de L'Estomac*. (These last two were afterward included in *The Miscellaneous Works* of Dr. Arbuthnot and may well have come from his pen.)[21] By May, the Woodward men had answered with *An Account of a Strange and Wonderful Dream Dedicated to Dr. Mead* and more soberly (the author claimed to be a "divine of the Church of England") in *An Appeal to Common Sense*. But before the end of the month they had to endure watching their hero become the chief character in a burlesque opera called *Harlequin-Hydaspes: or the Greshamite*, performed at Lincoln's Inn Fields and diversified by "the present humorous Dispute between some Members of the Faculty" (i.e., of Gresham College). In it the Doctor interferes with the marriage of his two daughters, but is outwitted and surrenders on the condition only that he be left undisturbed in his practice of administering oils and emetics. There is the usual fun with fossils, bilious salts, oils, clysters, and vomits.

Indeed it was time to call a halt. On June 5 the *Freethinker* (by Ambrose Phillips) found the whole controversy distasteful, and Richard Steele, a devoted patient of Woodward's, argued that the "Subject-matter in De-bate . . . has been wholly neglected, in order to turn the Person who gave rise to it, into Contempt and Ridicule."[22] Had not Woodward made observations and dissections and "search'd into the Body of Man with an inexpressible Diligence and Care," learning from the failures of others and building upon the foundation of nature herself? The Doctor, Steele claimed, had tried to rescue medicine from mystery and obscurity and reduce it to a science. Doubtless recalling his own successful treatment at Woodward's hands, Steele drew a picture of his physician in "constant Attention in the Cases of the Sick and Diseas'd . . . ever meditating on the Torments and Sorrows of those in Pain, when he is not at their Bedsides or Couches, observing the Adminis-tration or Operation of his Medicines, and abating or enlarging them as Events and Symptoms declare the momentary Condition of the Patients." It was a nice contrast to the very different and probably unfair recollection of Woodward's neighbor and fellow doctor, Dr. Daniel Turner, who remembered only "vomits and canthartics administered alternately, *de die in diem*, till the sick man grows tired, or being quite spent, is forced to give over."[23] Woodward's failures (if there were any) were naturally unable to speak out, but there were lots of patients who were eager to show their gratitude for his attention.[24]

The exchange was still in progress (with much more remaining) when the climactic event occurred, the personal confrontation between Woodward and Mead. It is a well-known tale, though told with a variety of endings. Mead and Woodward allegedly met before Gresham College and fell upon each other with their swords. Woodward slipped and lay at Mead's mercy. "Take your life!" exclaimed Mead. "Anything but your physic," replied Woodward. (In Austin Dobson's version, Mead bade Woodward beg for his life, to which he rejoined, "Never until I am your patient!")[25] It was variously reported in the newspapers of the day. The *Whitehall Post* heard that it was a duel with sword and cane between physicians who had already indulged in a paper war. "Both these Engagements were equally ridiculous, the former producing no Blood, and the latter no Argument." The *Post-Bag* claimed Mead had thoroughly whipped Woodward and generously spared his life, while the *Evening Post* had Woodward wounded in the hand and *The Weekly Journal* attributed his defeat to a fall.[26] In the end, Woodward himself was provoked to give his own account. Whatever its veracity, it quite surpassed the rest in vividness and detail. Here is the way the Doctor recalled it.[27]

There having been spread several false Reports of what lately happen'd between Dr. *Mead* and me, at *Gresham College*, I think my self oblig'd to give the Publick an Account of the Matter of Fact.

On the 10th Instant, about eight in the Evening, passing, on Foot, without a Servant, by the *Royal Exchange*, I there saw Dr. *Mead*'s Chariot, with him in it, and heard him bid his Footman open the Door. But Dr. *Mead* made no Sign to speak with me, nor did I in the least suspect that he would follow me. I walked so gently, that had he intended to have come up with me, he might have done that in less than twenty Paces. When I came to the College-Gate, which stood wide open, just as I turn'd to enter it, I received a Blow, grazing on the Side of my Head, (which was then uncover'd) and lighting on my Shoulder. As soon as I felt the Blow, I look'd back, and saw Dr. *Mead* who made a second Blow at me, and said, I *had abused him.* I told him, *That was false,* stepp'd back, and drew my Sword at the Instant; but offer'd to make no Pass at him 'till he had drawn; in doing which he was very slow. At the Moment that I saw he was ready, I made a Pass at him; upon which he retreated back about four Foot. I immediately made a Second, and retired as before. I still pressed on, making two or three more Passes; he constantly retiring, and keeping out of the Reach of my Sword; nor did he ever attempt to make so much as one single Pass at me. I had by this Time drawn him from the Street quite through the Gateway, almost to the Middle of the College-Yard; when, making another Pass, my right Foot was stopp'd by some Accident, so that I fell down flat on my Breast. In an Instant I felt Dr. *Mead* with his whole Weight, upon me. 'Twas then easy for him to wrest my Sword out of my hand, as he did; and after that, gave me very abusive Language, and bid me *ask my Life.* I told him, *I scorn'd to ask it of One who, through this whole Affair, had acted so like a Coward and Scoundril*; and, at the same Time, endeavour'd to lay hold of his Sword, but could not reach it. He again bid me *ask my Life.* I reply'd as before, *I scorn'd to do that*; adding Terms of

Reproach suitable to his Behaviour. By this time some Persons coming in, interpos'd and parted us. As I was getting up, I heard Dr. *Mead*, amidst a Crowd of People, now got together, exclaiming loudly against me for *refusing to ask my Life*. I told him, in Answer, *he had shewn himself a Coward; and 'twas owing wholly to Chance, and not to any Act of his, that I happen'd to be in his power*. I added, that *had he been to have given me any of his Physick, I would, rather than take it, have ask'd my Life of him; but for his Sword, it was very harmless; and I was ever far from being in any the least Apprehension of it.*

For all of Dr. Woodward's pique, the story rings true. The Doctor rarely avoided a quarrel and almost always had the last word. (Whoever had won the battle, he was absolutely certain that he had won the war.)[28] But whichever version of the fracas one prefers, the tale of the smallpox struggle nicely illuminates Dr. Woodward's character. At fifty, he was as stubborn and combative as he had been two decades before when he first startled the world with even bolder ideas about the formation of the earth. Although in neither case was he able to satisfy his critics, his ideas commanded respect[29] and were certainly not worse than those of his rivals. Perhaps they were premature, but as he claimed, he had constructed them painstakingly from observation and experiment, from wide reading and careful reflection. If they suffered from one fault, it was the apparent inability of the doctor to subject them to any criticism. But here it must be said that his rivals did not help. In the fashion of the day, they preferred satire and mockery to serious debate. But it is doubtful that even sober criticism would have touched our man very much. Dr. Woodward was too self-confident, too proud, too touchy, and too dogmatic. In this he was not unlike his old colleague and admired associate, Sir Isaac Newton. But alas, although he thought himself an equal, he had not the same genius, and where the achievements of the one remain to astonish us, it is somehow mostly the crotchets of the other that have left their residue.

II.

The Natural History of the Earth

The smallpox war went on, to the infinite amusement of London society. Woodward never conceded, but continued to collect the evidence for his opinions and to scribble replies to his opponents; Mead never forgave his rival, but carried his bitterness to the grave a score of years after Woodward had disappeared from the London scene. In the end it all proved of little consequence; the smallpox war continued unabated, and the quarrel gradually receded while new (and equally erroneous) theories were advanced, to be disputed in their turn.[1] It is time to let them rest, however, and to turn the clock back upon Woodward's early life and begin to account for the acerbity and vindictiveness of the quarrel. For it was Woodward's lifelong misfortune to have his views and his personality confounded, and it is the obligation of the historian to try to disentangle them and to do what justice is possible to each. Only then perhaps may we hope to sort out the complicated issues that surrounded the man — and his shield.

Unfortunately, Woodward's origins are obscure. He was born in a country village in Derbyshire, probably in 1665 but possibly three years later.[2] There he received the customary grammar school education: a good deal of Latin, a little less Greek, and scarcely anything else. He was then sent to London as an apprentice to a linen draper, although he was already recognized to be "a Man of very quick Parts, and having a genius to the Study of Natural Philosophy."[3] Now it *was* possible to remain in the trade and become a considerable scholar, but Woodward seems to have come to the notice of Dr. Peter Barwick, physician to the king, who promptly took him away from linens and prepared him for his own profession. In 1692 Dr. Barwick testified that Woodward had lived in his household for nearly four years and that the young man had "studied with so much industry and success that he hath made the greatest advance, not only in Physick, Anatomy, Botany, and other parts of naturall Philosophy, but likewise in History, Biography, Mathematics, Philology, and all other useful learning of any man I ever knew of his Age."

It was a tribute to Woodward's virtuosity, and not apparently undeserved. On the same occasion (Woodward's petition for a new post at Gresham College) he received endorsements from more than a dozen other well-known doctors. "He hath," wrote one, "attained to a very exact knowledge of the greatest part (if not all) good and usefull bookes in all faculties as well as Languages, Greek, Latin, Italian, and French . . . in Phylosophy, Physick, Mathematicks, History, Geography, Philology and whatever is necessary . . . to a Physitien or a Gentleman." It was perhaps an exalted view of the requirements of the profession, but it was not surprising that Woodward got the job he sought, Professor of Physick (i.e., Medicine) at Gresham College.[4]

One might well suspect the accuracy of such recommendations, except that there is other evidence to justify them. Woodward's range was uncommon, but it was not surprising in the waning years of the seventeenth century. Especially frequent was the combination of natural philosophy with historical and linguistic scholarship.[5] Among Woodward's acquaintances, for example, was William Wotton, a year younger but very much more precocious and erudite. Wotton had learned Latin and Greek beginning at the age of four years and six weeks, Hebrew only a little later, until at length he could lay some claim to almost every field of human accomplishment. He demonstrated his erudition in 1694 with his *Reflections on Ancient and Modern Learning*, a survey of the knowledge of his time as well as the beginning of the Battle of the Books. There he argued for the "moderns" against those who, like Sir William Temple, believed that the ancients had excelled in all things, specifically defending the achievements of the "men of Gresham" (a term which then applied both to Gresham College proper and to the Royal Society, which still met on its premises). But this meant more than praising and pursuing the natural sciences, for even the Royal Society mingled philological and antiquarian matters with "philosophy" in their meetings and *Transactions*, and there was hardly a member, not excepting Newton and Halley, who did not combine the two interests. Among Woodward's referees, Doctors Sloane, Martin Lister, and Tancred Robinson were thoroughly representative of the many physicians who seem then to have been equally attracted both to natural philosophy and to antiquities. And the great works of contemporary erudition frequently combined them also. Thus the new edition of Camden's *Britannia* (1695) had begun as history but added a wealth of scientific information, while Dr. Plot's *Natural History of Oxford-shire* (1677) began as philosophy but ended with a large dose of history.

The one thing that Woodward's many interests, like those of the other "virtuosi" of his time, had in common was that they were all "modern" — that is, unknown or undeveloped by the classical world. In the Battle of the Books, which was one of the distinctive intellectual landmarks of the time (it coincided almost exactly with Woodward's early career), the contest was between two antithetical forms of culture or education. Paradoxically, they had both originated in the Renaissance, where for a time they coexisted uneasily. On the one hand, there was the notion that the imitation of classical literature, especially the Latin works of the Augustan age, was the sole means to success in life and art; this was the conviction of the "ancients." The classics were the best; the sum of human achievement lay in the past. But this carried with it from the first an indifference, sometimes even a hostility, to classical philosophy and science. (Thus the quarrel between rhetoric and Aristotelian logic that persisted with the humanist ancients throughout and beyond the Renaissance.) No doubt this was due largely to the practical content of this form of education, a content that endeared it to the practical political men of the day, to statesmen like Sir William Temple, the champion of the "ancients," who had no difficulty identifying with, or at any rate trying to emulate, classical Romans like Cicero. No doubt too this is why classical letters (excluding classical philosophy and science) was the sole concern of the English grammar schools. On the other hand, the revival of antiquity had from the first meant also the restoration of classical culture in a much wider sense, and this carried with it two "modern" consequences, equally obvious to Wotton and to Woodward. It meant first the recovery of ancient thought as a whole, including science and philosophy, and hence one of the foundations on which was built the impressive new philosophical speculation of seventeenth-century Europe. It meant also, more immediately, the invention of classical scholarship, the recovery and elucidation of the ancient texts and the world in which they were created. Classical philology was thus a "modern" invention and it implied a new kind of history, the study of "antiquities."

Unfortunately neither the new science nor the new philology appealed to the "ancients"; they neither served the practical life of the statesman nor appealed to his esthetic sense, and indeed they seemed positively to impede the act of imitation. This was the argument of Temple in his reply to Wotton, as it was of Swift's account of the Battle of the Books. The footnote and the learned commentary (or the logical and mathematical argument) seemed impossible to reconcile with eloquence. And to complicate the matter further, the more the philologist came to know about classical literature and the more exact his learning, the more difficult did the possibility of imitation seem to be. (Here is the paradox of the Ciceronian controversy of the Renaissance that continued into Woodward's time: one could come to know too much about Cicero and the ancients to conceive of imitating them.) Philology might thus be an exalted historical discipline for Wotton, since it illuminated so much of the past; for Temple, only narrative could pretend to history – only the well-told story, told to classical standards of form and style.

Perhaps there was a time when Woodward puzzled over these questions. He seems, however, not to have hesitated in his choice: from the first he threw in his lot with the moderns.[6] Like most of them – like Wotton, for example – he did not deny the superiority of classical literature and art to anything modern, but preferred to limit the ancient achievement simply to those things. Woodward himself never pretended to "polite" literature, although he did develop a vigorous Latin and English style. His admirers generally applauded it for its clarity and strength – the very qualities which Woodward himself prized most.[7] Once, however, when he was criticized for a Latin work, he replied apologetically in a way which allows us to place him exactly in the quarrel. "There are certainly in it," he admitted, "Words not to be found in Cicero and the style is not Ciceronian." Apparently he had shown his work to two schoolmasters who had pointed out its defects. But that did not mean that he had not taken pains with it, as much as time allowed anyway. And those "who say *it has no style and 'tis not Latin* are too hard upon it."[8] Just so; Dr. Woodward's style in Latin and English was clear and presentable, although it was neither elegant nor polite.[9] In this, as in everything else, he was a modern. He was certainly never a part of the literary culture of London; the circles of the wits were entirely closed to him and, in the end, hostile. On the other hand, from the first that we meet him, he was actively engaged in the whole range of "modern" subjects. He was a thorough Baconian, for whom observation and experiment were the exclusive means to unlocking the world of nature; at the same time, he was an antiquary, for whom philology, archaeology, and a dozen other ancillary skills, all new (and all of them at odds with the culture of rhetoric), were the sole means to recovering the past.

Some letters are the first to throw light on Woodward's youth and to reveal the extent of his interests. In 1691 he was still uncertain about his career. He contemplated traveling abroad, perhaps as physician to an embassy or tutor to a young gentleman. He though himself not unfit for the post, "for besides Latin and French, I understand Italian and Spanish moderately well: nor have I been less diligent in acquiring the topography and noting such observables both of Nature and art as I could learn either from Books or Travellers."[10] For antiquities he had "a particular curiosity" and thought that he could give a competent account of all the architecture, museums, libraries, etc., in Europe, not to mention those men noted for learning in each country. He was anxious to hear of any opportunity. In the following year, while still with Barwick, he wrote another letter with an even more detailed account of his accomplishments, a rambling discussion on his favorite subjects, natural philosophy and antiquities. In it, he described the seashells that he had found inland and recounted in detail observations drawn not only from his own personal inspection of mines but from conversations with others, queries sent abroad, and a wide reading in Latin, French, English, and Spanish. He was not sure whether to publish his observations on fossils. "I have been already over the greatest part of England purposely and have at this time Queryes about those things in many parts of Asia, Africa, and America, as well as Europe, to some whereof I have already received Answers." But this was only part of his philosophical interest. "I have many things," he added, "in Anatomy and other parts of Physiology which would be more reputable to me as a Physitian which may hereafter be publick if I have leisure and encouragement." His correspondent had reported the case of an ear of rye being extracted from the side of a newborn infant. Woodward was very skeptical. "The supine credulity of some persons . . . and their fondness to relate and hear things unusual and strange, hath furnished the world a great many relations that might have been spared and some that are notorious Impostures." It was a subject on which he felt very strongly. In any case, anyone who had studied the structure of the human body and the formation of an infant in the womb (as had Woodward) would see the impossibility of the story. On supernatural grounds he found it equally implausible.[11]

About antiquities he had as much to say, and as broadly and confidently. He began with some remarks on an Irish version of Scripture that Robert Boyle had contemplated publishing, and the usefulness of the early languages to the study of ancient British history. "There hath been a vast deal wrote about our Antiquityes," he added as though he had somehow mastered it all, "yet . . . there is a vast scope still for anyone that shall undertake to write and we are in the Dark in many considerable things that I have good assurance are very retrievable." Woodward himself had several things in manuscript "in History, Philology, Antiquity, etc.," which he looked forward to publishing. In the meantime he wanted news of the Oxford booksellers. Had they anything in the way of classical or modern authors treating antiquities: of old buildings, statues, coins, gems, sculpture, etc.? Was there anything in the Ashmolean Museum or the College libraries that had not yet appeared in the standard antiquarian compilations? "I entreat you to enquire more especially after Egyptian amulets, Incunculae, or other Sculpture, or . . . Greek Antiquityes."

A little later, Woodward wrote to the same correspondent that he loved the study of antiquity only next to that of nature.[12] Within these vast domains there was not much that escaped his curiosity. Yet there was already something more than a little irritating about his tone. On every subject, the young man had an opinion, equally positive. When Woodward was only about twenty-five (in 1690), the naturalist Edward Lhwyd wrote to Dr. Lister to describe his new friend.[13] It was true, he admitted, that "He seems to have made a wonderfull progresse (considering his age) in several kinds of Knowledge." Woodward had for example given the famous Dr. Plot some useful specimens of coal slate and had made an impressive collection of fossils. But he had also told Lhwyd confidently that he knew *all the causes* of those objects which had so baffled philosophers for centuries, adding "that they seem soe plain, he wonders no body thought of it sooner." Since Woodward was so far unwilling to say more, the older man was naturally skeptical. He was doubtful whether his friend was "sufficiently experienced in such Observations as to be able to satisfie mens Curiosities in soe Nice a Phenomenon." But Lhwyd's suspicions had only begun.

[2]

For the natural philosopher in Woodward's day, the whole world seemed ripe for discovery. From the time that Bacon had urged upon his countrymen a fresh inquisition of Nature, the advancement of scientific learning had become the preoccupation of increasing numbers of Englishmen. The foundation of the Royal Society secured respectability for the new vocation and furnished a forum for its multifold activity. Newton's physics and Boyle's chemistry were its most ambitious achievements, but biological and geological speculation were equally evident. Woodward's youth was passed in an atmosphere of continuous and startling discovery, of passionate "philosophical" enthusiasm, but also of a lingering skepticism. For the new philosophy had so far called more in doubt than it had yet affirmed. The "ancients" remained suspicious of all natural speculation and were not impressed, though their composure had begun to be shaken, while the orthodox were too timid or frightened by what seemed to be menacing theological implications.

Woodward's first concern, besides medicine, was natural philosophy, not so much the mathematical speculations of Newton or Halley (both of whom he knew and admired, and tried in some ways to emulate) but more especially the description of the earth, i.e., natural history. "I began my Observations and Collections," he recalled, "in Gloucestershire where I was invited by Sir Ralph Dutton."[14] He had gone there about 1688 or 1689 along with Dr. Barwick, Lady Dutton's father. There he was able to pursue his many interests; he was furnished every sort of animal, bird, and fish that the country could provide. In London he also had the opportunity for botanical studies and could meet and talk with other naturalists, Sir George Wheler, Dr. Plukenet, and Mr. Doody. He was able to make frequent excursions to haunt the quarries and to examine the stone. Dr. Barwick (who was going blind) used Woodward as his amanuensis but often had to do without the young man's services because of his geological expeditions.[15] It was at Sherborne

(Dutton's seat) that Woodward made a discovery new to him and crucial for his career. He found shellfish lodged directly in the rock and, even more surprisingly, whole beds of shells in the plowed fields nearby. "This was a speculation new to me; and what I judg'd of so great moment that I resolved to pursue it through all the remoter parts of the Kingdom."[16] He traveled at once extensively, made close observations, and collected everything he thought interesting, sending his specimens on to London. (Boxes were still arriving two years afterward.) In an incredibly short time, even while pursuing his career, he became an acknowledged authority in the incipient sciences of geology and paleontology, an expert on fossils or "formed stones," as well as botany and comparative anatomy. And of course, he formed some definite ideas of his own.

As I have said, it was an exhilarating moment in the history of those disciplines. The seashells particularly had aroused new interest. Dr. Plot, for example, put the question thus in the *Natural History of Oxfordshire:* "Whether the stones we find in the forms of Shell-fish be *Lapides sui generis,* naturally produced by some extraordinary plastic vertue latent in the Earth or Quarries where they are found? Or whether they rather owe their form and figuration to the shells of the Fishes they represent, brought to the places where they are now found by a Deluge, Earthquake, or some other such means . . . in tract of time turned into stones?" For the first view there were the arguments of Dr. Lister; for the second, those of Robert Hooke and John Ray. Plot himself inclined to the idea that they were *lusus naturae* (sports of nature), and this indeed was the prevailing attitude when Woodward (who knew all but Ray personally) took up the question.[17]

If, however, fresh observations were furnishing fuel for speculation, there was another better-traveled avenue to the problems raised by fossils. For centuries theology had offered its opinions, founded upon the Mosaic testimony, about the creation of the world and all its creatures. Not only did the Old Testament appear authoritative; it seemed to furnish the *only* coherent account of the beginning of things. It may be hard to suppress a smile at the thought of fossil shells being especially created by a playful Providence to resemble living animals, and harder still to accept the Mosaic account of the origin of the world, or the story of Noah's Flood and the re-peopling of the world afterward. But what alternative was there in 1700? The immeasurable time spans required for evolution seemed inconceivable, while the thought of a prehistoric world in which thousands of strange plants and animals, long extinct, wandered over an unfamiliar terrain, was even more so. The sciences of geology and paleontology were hardly begun; the evidence for them lay largely still beneath the earth. How could anyone, from an odd shell or fossil bone, conjure up that lost universe? Meanwhile, there lay to hand a revelation that made all plain. In the Mosaic account there were answers to questions that seemed otherwise impossible, answers framed to the common experience of ordinary men. It was hard not to take them simply at face value. But the Mosaic creation fixed all the species in a limited time period, and so left the problem of the fossil shells still mysterious.

Meanwhile, in Woodward's youth, there appeared a work that stimulated further reflection on these problems, Thomas Burnet's *Sacred Theory of the Earth,* in Latin in 1681 and 1684, in English in 1684 and 1690. "My Book," the author reported gleefully to Sir Robert Southwell in 1688, "has at length passt the pikes at Lambeth" (i.e., the authority of the Archbishop of Canterbury) "and is now in the press. 'Tis concerning the conflagration of the world and the new heavens and earth that are to succeed."[18] It was very much a product of the new philosophy, a compound of naturalistic and theological speculation written in a particularly vivid and provocative style. Burnet purported to describe the creation of the world, and especially its form after the Flood. It was the result, he wrote, of "a particular curiosity to look back into the first Sources and Original of things," a motive of great appeal to the seventeenth century, equally important to both natural philosopher and historian. Indeed, the parallel was continuously in Burnet's mind. "What subject," he wanted to know, "can be more worthy the thoughts of any serious person than to view and consider the Rise and Fall, of all the Revolutions, not of Monarchy or an Empire of the Grecian or Roman State, but of an entire World?" The *Sacred Theory* was thus meant in the first place to be a foundation for universal history, and the argument was to rest on "Scripture, Reason, and the ancient Tradition."[19] The first and the last supplied the evidence of history, sacred and classical, which Burnet found nicely to concur. "Reason" was his equivalent for natural philosophy, but it was very deductive (some thought "Cartesian") and too little empirical for the Baconians. Put it all together, however, and Burnet was able to reconstruct the primeval earth before the Fall (egg-shaped and perfectly surfaced) and to describe, in larger detail, its transformation by the universal Flood into its present awesome and imperfect shape. Burnet's poetic description of mountains was especially stimulating to the imagination of his day.[20]

Burnet's tract signaled an immediate debate. It was reported at once to the Royal Society – and Newton, for one, found it plausible. To most of the naturalists, however, it was not persuasive. Above all, it failed to consider the evidence now rapidly accumulating: the new observations of shellfish and fossils, of rocks and ores and strata, of the actual configuration of the earth and its denizens. Burnet was too much the theologian, too little the natural philosopher. To most of the Royal Society it was entirely imaginary and fictitious or, as John Ray put it, "no more or better than a meer chimera or Romance."[21] This was exactly Dr. Woodward's view; as he complained in 1695, "how much soever it may relish of Wit and Invention, it hath no real foundation either in Nature or History."[22] But the theologians were equally unhappy. Among the several divines who replied to Burnet, Erasmus Warren pointed out that he had contradicted Scripture both in the letter and the spirit. Burnet's picture of the antediluvian earth, without mountains or streams or anything else of the topography of later times, flew directly in the face of the Mosaic description. Philosophy, he allowed, should be the hand-maiden of theology, but it could easily be overvalued. If one gave it "such a mighty influence as is conceited," he very much feared that it would be used to explain everything. What then would happen to God's miracles? Explaining the Flood simply by natural causes only "grievously impeaches Scripture." Or, as another churchman put it succinctly, "this way of philosophizing all from Natural Causes, I fear will make the whole World turn Scoffers."[23] Burnet was damned either way, by the naturalists for failing to make his case, by the theologians for making it too well! John Locke probably spoke for most when he wrote (privately) that he was unable to reconcile Burnet's work "either to philosophy, scripture, or itself."[24] Like everyone else, however, he found something to admire in it and much to stimulate further thought.

For the more scientifically inclined, for the naturalists who were scouring the countryside for clues to the primeval earth, the shellfish seemed to offer the most significant hints. There were other evidences, of course, but none more intriguing. How had they got to the tops of mountains and been buried deep within the earth? Why could living animals be found to resemble only some of the stone shells? And why were there so many sea remains and so few of any other kind? The honest John Ray, the most meticulous and patient of fieldworkers, probably the greatest naturalist of his time, could not make up his mind. "I have long fluctuated in my opinion concerning the Originall of these Stones," he wrote to Edward Lhwyd in 1690, "whether they be shells of fishes petrified, or primary productions of nature in imitation of shells. The arguments for and against both are so strong and pressing that they constrain me to settle in a middle opinion, that some are of one kind, some of another."[25] He had written about the problem as long ago as 1673; he remained equivocal in 1692 when he set down his thoughts once again in a work entitled *Miscellaneous Discourses concerning the Dissolution and Changes of the World.* He still could not decide between the arguments of Hooke and Plot, although he inclined as before to the notion that the fossils were once real. Unfortunately, the new observations of his friend Lhwyd, and of Lhwyd's friend Mr. Woodward of London, seemed (for the while anyway) only to have made the problem more complicated, although he was sure that a solution would eventually be found.[26] Apart from the paucity of the evidence, Ray saw a theological difficulty as well. If the petrifactions had once been real, he wrote, "there follows such a train of consequences as seem to shock the Scripture-History of the nativity of the World." The generally received opinion of divines and philosophers would be overthrown, "that since the first Creation there have been no species of Animals or Vegetables lost, no new ones produced."[27] He was not prepared to anticipate Darwin.

Yet not everyone was fazed by the problem. In 1686-87 Robert Hooke entertained the Royal Society with a series of lectures on the sea-fossils in which he forthrightly proclaimed their original vitality and accepted the consequences.[28] It was an eccentric opinion (though not without precedent) and must have startled his audience, but it was even more surprising to discover "that there have been many other Species of Creatures in former Ages of which we can find none at present and that 'tis not unlikely also but that there may be divers new kinds now, which have not been from the beginning." How had fossils gotten to the mountaintops and been buried within the earth? Hooke marshaled the evidence of observation and history, even of pagan mythology, to urge his conviction that it was earthquakes that had disturbed the earth and wrought its present configuration. Before that, he conceived a molten globe gradually congealing. Despite his ingenuity (and his prescience), however, there were few in England who were ready to believe him.[29]

[3]

Of those who were interested in these problems, the two most avid of the younger men were undoubtedly Woodward and Edward Lhwyd. The latter was a few years older and had begun his inquiries somewhat earlier, but the interests of the two men were remarkably parallel. Lhwyd took his degree at Oxford, tutored by Dr. Plot, whose assistant at the Ashmolean Museum he became in 1684. Among his first tasks was to draw up a catalogue of its shells, "it being the compleatest collection of any natural Body preserv'd there."[30] In 1690 he succeeded as Keeper, and it was about then that he met Woodward.[31] His official duties at the museum involved both natural philosophy and antiquities, and it was this common ground that drew the two men together. They soon began a correspondence and exchanged visits to London and Oxford. They examined each other's collections and exchanged gifts.

If their interests were similar, their temperaments were most unlike. Lhwyd resembled his very close friend John Ray; both were unselfish and ascetic spirits in a world of self-seeking and ambitious men. Like Ray he lived simply, even roughly, in order to pursue his ends, altogether careless of wealth or position. "He is a person of singular Modesty," wrote Thomas Hearne, who knew him well, "of good Nature and uncommon Industry. He lives a retired life, generally three or four miles from Oxford, is not at all ambitious of Preferment or Honour, and what he does is purely of Love to the Good of Learning and his Country." Ray thought he had discovered in his young friend "lesse of effectednesse, conceitednesse, pride, or vain-glory than in almost any man of my acquaintance."[32] Whatever Woodward's virtues, they were not these.

As we have seen, Woodward's positiveness soon nettled his friend. It was not long before they quarreled. Woodward was in a bookshop one day when he overheard a visitor named Beverland say he had been told by Lhwyd that "there was one Woodward at London who was setting up for a Collector but that he cryed down all men and run down every thing." Woodward was startled and wrote to Lhwyd at once. "Twas a character I was not over fond of and I think I do not deserve, especially from you." Lhwyd soon returned an answer. Beverland, he explained, had come to see the Museum as well as Lhwyd's own fossil collection. Lhwyd told him "that there was one Mr. Woodward a young Gentleman at London whom I thought a person extraordinary ingenious, and have made the best use of his time, of any that ever I was acquainted with, who had the most considerable collection of English fossils that I had any where seen, and all of his own gathering in their native places." Unfortunately, he had added that "you seemed a little fond of your own thoughts but that was a slender fault and would quickly wear off." Lhwyd owned that it was a slip (perhaps even a mistake) but hardly worth Woodward's expostulation. "However in regard you are of a very hot and passionat temper, and of a close conversation to me somewhat disagreeable and that I also have a greater shar than does one good of that Haste the common Proverbe bestows on my Countrymen, I think it our best course will be to let fall our Correspondence . . . ."[33]

In fact the correspondence continued. Woodward replied apologetically; he knew Beverland better now, the whole affair was not worth noticing. But the smart remained. "Softening our pens by degrees," Lhwyd wrote to Lister, "we became at last reconciled but how far from Friends I know not, nor am I very solicitous about it."[34] The truth was that temperamental difficulties were compounded by professional rivalry and a difference in philosophical outlook. Both men sought illumination to some ultimate questions in the fossil shells. But where Woodward had the answer almost from the start and could write to Lhwyd confidently on the subject, Lhwyd (though he had started earlier) remained equivocal and could see only the insufficiency of the evidence. As early as 1686 he wrote a letter summarizing the arguments on both sides for the origin of the strange stones. He inclined to believe them *Lapides sui generis,* yet he saw the difficulties and had to confess his own ignorance. Four years later, writing to the equally uncertain Ray, he remained dubious, though he inclined still to the notion that they were primary productions of nature. Of one thing only was he certain: "it seems an extraordinary delightful subject and worthy the inquiring of the most judicious philosophers."[35]

Indeed, on one point all were agreed, Hooke and Ray, Lhwyd and Woodward. "To find out the truth of this question, nothing would conduce more than a very copious collection of shells, of the skeletons of fish, of coral, pori, etc., and of these supposed petrifactions."[36] It was necessary to discover and to classify the evidence before proof was possible. Lhwyd set himself about the task. Inevitably Ray furnished the model; his great works of botanical classification offered an impressive standard. The older naturalist moreover directly encouraged his friend. "Your design of publishing a Catalogue of formed stones," he wrote in 1691, "I doe very much like and approve of. Only I would not confine yourself to so narrow a compasse as the neighborhood of Oxford but take in all of your knowledge that are found in England." Lhwyd took the advice. While Woodward was promising the world a complete explanation, he would attempt something more modest. His erstwhile friend need not worry at their writing on the same subject, "his work being philosophical discourses and in English, but myne onely a classical enumeration with short descriptions of such of our English stones as have been hitherto discovered, in Latin."[37]

Woodward does not seem to have been disturbed. From the first he offered to help Lhwyd, and through the years sent opinions and descriptions and offered advice. "I have had a real kindness for him from the first that I became acquainted with him," he wrote to a friend – chiefly, as he explained, because of their "like Studyes and Searches."[38] In the end, it was Lhwyd who was concerned, despite himself. The longer Woodward's promise was suspended (and all sorts of business interfered with its resolution), the more curious Lhwyd and his friends became. "How forward is Mr. Woodward's book? How many cuts has he and what hypothesis does he go upon?"[39] And the longer they waited, the more extravagant was the promise. "'Twill be of considerable extent," Woodward wrote Lhwyd in 1693 of his intended work, "for I do not confine myself only to the shells and figured stones." These had started him on his inquiry, but they had led him necessarily to investigate metals, minerals, ores, etc. "These therefore I treat of, and of the origins of Springs with other things about the Earth." His main conviction at least was clear, if little else. "I begin with my Fossil Shells which I prove to be real and shew by what means they became buryed in the earth and lodged in Stone." In this matter, he saw that he would have to disagree with Drs. Lister and Plot, among others, friends for whom he had great respect.[40] He seems not to have noticed that he would also be at odds with his correspondent, Lhwyd – though of course the Welshman saw it only

too plainly. About this time, Woodward could report his first lectures at
Gresham College: his subject was the fossil shells. He was able to show their
real origins, so he wrote, "to the full satisfaction of the Auditory."[41]

[4]

By the winter of 1693 Woodward had settled into Gresham College, assisted
apparently by his fellow professor Robert Hooke. Moving in with his collections was a time-consuming undertaking and interfered with his work. But he
had ample room and could even let part of his lodgings to increase his
income.[42] Altogether it was a comfortable situation, and Woodward spent
the rest of his life lodged happily in his new quarters. In November he was
proposed by Hooke for the Royal Society, and soon elected and admitted.[43]
By the end of the year he was back at work. "I have now (I thank God)
settled my affairs," he wrote to Lhwyd, and he hoped soon "to fall upon a
review of my papers in order to the publication of somewhat concerning
those shells, plants, etc., in stone." He did not know yet where to start,
however. Lhwyd's response (in a letter to Ray) was characteristic: "So far
Mr. Wdwd; to which I add let him have never so much to say; he'll scarce
outdoe his promises." The Welshman found his friend's letters now "so full of
pride and conceit, that I find he rather improves than corrects the fault I
observ'd at our first acquaintance."[44] Lhwyd's irritation was naturally in-
creased by growing envy. At least one of his friends, William Nicolson of
Carlisle, noticed it. "I see you have not quite laid aside your jealousies of
being supplanted in the Discoveries you have made (in the Secret of form'd
Stones) by your friend Mr. W." He was sympathetic, and advised Lhwyd to
hasten with his own work.[45] Scientific rivalry had obviously not improved an
uneasy friendship, though it does not seem that Woodward was as yet
concerned.

For the while, however, his book, like Lhwyd's, remained a promise. Both
men were diverted by other tasks. Lhwyd turned for a time to Welsh
antiquities for a new edition of the *Britannia.* Woodward was preoccupied
with other scientific interests and with his career, now beginning to blossom.
He had all along been busy experimenting and recording observations and
trying to advance his medical career. He was not idly boasting when he wrote
to Lhwyd (deferring an answer to an inquiry about shells) that "the Truth is I
have been . . . engaged in other Natural Enquiries, Anatomy, Pharmacy, Vegetation of Plants, etc."[46] Some of his experiments were to bear fruit later,
particularly his remarkable series on vegetation. Woodward was apparently
the first to discover the principle of transpiration in plants, a discovery that
was much appreciated then and now.[47] Others, like those about the weather,

have disappeared. (Among his schemes was a design for a "Cannon of the quantity of vapours that rise"; he sent Lhwyd a careful description of his method and corresponded widely about it, but his conclusions do not seem to have survived.)[48] Meanwhile his progress to a career was crowned with success. On the fourth of February 1695 he was awarded a doctorate in medicine by the Archbishop of Canterbury, a "Lambeth degree" that was confirmed by another (not "over-regular" according to Thomas Baker)[49] at Cambridge about six months later. Thus prepared, Woodward began to attract a broad and busy practice. Soon he was elected to the College of Physicians, where (as we have seen) he was ultimately to read his lectures on the bile. Under these circumstances it was something of an accomplishment simply to complete his manuscript, even though it was not quite the promised performance.

Woodward had pledged a full explanation of the nature and origin of fossil shells, as also metals and minerals, springs and strata. In the end he had to confine his ambitions somewhat. "'Twill be about 300 pages in octavo," he wrote in the autumn of 1694. But it would be as concise as possible, an outline only of his conclusions; he would reserve for a second edition the full proofs and illustrations. Nevertheless, he was insistent on the authority of his argument. "The method of my proving things," he wrote, "is all along by inductions from my observations, upon which I ground everything that I advance." Thus he expected (unlike the deductive Burnet) to treat the antediluvian earth only at the end of his work, coming to it with the ground appropriately prepared. He asked Lhwyd to read the manuscript but the latter declined, against the advice of Ray. Apparently the Welshman wanted to be free from all complicity in it; free also perhaps to reply in public.[50]

When it appeared early in 1695, the book was entitled modestly *An Essay Toward a Natural History of the Earth,* though its subtitle declared less humbly its intention to deal with "Terrestrial Bodies, especially Minerals, as also ~f the Seas, Rivers, and Springs; With an account of the Universal Deluge And of the Effects that it had upon the Earth."[51] At the outset Woodward made clear that the work was only preliminary; more was promised for the future. It was done in haste and "rather too short". Woodward was painfully aware of its limitations. However, he wanted at least to state the argument and outline the evidence. He was aware that his thesis could shock some readers, but he was confident of its truth.

For the modern reader the *Essay* is notable for its persuasive argument in favor of fossils as once-living creatures. For the contemporary reader, the most interesting part was doubtless where Woodward reconstructed the history of the antediluvian earth and its dissolution at the Flood. Here was a dramatic alternative to Burnet. Woodward built his argument carefully. He intended to be guided simply by observations, "the only sure Grounds whereon to build a lasting and substantial Philosophy." He introduced his hypothesis therefore with a scrupulous account of the way in which he had accumulated the "Matter of Fact" that supported it. He thus recounted at some length his travels, inquiries, and collections, and the lists of queries which he had systematically sent abroad. In this way he reassured the reader that he had tried by every available means to secure reliable information; his confidence throughout the *Essay* was largely a result of his thorough commitment to the evidence of nature.

The point is worth remarking. Woodward was addressing a problem that had been considered for centuries. But he was convinced that his solution alone was original and true, and that its persuasive character lay in its self-conscious concern with "modern" scientific method — that is, the method of the experimental philosophers of the seventeenth century. A proper natural history of the earth must be painstakingly reconstructed from those visible remains of its antiquity which might be made to yield up their secrets. As with Hooke and others, it was the fossil shells especially, but geological formations more generally, which were the clues to ancient natural history. They had to be systematically studied, collected, classified. When observed sufficiently, they would tell their tale.

In this respect, the fossil shells were exactly analogous to the new kinds of evidence which were then being examined by archaeologists and antiquaries for the clues to early human history — especially the coins of ancient Greece and Rome. They were, in Hooke's words, "Records of Antiquity," just like the ancient monuments and hieroglyphics, "transactions of the Body of the Earth, which are infinitely more evident and certain tokens than any thing of Antiquity that can be fetched out of Coins or Medals." The pursuit of shells was not a trifle; it was a valuable and important discipline, for it furnished the words and characters of the history of the world before Noah's flood. "Methinks," Hooke wrote elsewhere, "Providence does seem to have design'd these permanent shapes, as Monuments and Records to instruct succeeding Ages of what past in preceding."[52] To Woodward's contemporaries and for a hundred years afterward they seemed the "medals of Creation," and like the medals of Antiquity were examined, collected, and studied for much the same purposes and in much the same way, often by the same people. "As an accomplished numismatist," an English geologist wrote in the nineteenth century, "even when the inscription of an ancient and unknown coin is illegible, can from the half-obliterated effigy, and from the style of art, determine with precision the people by whom and the period when it was struck; in like manner the geologist can decipher those natural memorials, interpret the hieroglyphics with which they are inscribed and ... trace the history of beings of whom no other records are extant."[53]

Woodward shared this conviction exactly. In an amusing passage written later, he argued against the theory that fossils were sports of nature by the analogy of antiquities. Suppose, he wrote, that someone was to argue that the medals and coins found beneath the earth — or indeed the Roman urns, *paterae,* and other ancient objects, which were discovered from time to time — were also *lusus.* Was it not just as easy to imagine God playfully implanting these objects (so remarkably like the real ones) in the earth just as, allegedly, He had the fossils? But if no one was willing to believe this of antiquities any longer, it was surely time now to give up that thought about fossils also, and for the same reason. Now that they too were being examined and classified with the same thoroughness and system as the human medals which had so long filled the cabinets of collectors, now that they too could be compared with real objects — i.e., living animals and fish — the truth about their origin had become equally apparent. Fossils, like human antiquities, were "so many standing Monuments" to the history of the world.[54]

Indeed, Woodward never stopped insisting upon the importance of reliable evidence. Almost at once after publishing the *Essay,* he brought out a small pamphlet entitled *Brief Instructions for making Observations in all Parts of the World, as also for Collecting, Preserving, and Sending over Natural Things.* It was intended, Woodward urged, as an "Attempt to settle an Universall Correspondence for the Advancement of Knowledge both Natural and Civil." The parallel between philosophical and civil knowledge, between natural and human history, was thus explicit again, although it was not elaborated any further here. As with Bacon, the advancement of learning by observation and experiment was Woodward's profoundest conviction. The *Instructions* were detailed and meticulous, even to accounting for the requisite supplies and "A List of such Instruments, as may be serviceable to those Persons who make Observations and Collections of Natural things." For the modern geologist they still appear to contain just about everything essential for a serious collector.[55]

Before developing his own hypothesis, however, Woodward saw the need to clear the ground, to dismiss the opinions of his predecessors and his contemporaries. To be a "modern" required an intellectual arrogance that was only too apparent in the young man. Yet his was no more than the old Baconian view, that "over-great deference to the Dictates of Antiquity" had betrayed them all. Fortunately, Woodward thought, "we have at this time of day better and more certain means of Information" than any of the ancients. Here, however, he introduced an *ad hominem* argument which he might well have avoided. He explained the ignorance of his contemporaries not only by their over-deference to the classics but also by their "slothfulness." "Those who can content themselves with a Superficial View of Things, who are satisfied with contemplating them in gross; and can acquiesce in a general and less nice Examination of them; whose thoughts are narrow and bounded, and their Prospects of Nature scanty and by piecemeal, must need make very short and defective Judgments and oftentimes very erroneous and wide of Truth." Nor would he let the matter rest there; he continued contemptuously to list some of the heresies and promised more for the future: an historical account of the labors of his predecessors (including expressly Mr. John Ray), "shewing what they have already done, wherein they failed, and what remains to be done." Not one of these, he thought, had abided the main test; not one had had "due warrant from Observation."[56]

Woodward's own view rested, he believed, on several indubitable propositions. In the first place there were the large quantities of marine remains found embedded in or composed of stone. Next there was the fact that these were reposited from the surface of the earth to the greatest depths known. Then there was his observation that terrestrial matter was arranged in strata or layers. So far so good, but finally he insisted that these marine bodies were lodged in the strata "according to the Order of their Gravity, those which are heaviest lying deepest in the Earth, and the lyghter sorts . . . shallower." If these observations were allowed, his main hypothesis, he was certain, followed, "deduced" exactly in the fashion of his conclusions about smallpox. His explanation was that the whole terrestrial world had been dissolved at the Deluge, along with everything in it, and that the present earth had been formed out of that "promiscuous Mass of Sand, Earth, Shells, and the rest, falling down again [according to their gravity] and subsiding from the Water." Noah's Flood was thus a universal deluge, and it accounted for all the features, even the strange fossil shells and formed stones, of the present earth.[57]

If Woodward's "hypothesis" thus purported to account for the observed phenomena, it had another virtue: it exactly accorded with the Mosaic narrative. In the words of a reviewer, "He has given full Evidence of the Certainty of every Single Natural Proposition that Moses has laid down." [58] Woodward's history was thus both natural and providential at one and the same time. He had the satisfaction of vindicating together both the new science and the ancient Scripture, all with a single hypothesis. He was only surprised that no one had seen it before. On the other hand, he was only too conscious of the limits of his work; to remove every difficulty required a greater undertaking for which this was only a model or "platform." Eventually he would make amends for the omissions here; he would show why the stone and other terrestrial matter was dissolved but shells, teeth, bones, etc., were not. He would describe how the water was raised at the deluge from the center of the earth and how it had subsided, leaving the earth surrounding it. He would show that the arguments for the continuing change of the earth's surface were erroneous. And he would consider fossil woods and vegetable remains, showing how they too could be made to fit into his hypothesis. Perhaps most intriguingly, he intended to describe in a separate work the re-peopling of the earth after Noah, the beginning of human history. There was not much about the origins of the world and man that Woodward thought beyond his grasp.

[5]

Lhwyd could not restrain his curiosity. What, he asked in March of 1695, did Lister think of Woodward's book, and what was its reception? "It has acquired a great name in this Town," he reported ruefully from Oxford, "partly because he undertakes to confirm by his Observations the History of Moses and partly because he bylds upon experiments." It was also, he remarked slightingly, in *English.* [59] In any case, it was a great success. Lhwyd's friend John Archer found it more in vogue in London than any similar book, and Woodward was "look'd upon to be one of the most ingenious men in the whole world." Woodward was not in town, he continued, "though I have not yet seen his Worship, and was yesterday admitted MD according to custome being created by the Archbishop." [60] The two events may not have been unconnected. Woodward presented a copy of the *Essay* to the Royal Society, and a summary duly appeared in the *Philosophical Transactions.* [61] His first published work had brought him fame and celebrity.

It also brought him notoriety; the *ad hominem* argument was at once returned, to be followed soon by more serious criticism and a veritable explosion of pamphlets. The Doctor for the first time was engulfed in controversy. Needless to say, Lhwyd and Ray were immediately offended, as on cooler intellectual grounds they remained skeptical. A few months earlier Ray had been optimistic, happy (if a little surprised) to notice "a straiter tye of friendship and confidence" between Lhwyd and Woodward than he had thought. Between the two young men, he believed, the difficulties of the fossil shells could be untangled and all reconciled with the Old Testament history. On hearing Woodward's hypothesis a month later he was less sure (with Lhwýd) that it could be defended. When at last the *Essay* appeared, with its "tacit slight" at Ray, he could only lament that it had not been more "modestly propounded." It was, after all, a plausible conjecture, though no more. "But to goe about so magesterially to impose it upon our belief, is too arrogant and usurping. I cannot but wonder to find such a strain of confidence and presumption running through his whole book." [62] It was, alas, too true; Woodward's arrogance had become apparent to the world.

To the naturalists, the *Essay* had more serious weaknesses. Woodward himself had noticed some difficulties but had tried to forestall criticism by promising explanation in a further work. "'Tis wonderful," wrote Nicolson from Carlisle, "that all Solids (Marbles, Mettals, and Minerals) should be dissolved in the universal Deluge and yet several species of Shells still keep their primitive figure." He was also disturbed "that the several Strata of Mines and Minerals have continu'd in the state they were first settled in without the least change or alteration and yet our Miners do confidently report that Coal as certainly grows in England as Copper does in Cyprus." It was hard to be satisfied with a work when there were "forty more such choak-pears in the Book." Ray noticed that he had seen the heaviest fossil shells on the surface of the earth. This appeared to make nonsense of Woodward's notion of gravity as an explanation for their location. For Tancred Robinson the idea was ridiculous, "though he tells me he hath written 40 sheets on it." "He pretends [also] to have compared the Old and New World in every particular," Robinson continued, in order to prove the universality of the Flood, "but in discoursing with him I discover his ignorance of the History of both, especially Asia and America, whence he affirms the Animals are the same and the migration out of the first into the latter demonstrable, though he knows not when nor which way." A host of other objections occurred to Ray.[63] Ironically, it was in terms of evidence first that Woodward had failed to convince. In the absence of further proofs, his hypothesis seemed no better than Burnet's, another "romantic theory."[64] The closer the naturalists looked, the less they liked it; Woodward's imagination had quite outrun his observations.

There were other problems. Even the attempt to support Moses seemed weak. Woodward had insisted on placing the Deluge in May, because of fossil leaves discovered in the Earth, and intimated that the dissolution of the Earth must have occurred within a fortnight; "whereas Moses," Lhwyd noticed, "tells us the Waters were increasing on the Earth 150 days . . . ."[65] Nor was Woodward's theory as original as he had claimed; Tancred Robinson believed it to have been articulated first in the work of Agostino Scilla, and thought Woodward little better than a plagiary. Thus, within months of its first inception, the battle-lines were drawn. Arrayed against the brash young doctor were some of the ablest naturalists of his day.

Woodward was disturbed, and characteristically thought it a conspiracy against him. He was not entirely mistaken, for the world of learning in his day was small and close-knit and the slights to his contemporaries offended the sensibilities of the whole circle. "Woodward is groane most prodigious impudent in the Coffee Houses as I am told and troublesome to all mankind," wrote Lister to Lhwyd (his own ideas had of course been roughly treated in the *Essay*). "I have had noe conversation with him in the last 12 months for Mr. Ray's sake whom he vilified."[66] Ray and Lhwyd, Lister and Robinson, were all close friends and were all arrayed against him. Ensuing events made their alliance more real and (to Woodward anyway) entirely personal. But however much he may have regretted their hostility, he saw at least one advantage. As in the case of other controversial works, the uproar had made Woodward's name and reputation. "The Truth is their opposition is of real advantage to me, and hath made me appear more considerable to some than perhaps I should otherwise have done."[67] It never occurred to him that there might be something substantive to their objections.

[6]

The first publicly to enter the field against Woodward was one L.P., author of two brief essays, "The First concerning some Errors about the Creation, general Flood and the Peopling of the World." It is tempting to ascribe the work because of its initials to Lister and Plot whose fossil theories had both been mocked by Dr. Woodward, but he preferred to believe it the work of Tancred Robinson, a fellow doctor and member of the Royal Society. Just who wrote the *Essays* I cannot say, although Robinson admitted publicly to assisting the author and Lister remains a likely candidate.[68] (How it must have rankled with them, and Plot also, that they had all recommended Woodward for his Gresham College post, Lister even predicting then that he would some day "oblige the world with new Discoveries"!)[69] "I am not afraid to own ingenuously," Robinson wrote, "that the Composer of them shew'd me Part of those Essays, I mean those relating to America, which I did touch here and there and [that I] furnish'd him with some Books and Letters." His impulse, he explained, was nothing more than, "a Private Disgust at D.W. who (as I was inform'd . . .) had spoke very reflecting Words upon Mr. Ray."[70] Whatever his responsibility, it was Robinson who seems to have organized the opposition at this point, writing maliciously against Woodward in his letters to Ray and Lhwyd and trying to stimulate a public opposition.[71] L.P.'s essays were brief but suggestive, especially the detailed postscript which raised some specific objections. Why were vegetable remains not far more plentiful today if they, like shells, had survived Woodward's flood? Why were so many sea animals destroyed "by a deluge of their own element?" The author was confident that the landscape had never ceased changing, and he was willing to reaffirm that the fossils were independent natural phenomena. On the Mosaic testimony, the author was more daring than either Ray or Lhwyd. He could see no reason to accept the Mosaic account literally at all; repeating an old argument, he suggested that Moses had simply adopted a "fable" to persuade a rude and ignorant people of an underlying truth.

When Woodward read the pamphlet, he at once suspected a plot. Apart from Robinson's complicity, one passage in the *Essay* had in fact been taken from a letter by Lhwyd in which Ray was unwittingly the intermediary. (Robinson, however, was again the culprit.) The "hot and passionate temper" that Lhwyd had spotted in his rival was inflamed. "Mr. Woodward is your mortall enemy," Ray at last could report to Lhwyd (on Robinson's authority, however) "and doth all the mischief he is capable of." He doubted that his friend had much to worry about, "for the Doctor begins to be contemned in all places for his conceited and insolent behaviour." [72] Nevertheless, Lhwyd *was* concerned; Woodward, he felt sure, retained his influence with the clergy (who doubtless relished his orthodoxy), if not with the naturalists, and Lhwyd still hoped for their patronage. In fact it was a realistic worry and according to his friend John Morton, Robinson was quite wrong to say that there were very few to support the Doctor. "The Archbishop of Canterbury is, I know, his great friend and so are most of the bishops and other very great men. When he was at Cambridge with us, he met with very honorable entertainment at the Vice chancellor's hands and others, the great men of the university." Woodward, he thought, could either do Lhwyd great good or great harm. [73] According to Lister, however, Woodward was soon spreading the rumor that the Welshman would never complete his own work on fossils and had given up hope of publishing it. Woodward rejected the accusation indignantly, but all pretense of friendship was at last given over; the two naturalists were now "implacable enemies." [74]

"There's a great fray betwixt the Virtuosi," wrote an observer after several years of controversy. [75] To recount the whole of it, however, would be too tedious. It had too many levels, too many contributors, too many irrelevancies, to occupy us fully. It remained personal, perhaps increasingly so, though it continued to involve some large questions of philosophy and method. In a year or two the young Newtonian disciple, William Whiston, for example, compounded the discussion further with still another theory about the Creation and Flood, inspired by Woodward's work but this time with a precisely dated comet's tail as cause. [76] (In fact, something very like this notion had earlier been suggested to the Royal Society by Edward Halley.) [77] The clergyman Nicolson, in Carlisle, watched each episode in the quarrel with consuming interest but in the end with growing dissatisfaction. "Our late refiners upon the Creation and the Deluge are unanimously agreed, that the old interpretors of Moses were all Blockheads . . . Whether Dr. Burnet's roasted egg, Dr. Woodward's hasty pudding, or Mr. Whiston's snuff of a

Comet will carry the day I cannot foresee." [78] It was indeed difficult to choose among them.

Woodward had his defenders. John Harris, for example, turned upon Woodward's critics, especially L.P., and bombarded him with an effective combination of satire and geological learning borrowed from his master. [79] L.P. (whom Harris insisted was Tancred Robinson) had left himself open to criticism on two main points. In the first place, he knew little directly about the fossil remains and it was easy to contrast his ignorance with Dr. Woodward's painstaking study of the evidence. If (like Ray before him) L.P. did not know whether the *cornu ammonis* was once a real animal, Dr. Woodward "hath real Sea-Shells of that kind now by him, which I've more than once seen and compared with the fossil ones." [80] L.P. was clearly no match for a man who had surrounded himself with a "vast variety of Shells, Teeth, Bones and other marine Bodies . . . Fossils of all kinds, as well as shells fill'd with all sorts of Natural Minerals, Spars, Flints, etc." Dr. Woodward's collection, Harris claimed with some justification, "for Number, Variety, and Excellency, is not to be match'd by any of this kind in all Europe." [81] In the second place, L.P. had been complelled to argue against Woodward that the Scriptural account had been adopted by Moses for a rude and simple people, that it was in effect a lie. His critic thought that he would find little to support that notion outside the works of the deists — or worse.

On the whole, then, Harris' tract did not add a great deal substantively to the controversy, but the author did borrow at least one passage from Dr. Woodward's "Larger Work" in which he explained why the heaviest bodies were not always found deepest in the earth. (In the intermixture at the dissolution, some of the light bodies happening to be near the bottom arrived and settled there first, despite a lesser specific gravity. Dr. Woodward had meant to describe only a *general* condition which admitted of exceptions.) Harris was also able to qualify some of the Doctor's statements, noticing, for example, that Woodward had not intended to deny *all* change on the Earth since the dissolution. Finally, to clear his friend from the charge of plagiarism, Harris reviewed the whole progress of the new science from Fabius Columna in the sixteenth century to Dr. Woodward. (It was, however, only pouring fuel on the fire to contrast Ray here as drawing his materials exclusively from other writers, while Dr. Woodward had relied "solely upon Observations of matter of fact.") [82] We also learn that Dr. Woodward still intended a *Discourse concerning the Migration of Nations and the re-peopling of the World after the Deluge by the Posterity of Noah,* in which Moses was to be vindicated again as an historian and "several things in Ancient History that were not known will be made out, and many that were perplext and uncertain will be effectively cleared." Robinson's views (for here he admitted helping L.P.) about the improbability of Blacks and Whites descending from the same stock would be shown to be fallacious, and the unity of the human race reaffirmed.[83]

Harris' work was written with gusto and left its mark.[84] Ray and Robinson were both badly stung; others like the Newcastle virtuoso Jabez Cay, or the Bristol geologist William Cole, thought that Harris had come off rather well. But the only demonstrable effect of his tract was to embitter the combatants further and to raise the quarrel to a new pitch of personal intensity. Some – like William Nicolson, who tried to mediate, or John Evelyn– thought the acrimony regrettable. "Methinks philosophers should not fall out about Shells and pebbles."[85] Certainly it did not stop the attacks on Dr. Woodward. Most damaging of these perhaps was the collaboration of Dr. Arbuthnot with William Wotton in *An Examination of Dr. Woodward's Account of the Deluge* (1697). Wotton's contribution, however was slight, an abstract of Agostino Scilla's earlier book on fossils and Flood, the alleged source for Woodward. He claimed to print it simply for its intrinsic interest, not necessarily to criticize the Doctor, although Woodward saw it otherwise and took offence.[86]

It was Arbuthnot who presented the more formidable obstacle. His career as satirist and friend of Pope and Gay still lay in the future. His reputation now (as Lhwyd noted) was as an "excellent Mathematician; and one very well skilled in experimental philosophy."[87] True, he knew little about the figured stones, but he concentrated his attack upon the "philosophical" part of the book. Arbuthnot saw at once that the chief difficulty with Woodward's hypothesis was not its evidence but its lack of explanation. How had the water come from the earth and what had replaced it? Why did material substances not return with the fluid to the center of the earth at the end of the flood? How had the strata attained their solidity in the first place? And so on. Arbuthnot, who combined medicine with his mathematics, was able to train his special weapons and satirical gifts against Woodward's thesis. How much water was required to dissolve the dry solids proposed by Woodward? As with everything else, the Doctor had left the answer for the future. But he should at least, Arbuthnot thought, "have calculated the Proportions of his Drugs before he mix'd them." He was ready to work it out himself, giving Woodward the benefit of the doubt and allowing a mixture for the dissolution of the earthly matter of equal parts of water and solids. He concluded that to have enough, it would have been necessary to raise the water 450 miles above the earth![88] (Woodward, following Moses, had allowed but fifteen cubits above the mountaintops.) In a similar vein, Arbuthnot pursued his quarry through other objections, haphazardly but amusingly argued and always with good humor. Nicolson — who, as a divine, was much taken by Woodward's arguments — at once understood Arbuthnot's main thrust; he was quite right to show "that a successful theory must be built upon many nice enquiries and not forwardly advanced on the encouragement of a few likely phenomena."[89] The mathematician, even better than the naturalists, had exposed the weakness of Woodward's philosophical method, or so it seemed. Yet not everyone was persuaded, and Nicolson himself was confident that the Doctor would find an answer.[90]

[7]

Woodward was not discouraged. Although the attacks continued and the debate spread abroad, there were many who rallied to his opinions and his work was translated eventually into Latin, French, Italian, and German. His most enthusiastic supporter was undoubtedly his Swiss friend Johann Jacob Scheuchzer. The two men were almost exact contemporaries and, like Woodward, Scheuchzer was a physician with a taste for natural philosophy and antiquities. He was described to Hans Sloane in 1705 as "a good man and a most carefull and able physitian. From what I find by his conversation, and the accounts I have from others, he is indefatigable in his inquiries after all the parts of Natural Historie and he is now particularly employed in compiling the natural historie of Swisse."[91] He first published a paper on the figured stones in 1697, holding there, however, and for some years afterward, to the traditional belief that the fossils were *lusus naturae.* In 1701 the two men began a correspondence, exchanging information and books on that great range of subjects that united them, as well as specimens of fossils and antiquities that they each pursued so avidly.[92] From the beginning the Englishman was the dominant partner, holding out hopes for patronage and the sponsorship of the Royal Society to Scheuchzer and his brother, giving him books and medals, and trying to find him a place in England, at Leyden, and elsewhere. In return, he was eager to learn about Swiss geology; the Alps were especially alluring for a naturalist, and Scheuchzer was already their master. It was a correspondence and a friendship that lasted (although the two men never met) until Woodward's death, and it is the best single source of information about the Doctor's life and opinions after 1700. In one letter Woodward persuasively rehearsed the evidence that the fossils had once been real shells, while throughout he reiterated his conviction that the deluge explained everything. Unfortunately he had to add about his great work that the materials were so plentiful, and business so intrusive, "that God knows when I shall be able to finish it."[93] Gradually Scheuchzer came under the spell of Woodward's ideas and adopted them for his own; he soon became their most formidable exponent on the Continent. Already in 1701 the two naturalists had discussed a translation of the *Essay* into German; in 1704 it appeared under Scheuchzer's name, with a eulogy for Woodward, the beginning of a long and fruitful if somewhat one-sided partnership.

Meanwhile controversy continued: the Tuscan geologists, for example, objected to Woodward's theories, while the Bolognese supported it.[94] The Scots followed the quarrel with growing interest. In Germany, particularly, Woodward found many followers.[95] More and more, his view (and Hooke's) that the fossil shells had once been real came to prevail, though he had less success with his hypothesis about the dissolution. Even Ray came around distinctly to Woodward's notion that the fossil plants represented real species that had once lived in Britain, although he could not accept the rest. Woodward's correspondence widened and his collections broadened. Gifts of fossils came from all over the earth, wherever England traded or colonized. Through the years Woodward continued to emphasize the importance of accumulating the evidence without which, he insisted, all natural speculation was reduced to "Amusements." "They who would hereafter write well and justly must be very cautious how far they rely upon a vast number of both Observations and Reasonings that some of late have set forth to Light." [96] Cotton Mather was one American who helped him; from Africa, from the East and West Indies, from Arabia, Persia, and especially from Europe, he won new specimens. Among his greatest prizes was the entire collection of Agostino Scilla, along with his original drawings.[97] Nearer home, his contributors included both Newton and Locke as well as William Nicolson. Never once did he falter in his views of 1690. He was not very good at self-criticism; his collections and new observations apparently served but one purpose, to bolster his original opinions. He ignored L.P. and Arbuthnot, at least in public;[98] he tried to persuade Nicolson by letter, but in the end grew angry at the obstinacy of his friend in Carlisle. Eventually he printed a Latin treatise to quiet one of his opponents, but he disappointed everyone by failing to produce his *magnum opus*.

One of his assistants was a young man, "brought up from his Youth in Mines," named John Hutchinson. As steward to the Duke of Somerset, he traveled to London about 1700 and fell under the spell of Woodward's ideas. He soon began to assist the Doctor in his collections and gathered materials for a pamphlet of his own on the observations he had made in the year 1706.[99] He thought he had discovered plentiful evidence to confirm Woodward's theories, even the most suspect of them: the Doctor's notion that gravity had disposed the fossils and arranged the strata in their present positions in the earth. Woodward in turn showed him the book which he had promised the world, a large folio which he kept on an upper shelf in his study but which he would not let Hutchinson read until it was completed or at least polished. Hutchinson grew suspicious and determined to see if somehow he couldn't get a peep into the mysterious tome. He tried assisting the Doctor at those hours when he was likely to be abroad on business, but the Doctor kept too wary an eye on the young man. "One day whilst the Doctor and Mr. H. were together in the study, a servant came hastily in with a message, upon which the Doctor went out in a hurry, and inadvertently left Mr. H. alone, who did not slip the opportunity but immediately seized and opened the Book." His worst suspicions were confirmed; he found (so he wrote) only a few heads of chapters and many blank pages.[100]

Disillusionment followed at once. Hutchinson demanded his fossils back; he was unsuccessful. Eventually he altered his views; Woodward had been wrong. His resentment lasted a lifetime. A vast collection of materials had been made, he recalled bitterly, "not for a Raree-shew," and a vast series of observations had been accumulated, "not to support Mistakes." In the end they proved, "tho they were twined and bended as much as possible," that every article Woodward had advanced was wrong. "He had begun wrong . . . could not go backward, so could not go forward."[101] Unfortunately Hutchinson was as opinionated and ornery as Woodward, and his personal animus mars his testimony and makes him often incoherent. Whatever the origins of the dispute, it was embittered by Hutchinson's discovery of views of his own on the fossils and on Scripture. (He was much more literalist about Moses than Dr. Woodward.) He had "shaken off the Miner and started up into a Philosopher."[102] After his death and Woodward's, his ideas lived on in a veritable "Hutchinsonian party," and his collected works in twelve volumes were piously edited by his followers.

Hutchinson's story, reflected in anguish many years afterwards, rings only partly true. It is certain that Woodward continued in fact to work at his theories, and that his plans for a great treatise were honestly intended. Nicolson for one was shown the great work in those years and was impressed.[103] When Woodward brought out a second edition of the *Essay* in 1702, it was however only slightly revised. "I may enlarge upon it hereafter," he wrote to Scheuchzer, "but can never make the thing more plain and evident." He saw that he should say something more about the *causes* of the dissolution of the earth but for the fact, that "'twas very certain" and beyond dispute. He believed that he had put the whole study of fossils on a new foundation. "Indeed till my *Natural History of the Earth* came forward the Fossil Shells etc. past universally in England for Stones and Lussus's in Nature. But that Discourse . . . put the Study of the whole Mineral Kingdome upon a new Bottom." He continued to dismiss all criticism and any alternatives as nonsense. "Dr Langius's work," he wrote characteristically, "has set

us all laughing here as [much as] Mr. Lhwyd's ever did." The Frenchman,
Tournefort, "hath been very defective in his Observations relating to Fossils.
He hath very wild Fancyes concerning them." To the end, he was still
dismissing the "childish errors of Buttner in his *Corallographia Subterranea*
and that "rattle-headed Fellow Spreckelstein of Hamburg." Meanwhile, busi-
ness of every kind kept interfering with his own progress. "What am I doing?
I must answer — *Nothing*. And yet methinks I am alwaies much employed.
But I see too much Company: I have too little Retirement. Then the little
Business that I have in Physick takes up Part of my Time. The rest I dedicate
to the Serving my Friends and doing them good offices . . . Upon the whole
I'm allow'd little Leisure for Study." [104] When he was attacked by a German
doctor, Elias Camerarius, in 1712, he at last sent forth a rebuttal, in Latin,
which made some effort to develop his argument. [105] Even so, it was no
substitute for the great work which he had promised.

As Dr. Woodward grew older and his *magnum opus* seemed more elusive
than ever, it occurred to him that he might get some assistance to put his
collections in order. Curiously, it was a future Hutchinsonian, the young
divine Benjamin Holloway who agreed to assist Woodward in these last labors,
and even more strangely perhaps it was William Wotton (now living in
retirement) who also offered his help. Holloway and Woodward had struck up
a correspondence about geological matters as early as 1713; in 1723 Holloway sent a letter to the Doctor describing some geological observations he
had made near Woburn, an important corroboration of Woodward's notions
about strata, which was then printed in the *Philosophical Transactions*. [106] At
Sloane's recommendation the young man was admitted to the Royal Society
and shortly afterward translated Woodward's Latin tract against Camerarius,
adding to it a long introductory account of the Doctor's subsequent efforts
and several other pieces. The new work was popular enough to be translated
in its turn into both French and Italian. [107]

Among other things, Holloway included with his translation some letters
that had passed between Woodward and Sir Robert Southwell (to whom, as
President of the Royal Society, he had dedicated the *Essay*), written sometime before 1702, and clarifying his views on a number of important matters.
Woodward enlarged there, for example, on his ideas about the cyclical nature
of moisture in the atmosphere, which he related both to his experiments on
vegetation and to his theory of the abyss. And he explained at last how the
dissolution of the earth had actually occurred and why the shells were not
dissolved during the Flood like the stones and other terrestrial materials.
Woodward now made clear that it was not the water which had caused the
dissolution, but rather a Providential suspension of gravity during the Flood
which, as soon as it was restored to its natural role, had re-created the earth in

its present condition. "Should Gravity once cease," Woodward had already hinted in the *Essay*, "or be withdrawn, [the World] would instantly shiver into millions of Atoms and relapse into its primitive Confusion." [108] Since the structure of organic matter was dependent upon fibers rather than upon grains, it was not, he thought, like the inorganic materials, subject to dissolution by the suspension of gravity, and therefore remained as witness to the antediluvian world. Apparently Woodward's views persuaded Southwell, for in a manuscript letter of 1702 he replied that he agreed to what he thought were the Doctor's three main conclusions: that Scripture should best be read in a "finite and limited sense where the Expression is generall"; that the dissolution of matter must have occurred after the waters had covered the earth; and that the laws of gravity must indeed have been suspended by God. Why then had not all the matter gone to the center of the earth? Because, retorted Woodward, the suspension of matter need not have lasted long enough to allow "the melted matter to goe lower than the former Bounds." It seemed a plausible answer to Southwell, who had spent a lifetime in the Royal Society considering the problem. [109]

The reply to Camerarius which formed the bulk of Holloway's work did not go much further, however, although we do learn another reason why the heaviest bodies were not always found deepest in the earth — a fact that had baffled more than one of Woodward's critics. It appears that in falling down after the dissolution there would have been much collision and obstruction, which must have prevented the natural play of gravity. Here also Dr. Woodward accumulated the ancient non-Biblical testimony in support of the historical reality of the Flood. On the whole, however, Woodward in 1713 was still hoping to treat all these subjects "more at large," if God should give him leisure. Was it this threat that caused Camerarius finally to submit? [110] Meanwhile, Holloway referred again teasingly to the work that Harris had described still in manuscript, entitled *A Representation of the State of Mankind in the First Age after the Deluge*, in which Woodward treated the subsequent history of the world. According to Holloway, he dealt there with nothing less than "the Manners, Customs, Opinions, and Traditions, as also the Arts, Utensils, Instruments and Weapons, of all the most Antient Nations." That the Doctor entertained original opinions on these matters is quite certain, although his manuscripts and notes on the subject, with one exception, have vanished. Among other things, he believed that he could trace all the different races in the world to a common stock by way of Noah, and explain their skin color and other variations in terms of their adaptation to climate and geology. [111]

However imperfect, Holloway's collection was an impressive performance. Although it remained largely in fragments, it is clear that Dr. Woodward was aiming at a vast synthesis in which his views about geology, medicine, and history (each very broadly conceived) were to be combined and interrelated. If only more had survived of the manuscripts which he left at his death! But even the mere enumeration by his executors suggests the range and ambition of Woodward's system. No wonder that he was unable to complete his work or to surmount all the difficulties.

One other contribution Holloway was able to make. He included in his miscellany, by way of introduction, several long extracts from the mysterious manuscript of Dr. Woodward's *magnum opus.* [112] Holloway was especially impressed with the Doctor's argument that the evidence of nature exactly confirmed Holy Writ, and he selected some large passages which played upon that theme. Apparently Woodward, provoked by the challenge of skeptics and free-thinkers, had given further thought to the theological implications of his natural philosophy and had reached the interesting conclusion that Nature could provide direct confirmation for Divine Revelation — rather than the other way around. While this was soon to become commonplace, it was a rather novel idea at the beginning of the eighteenth century. That it should be Nature now that seemed secure, and Scripture in need of defense, seemed to some at first sight paradoxical. What Woodward showed was that Moses had reached correct conclusions about the history of the earth (conclusions, that is to say, identical with his own views) without having had access to any of the records, traditions, collections from nature, or personal observations which were necessary to discover it. Since it was impossible that he could have hit upon this by chance, there was left but one explanation: Moses could only have known the truth by divine revelation. So, for example, it was only possible recently to know the universality of the Flood by inquiring into the fossils and strata of the entire globe, but obviously neither Moses nor any other ancient could have done this. [113] In the same way, even to the smallest details, Moses' narrative anticipated the results of modern science, but without its method. Indeed, Woodward was at pains to show that medicine equally with geology confirmed the point, and Holloway recited a long series of dissections that Woodward had made between 1698 and 1714, on a host of different animals and insects, to prove that Moses was right (and the Cartesians and others wrong) when he claimed that "The Blood is the Life of the Flesh." [114] Woodward denied that either heart, or brain, or pineal gland entirely controlled sensation, which could be shown by experiments (here recorded) to continue quite apart from the vital organs. "The Life of the whole Animal and its power of Sense, Action, and answering the Ends of Life . . . is exactly commensurate to the Quality of rightly constituted Blood in it." With this Woodward seems to have meant to recall his theory of the bilious salts, but the manuscript in which he developed the idea has disappeared. On the general matter of the Mosaic testimony, Woodward apparently had much more to say also, in still other manuscripts which Holloway saw but which have likewise vanished.[115] There was apparently no implication – nor indeed any obstacle – to his theory that the Doctor had failed to consider.

III.

Science and Religion

How seriously should we take Dr. Woodward's ideas? There is not much difficulty about his ingenuity or the range of his views, but what are we to make of his speculations? There is a real temptation to dismiss them out of hand simply because they appear to be so wrong. Their dogmatic air does not help, nor does the eccentricity of Dr. Woodward's character. Above all, there is the Doctor's apparent confusion between science and religion, his confounding the evidence of nature with the authority of Scripture. Yet after all, there is much to be said in his defense.

In the first place, it is clear enough that no one then knew, nor could know, the conclusions of modern science about the subjects that interested Dr. Woodward. On these matters all were wrong – Hooke, Newton, Halley, and the rest. It is true that some speculations seem more nearly to anticipate modern conclusions than others, and here Dr. Woodward was not always fortunate. But science proceeds by proofs, and the proofs for most of the modern conclusions of geology or medicine were simply not then available. It is not only that most of the evidence for them still lay in the ground, or even above the ground, unobserved, but the techniques for examining it – microscopy, chemical analysis, comparative anatomy, etc. – were still far too imperfect to be very helpful. More than most of his contemporaries, Dr. Woodward at least tried to employ them where he could. If the Doctor was seriously at fault, it was in his haste to devise explanations for which there was as yet no conclusive evidence – but in this of course he was at one with most of his critics.

Take for example Dr. Woodward's most dogmatic and useful conclusion, his insistence that the fossil shells were once real living creatures and not simply sports of nature. Here Dr. Woodward was right by the terms of modern science; here his arguments seemed most secure, as they were certainly most influential. Yet even so, the evidence was hardly enough to support them. What were the alternatives? On the one side, as we have seen, there

48

were those like Lister and Plot who claimed that the petrified shells were merely sports of nature due somehow to a "plastick power" latent in the earth; on the other side there was Hooke, who argued that they were real but represented among them many species now extinct. How was one to know?

Obviously, proof rested (among other things) on an exhaustive comparison between the fossil shells and all those living species that could then be observed. Now Lister had just published at his own expense one of the great descriptive treatises in the history of conchology (a work, incidentally, very useful to Dr. Woodward), his *Historia sive Synopsis Methodica Conchyliorium* (1685-92) — it had taken him ten years and two thousand pounds to produce — and he knew more about living shellfish and as much about the petrifactions as anyone in his time.[1] But it was precisely because of what he knew and what he could not yet know that he remained skeptical of the figured stones. For he saw that in a great number of cases there were no living analogies to many of the fossils, and he took this naturally enough to be conclusive proof that they had never actually existed. It is only fair to add that (according to the modern scientist) there are indeed figured stones in nature which look almost exactly like living species but which are not fossils — accidents if you like, but *lusus naturae* beyond a doubt![2] Nor was it easy to convince Lister or Plot without an explanation of just how shell could turn into stone.

Thus Lister's view was not implausible, although in the end it was inadequate.[3] The central difficulty was that it was still impossible to know to what extent there *were* living analogies in 1700, when every day new species were being discovered in different parts of the world. (Perhaps the most characteristic activity of the naturalists in the Royal Society was the race to discover new animals and plants, although almost everyone still woefully underestimated their number and variety.)[4] Dr. Woodward could thus reply that, while all the fossil shells had not yet been matched with living species, it was only a matter of time before they would indeed be discovered. He saw that this must be a "great Undertaking," for after all, even "the shores of England have not been yet so carefully search'd but that we now daily find shells that have escaped the notice of former Naturalists." There were also distant lands, and he was sure that there were even greater numbers at the bottom of the seas. (As early as 1692, Ray had anticipated this very argument against Hooke.)[5] But if one gave up the conviction that all fossils could be matched with living animals — impossible to know in Dr. Woodward's time — one was forced back upon Hooke's view that many species had been lost out of the world, and this too, was equally undemonstrable as well as raising further difficulties. It was, as Ray noticed, difficult to accept philosophically, because it implied an imperfect world; and even harder to accept empirically, because it meant not

only that some species had disappeared but that whole genuses once represented by countless individuals must have perished altogether without leaving
any survivors. Indeed, it was necessary to imagine that an entire world of
animals and plants had quite vanished except for their fossil remains. Since
Hooke (like everyone else) refused to allow more than a few thousand years
for this to occur — and all geological change to take place — his conclusions
were, in the words of a later writer, "only less extravagant than the world
visions of Burnet and built upon foundations equally insecure."[6]

Thus, paradoxically, the more one knew about the petrifactions the harder
it was to accept extinction — at least up to that critical point when the
balance tipped and no other conclusion could reasonably be held. Only when
the world of living animals and of stone relics had been largely classified and
catalogued could a conclusion be safely drawn. Before then, Dr. Woodward's
view was reasonable enough, although it could hardly be called definitive. As
he worked away at the problem, built his collections and attempted his
comparisons, he discovered in fact many more analogies than had been
known before. If he could not always match the shells with the petrifactions,
he could, as Harris remarked, "pair several Fossil Shells with the Sea ones
which were pronounced by some inquisitive Gentlemen to be absolutely
unlike any that the Sea produceth."[7] It was not unreasonable therefore,
although in the end it was quite wrong, for him to assume that they all could
be made some day to match. Moreover *he* at least had tried to explain how
the process of petrifaction could take place.[8] Of the various conjectures that
were then advanced, wasn't his then, with all its flaws, the most plausible?

A parallel problem arose from those even more disconcerting fossil remains, the bones of various large animals which were now and again being
unearthed in different parts of the world. "It is certain," Lhwyd wrote to
Ray in 1692, "they differ totally from the bones of any land animals at
present in the island." Hooke reminded the Royal Society, in one of his fossil
lectures, of evidence that a large number of animals had once lived in Britain
which were no longer to be seen. There were for example bones from a hippo
found at Chartham in Kent and sent to the society by the Archbishop of
Canterbury; there was the great thighbone of an elephant sent by Thomas
Browne (of the *Religio Medici*) from Norfolk; and there were the large deer
horns from Ireland — too large by far to fit any modern animal. Hooke's
fruitful mind found much to question here. "Whether some species of animal
substances might not be lost, they not being to be found? Whether the
latitude of places were not changed so elephants might have been here
inhabitants in former ages? Whether the bottom of the sea might have been
dry land and what is now land might not have been sea?"[9] From abroad,
Leibniz made another suggestion. Might not the elephant bones "be of some

Creature which might differ from what wee see in the world by being somewhat changed by the space of time since that animal had been alive?" [10]

The deer horns particularly captured the imagination of the Royal Society; when a full report was transmitted from Dublin by Dr. Thomas Molyneux, it was found "extremely pleasing" and duly printed in the *Transactions.* [11] Molyneux noticed that to his own knowledge some thirty pairs of horns had been unearthed within the past twenty years in Ireland (all by accident), showing that the creature had once been very common indeed. Yet there were no deer anywhere in the British Isles that even closely approximated their size. Had they then disappeared completely from the world? Not according to Molyneux. "That no real Species of Living Creatures is so utterly extinct as to be lost entirely out of the World, since it was first Created, is the Opinion of many Naturalists; and 'tis grounded on so good a Principle of Providence taking Care in general of all its Animal Productions, that it deserves our Assent." Exactly had Ray written a few years before. [12] But the reply to Hooke did not have to be only "theological." Could it not be argued as well that the missing animals had simply changed their location rather than become extinct? In fact there was one animal, still very much alive, whose horns compared in size and shape with the unlucky Irish deer, and these furnished a convenient analogy. In America there was the moose, news of which had recently crossed the ocean to impress the naturalists. Needless to say, Dr. Woodward, who had his own numerous American sources, at once adopted the suggestion; if one only looked hard enough, there was always a living counterpart. He was much encouraged by his friend, William Nicolson, who was able to supply him with a fossil horn and some arguments against Molyneux's views; Nicolson was certain that they had been brought to Ireland "at the same time with the Spoyls of other American Animals; and that they are a further Illustration of your Hypothesis." [13] And so when Molyneux visited Dr. Woodward and surveyed his collections, he was told with typical "confidence" that "Your great Mouse deers Horns found in Ireland were transported thence by the Deluge from America and never belong'd to any Creature that had life there." What was needed of course were some actual moose horns to compare with the fossils, but these were not so easy to obtain despite the best efforts of Molyneux, Lhwyd, and Dr. Woodward, and for the while everyone had to rest content with such imperfect descriptions as John Josslyn's in *The New England Rarities* or Cotton Mather's in a letter to Dr. Woodward which was published in the *Transactions.* [14]

So Dr. Woodward was right in his conclusions (that the fossils represented once living creatures) but wrong in his evidence (for example, that the Irish deer were American moose). More importantly, perhaps, he was on the right track. Only systematic collection and comparison of the evidence, living and dead, could ever settle the question, and Woodward's analogies were more likely to be helpful and illuminating than not. Thus the Doctor welcomed all new fossils whatever and he cheerfully added to Hooke's list a variety of further examples: crocodile skeletons from Germany, for example, as well as great numbers of trees found all over England "where no such, not only in the Memory of Man but in the Records of any History have been known to grow." (Woodward was thinking how Caesar had expressly denied seeing the great pines and firs that were so common underground.)[15] The more he thought about it, the more persuaded he was that only the deluge could account for them.

Indeed, the only way to avoid Dr. Woodward's flood was to allow for time, immensely long periods of time in which the earth could change its surface – and its animal and vegetable life – without recourse to a single great catastrophe. Time was necessary, for there was nothing in the observations of men, or the record of human history, that suggested that natural, continuing processes could possibly account for the vast changes that the fossil remains seemed to import. How could mountains be raised and seas disappear through natural processes like the wearing away of the seashore, or the sedimentation that slowly built up islands, or even the more spectacular earthquakes that were Hooke's favorite, unless a vast period of time was imagined in which they could operate? Woodward specifically denied these possibilities in his *Essay*. He allowed that some change in the earth's surface had taken place after the Flood, but he was adamant that these changes were insufficient to account for the size of the alterations which he wished to explain.[16] There simply was not enough time.

As we have seen, not even Hooke was ready to imagine sufficient time for his theories, for there were great difficulties in extending the history of the world to that immensity which has become familiar to us. In a way it is wrong to ask why men believed in a relatively short life for the planet; *that* was natural enough. The question should be rather how it is that we have come to believe in something so far beyond the possibilities of sense and imagination as time-spans of hundreds of millions of years. Needless to say, conclusions drawn from radiocarbon, solar radiation, or the tree-rings of California redwoods are neither common-sensical nor necessarily consistent with one another, and it is doubtful that there are many literate men today who could adequately explain their convictions about the longevity of the earth.[17] True, Aristotle had argued for the eternity of the world, but barring that extreme possibility it was hardly conceivable in 1700 that the earth's beginnings could be more than the six thousand years or so that the Bible seemed to suggest. (Thus the Greeks had not thought the world much older, though they were not inhibited by the Bible.) What reason was there to think so, apart from the possibility of extinct species? There was only the notion that processes still apparent in the world, like the slow wearing away of the seaside by waves, could be read backward to explain extraordinary changes in the earth's surface. It is true that this idea was glimpsed more than once during this period. Edward Halley, for example, read a paper to the Royal Society on the salination of the seas in which he calculated that much more time was needed to account for the manufacture of the oceans' salts than the usual history of the world allowed. But even he was willing to extend it only so far as to allow the six days of Moses' creation to stand for six thousand years. (The idea seemed bold enough at that.)[18] Lhwyd was puzzled by the sight of great numbers of boulders strewn in the valleys of Wales, though only two or three had ever fallen within the memory of living men. Unless one allowed some catastrophe — a universal deluge, for example — "in the ordinary course of nature we shall be compelled to allow the rest many thousands of years more than the age of the world." But one or two such observations were insufficient by themselves, however suggestive. (The boulders, for example, might have been tumbled down upon an enemy by an army, as in Switzerland against Duke Leopold of Austria, or Alexander the Great.)[19] It took a century or more and a vastly extended knowledge of the natural world before the modern notion of geological time could become acceptable and be reconciled in its own way with the Scriptural story.

In other words, belief in the Bible was not the *cause* of the idea that the life of the world was relatively brief. Scripture merely confirmed what was already expected, though with a helpful authority and precision which made it in the end harder to escape. Natural history was (as Bacon had explained) one branch of history in general, and drew its testimony from the records of men as well as stones. What else were the narratives of the *Pentateuch* than a kind of historical testimony about the making of the world? No one yet conceived that men might not have been around in the first place to witness and to record the early events in the history of nature — nor was there yet any reason to suspect it. The trouble was that no matter how one tried to resolve the problem, the extension of time brought with it at least as many problems as it solved.

[2]

When the nineteenth century came to consider Dr. Woodward, it was ambivalent. For a hundred years or so his ideas had been taken very seriously, even by his critics. It was recognized, as Buffon put it, that Dr. Woodward "was better acquainted with the materials of which the globe is composed than those who preceded him." If his system was incorrect it had nevertheless "dazzled many people."[20] But the geological "revolution" of Charles Lyell and his fellows had to be accomplished in the teeth of a Biblical resistance. To create the modern science they had to cast aside Moses; Genesis and geology were torn asunder, and the evangelical reaction was strong. They could not therefore help but look back upon Dr. Woodward with mixed feelings. They admired him for his single-minded devotion to their subject; they applauded his observations and collections; they even approved some of his particular judgments. But they lamented his explanations, which they believed must have been a result purely and simply of his bibliolatry and which they insisted had retarded his understanding, as in general it had retarded the whole progress of the science. "It is singular to observe how acutely the soundest reasoners argued while confining themselves to matters of observation and how suddenly they wandered from the truth when the narrow theology of the time barred their way to further progress."[21] They could not understand how reasonable men had failed to anticipate their conclusions, unless willfully — forgetting how much had transpired in a hundred years and seeing only their own immediate obstacles. Perhaps if they had noticed how easily they won their way, despite much opposition, and found (besides a new history of the earth) new ways to make the Bible acceptable, they would have tempered their judgments. Unfortunately it is their views, and their prejudices, which have lingered in the literature, despite a few honorable exceptions. Thus one of the very latest students of the subject is astonished at the "tenacity" with which even the ablest of the seventeenth- and eighteenth-century scientists clung to the belief that *Genesis* provided a reliable account of the Earth's early history. Woodward, he allows, was probably the best geologist of his time, but "in our day . . . it seems almost inconceivable that scholars of earlier generations should have implicitly accepted the infallibility of Scripture and based a lifetime's scientific work upon this premise. But so it was."[22]

Well, not exactly. Certainly the charge would have surprised Dr. Woodward. He had been careful, he says again and again, to construct his theories solely upon the basis of observation and experiment, independent of any preconceptions. He professed himself a rigid empiricist and scorned any hypothesis that strayed from the evidence. Truth in geology could only be established by the methods of natural philosophy. Moreover Woodward was not — like Ray, or Burnet, or Nicolson — a cleric. His own view had been set out in the preface to his *Essay* in 1695. He had included there, he says, a few remarks about the fidelity and exactness of the Mosaic narrative, although some might stumble at it, thinking it strange of him to meddle with such matters in a physical discourse. (It was his model Bacon, after all, who had been among the first to try to separate natural philosophy from religion.) But since Moses had written of similar matters, "I was bound to allow him the same Plea that I do other Writers and to consider what he hath delivered." In order to accomplish this, he assures the reader, he "set aside every thing that might byass my Mind, over-awe, or mislead me in the Scrutiny, having regard to him only as a Historian. I freely bring what he hath related to the Test, comparing it with Things as they now stand." In short, Woodward had meant to treat Moses as he would any "common historian," no different say than Herodotus or Livy. It was only coincidental that Moses had passed the test and turned out to be right.

Now it will be said that Woodward labored under an illusion and that in fact, whether deliberate or not, he allowed a belief in the literal meaning of Scripture to color all his philosophical activities. No doubt that is so, for it has become obvious that even the most dispassionate of scientists brings a mind filled with ideas and preconceptions to his work. Yet it must be allowed that in Woodward's time there were a number of respectable possibilities open to believers in Scripture which would encourage them either to reinterpret or to bypass *Genesis* in scientific matters. Both Burnet and Newton, not to mention L.P., for example, agreed that the Mosaic account should not be taken literally about the history of the earth, and the whole experience of Copernican astronomy lent weight to that proposition.[23] Did Dr. Woodward then take Scripture literally because of some theological preconception or did he, as he himself imagined, believe in it because he found it to agree with his science?

It seems to me that there is something to both of these possibilities. What I have been at pains to suggest in these pages is that the resort to Scripture for "scientific" or historical information was in part the result of its appearing to furnish a plausible explanation for natural or human phenomena. Not that its authority rested solely on its plausibility, but that its plausibility reinforced its authority. As long as that appeared to be true, it was bound to influence the course of all speculation on such matters, whether conscious or not. Dr. Woodward chose to treat the Mosaic account deliberately as an hypothesis – in his own words, as something which might "solve the appearances" – and he tried hard to see if it would work, of course entirely predisposed in its favor. And it did work; although how well it is for others, more expert than I in the problems of natural history, to determine. "Mine is the only Hypothesis," he claimed, "that answers Nature, and shows all the Phenomena observable in the Earth in an easy and Geometrical Manner." With this at least one modern student seems to agree.[24] What is clear is that it will not do to dismiss the Doctor simply because his conclusions were wrong in the light of what we now know. It is doubtful if they would have been more correct had he never heard of *Genesis*. Two and a half centuries of observation (and the elimination of many another worthy theory) have intervened between Dr. Woodward and ourselves and revealed a world of information not only in geology but in

chemistry, physics, and biology, all bearing upon his problems but of which
he could have had only the faintest inkling. Nor is it right to detach Dr.
Woodward's explanations from his observations and collections, as many have
done, to dismiss the one and to applaud the other. For clearly they went
hand in hand, his hypotheses guiding and stimulating his empirical researches.
If they were eventually proved wrong, they were in their own time fruitful;
they helped to lead him and others to discoveries about strata and fossils that
were of permanent importance. If they failed to solve all problems and to
anticipate the conclusions of the future, they shared in that the common
destiny of all scientific theories. The fact is that they led a long and useful life
in the eighteenth century.[25] And Dr. Woodward's belief that he had demonstrated the truth of Revelation as history, howsoever it might be disputed in
his own time or ours, was neither unsophisticated nor "unscientific." It was
just wrong.

[3]

Whatever his initial inspiration, then, Dr. Woodward confronted the theological challenge directly. It was almost impossible to do otherwise. The fact is
that the relationship between philosophy and religion, science and Scripture,
was one of the liveliest issues of his time, a backdrop to all contemporary
speculations about either nature or history. Much of the excitement seems to
have been generated by attacks, real or imagined, upon religion by atheists
and free-thinkers, deists and libertines. In reality it was more likely the
reflection of a gradual and subtle shift of ground in which philosophy and
science, like the practical disciplines beneath them (economics, for example,
or politics), were becoming secularized. In short, it was the beginning of the
"enlightenment." "Rationalism," wrote Mark Pattison, "was not an anti-
Christian sect outside the Church making war against religion. It was a habit
of thought ruling all minds. . . . The Churchman differed from the Socinian,
and the anti-Socinian from the Deist . . . but all alike consented to test their
belief by the rational evidence for it."[26] Had there been no atheists they
would have had to be invented. For Christianity had to be defended, and it
was better to throw up bulwarks against an imagined enemy without than to
recognize a growing cancer within.

   "With some trifling exceptions," therefore, "the whole of religious literature was drawn into the endeavor to 'prove the truth' of Christianity."
Pattison had read it all, not only the theological works but the sermons, the
learned treatises and philosophical disquisitions, Addison and Bentley, even
the great Newton, and found them all set upon the same task.[27] It is not hard
to place Dr. Woodward here. In his *magnum opus* he too meant to tackle the

critics of Christianity. For those who sought to undermine Moses, he offered the argument of science. The Pentateuch was authentic because it agreed with the findings of observation and experiment. Like his great predecessor in the Royal Society, Robert Boyle, Woodward had not the slightest doubt that "being addicted to Experimental Philosophy, a Man is rather assisted than Indisposed, to be a good Christian." And no more than Boyle was he ready to face the possibility that the conclusions of the one might in the end be found to differ from the other and thus force an unpleasant choice. "A great Esteem of Experience," Boyle had written, "and a high Veneration for Religion should be compatible in the same person."[28]

It does not appear that Woodward ever doubted this. Here as elsewhere he had settled his convictions early. Perhaps the most representative of the deists in his youth was Charles Blount, who scandalized the London world (not for the first time) with a posthumous publication in 1693 entitled the *Oracles of Reason*. Blount employed the arguments of several others to throw doubt on the authorship and thus on the authority of the *Pentateuch*. He quoted Burnet, for example, who in a recent work, the *Archaeologiae Philosophicae* (1692), "seems to prove that many parts of the Mosaic History of the Creation appear inconsistent with Reason." Whether this must destroy belief in a universal Flood or a natural history of six thousand years was a little hard to say; Blount was circumspect on these matters. In any case, he was quite persuaded that Deism was "a good manuring of a Man's conscience" and that if it was *joined* with Christianity it would produce a "most profitable Crop." Such views (but I have not tried to state them fully) were, when put this way, surprisingly timid, but they called forth a storm of indignation anyway and it was probably as well for Blount that he was dead by then and did not have to face the consequences. (Burnet did not get off so easily.)[29] For Dr. Woodward it was all simple enough. "I doubt not but you have seen Mr. Blount's late Book," he wrote to the Bishop of Norwich soon after its appearance; "it is not licensed but handed about privately and makes a noise amongst some People. For my part I see nothing in it but what we have had long before from Vaninus, Spinoza, etc., and what hath been over and over refuted and exploded."[30]

Still the clamor was growing, and the deists were becoming more provocative with every passing year. Perhaps this was why Dr. Woodward returned to the subject in his *magnum opus* in the passages I have quoted. Nevertheless, there was a stimulus both more provocative and closer to home; for the Doctor's problem was not so much his decision to defend orthodoxy, which was generally appreciated, but his failure (for most of his readers) to settle the issue. He had thereby, however inadvertently, added fuel to the argument of the unbelievers, since the proliferation of new theories could only serve to spread confusion and doubt and to implicate natural philosophy in the decline of religious conviction. What was one to believe when everyone seemed to have a different opinion?

Apparently this was the view expressed in a small book that followed anonymously upon Dr. Woodward's *Essay*. It was entitled *Reflections upon Learning* and it was by Thomas Baker who, however, steadfastly refused to acknowledge its authorship. The work was related to the controversy between the ancients and the moderns, but it sought to make another point. Baker was concerned to show the "insufficiency" of *all* learning, in order, as he put it "to evince the Usefulness and Necessity of Revelation." [31] He was of course using a very old device to shore up the claims of faith as opposed to reason, and in truth the work was not a very impressive performance although its scope was wide and its manner modest and unassuming. It was, nevertheless, popular enough, passing swiftly through many editions, and Boswell was still enjoying it half a century later. [32]

Baker was a non-juror who had been turned out of a Durham living and returned to his Cambridge College, where he lived out his days in scholarly retirement, helping all who called upon him in matters of scholarship and earning universal praise for his honesty and humility. In the *Reflections* he naturally considered the claims of philosophy to learning, especially the various hypotheses and theories of the earth that had been recently advanced. He thought the most plausible to be Burnet's but noticed several weaknesses, particularly his failure to provide for antediluvian waters, springs, and rivers. He admired Whiston's "vast reach and depth of Contrivance," as also his explanation of the cause of the deluge, but found his "Paradisiacal days" too long, and the quantity of water in the air insufficient for his needs. Finally, he thought Dr. Woodward's theory so natural "as to induce a good man to wish it true." But like everyone else, he believed it needed proof and doubted that Woodward would find it. "I am afraid the Dr. will never be able to prove it for want (amongst other things) of such a universal *Menstruum* as shall dissolve all things except his Shells." In the end he saw nothing to distinguish the different hypotheses and theories and despaired of ever seeing a good one, "so many Observations and Experiments [being] required." [33]

As I have said, it was not a very impressive performance; had Baker known anything about natural history (had he any of the weapons of Lhwyd or Ray), he could have done immensely better. But, typically, it irritated Woodward anyway. Apparently one Mr. Gualter called the offensive passage to Woodward's attention and offered to reply. The doctor encouraged him and this began a correspondence which Woodward continued, apparently with thoughts of eventual publication. When it was over, he summarized it in a brief note. "I carry'd on the controversy," he explained, "to see if I could drive Mr. Baker out of his Sophistry and Evasions; in which with much Difficulty, I succeeded." It does not appear that Baker understood the results quite this way; but then Dr. Woodward's opponents somehow never seemed to understand defeat.[34]

Gualter's letter, written under Woodward's supervision, made short work of Baker's "slurring" remarks. He pointed out that Baker, like so many others, had misunderstood the theory. Woodward had not argued that water was the agent of the dissolution; therefore there was no need, as Harris had already shown, for a great *menstruum.* Nor had he said that all was dissolved except the shells, preferring (as we have seen) to distinguish instead between organic and inorganic bodies. What he had done from the direct evidence of his observations was to prove that a dissolution had occurred; he hoped in time to show both how and why. Meanwhile, it was paradoxical for Baker to say that Woodward's theory was "natural" and yet "not true." If it was conformable to nature, then it *must* be true. Here was a work which pretended to be written to show the necessity of Revelation, yet slighting a treatise which was everywhere "acknowledged to give incontestible Evidence from Nature of the Certainty and Truth of so many considerable and important Passages." It was all extremely annoying.[35]

Baker's reply was characteristically cautious. He had meant no injury. It still looked to him as though Woodward's theory required a *menstruum,* as it certainly needed an explanation for why the shells were not dissolved. "I do verily believe your design in that book was pious," he conceded, "yet I am afraid your account of the dissolution of the Earth will upon Examination be found not so reconcilable to the Text." None of the theories of philosophy had helped the cause of religion, because if they failed in any single thing they failed altogether. He promised however to use Woodward's own words to describe his theory in any future edition. Gualter was baffled. "Strange that a Man that has wrote against the Deists, battled and defeated all their Arguments and clearly made good the Cause of Christianity in opposition to them should for this be stiled a Deist!" How could men on the same side, like Baker and Woodward, be found to differ so fundamentally? Just where was Woodward's alleged discrepancy from Holy Writ? Meanwhile Gualter could only promise an answer to Baker from "those Papers" still in manuscript, "which have satisfy'd all that have perus'd them [that] he hath pitch'd upon the true Course."[36]

In the face of this reticence the argument was not likely to advance, and a further exchange between Woodward and Baker led only to a promise on Baker's part to leave out the offensive passage in future editions. He was, after all, only "a very inconsiderable Author scarce worth your notice," and he wondered "to see you so much concerned at so feeble an attack." There *was* clearly something disproportionate in Woodward's response to Baker's brief remarks, which nevertheless continued to grow hotter through several further exchanges. The issue was not clarified by Baker's unwavering insistence that he had not written the original but was only a friend of the author's or by Woodward's stubborn attempt to pin it on him anyway. In one way, however, the argument did expand. Woodward insisted that Baker was no better than a skeptic whose arguments endangered Scripture more than they had served it. He pointed out that Baker had said that all historical knowledge was imperfect and that this could only undermine Holy Writ. But Baker refused to concede. It was unquestionable, he thought, that all profane history was subject to error, that even the best historians were liable to mistakes or moved by prejudice or interest. This did not mean that he was ready to reject it altogether; there was "certainly enough in these matters, tho much short of the assurance we have from sacred Story." He thought he had as much confidence in Caesar's commentaries or the foundations of Rome as did Dr. Woodward.

But Baker had missed the point. Woodward replied at once in a long and carefully composed letter.[37] His critic had tried to show that all history except Scripture was subject to error. From that fact he had pretended to prove the necessity of Revelation. "Now my Arguement was, if History be insufficient and uncertain, we have no way left to prove Revelation is really Revelation. The best we could do," Woodward insisted, "would be to have recourse either to Rome or Enthusiasm and take our information from Oral Tradition or the light within."

Take away the Certainty of History and natural Philosophy, which likewise he attempts and you destroy all means of evincing his Existence. In particular, the Study and contemplation of Nature carryes us directly to the Knowledge not only of his Being but of his Power and his Providence. And therefore the Prophets of old and the Apostles all along refer to it. In brief without it we could know nothing of him, unless he discover'd himself to us by Miracle.

Baker was taken aback but tried to reply. On one point he wavered a little, "the mistakes he had observ'd in history are all of them innocent and such as are proper for an Essay without entrenching upon Religion or good manners. . . ." He had deliberately forborne reflecting upon ecclesiastical history, "from which the Canon of Scripture may be prov'd by as universal a consent as can be had for any Truth." (The French and Dutch, he added, had other rational arguments, like the evidence of the books themselves.) As for natural philosophy, he was more forthright. There were a whole series of ways in which Dr. Woodward's theory seemed to him to depart from the Scriptural narrative. If there was a general dissolution at the time of the Flood, how were the ark and all its materials preserved? Did not the *Pentateuch* refer to the same places (like the Euphrates River) both before and after the Flood? And so on. Above all, Dr. Woodward had *added* some things that were nowhere suggested by Moses. Take it together, and Dr. Woodward's account plainly contradicted the Biblical story, as though "Revelation taught one thing and Nature says another." It was a predicament that neither the Doctor nor his fellows in the Royal Society were likely to admit. But for the fideist (for Baker was not really a skeptic) there was no problem. If the choice came to nature or to Scripture, "I must think it the safest way to adhere to Revelation tho I could not account for all the Phenomena in nature which I should rather believe to have happen'd at another time or in another manner." "And I will confess to you," Baker concluded, "that nothing shocks my Faith more, than the many new opinions in Philosophy which are vented with as much assurance as if they were Gospell and yet vary every day."[38]

Woodward could hardly reply without reiterating everything that he had already said. *He* saw no contradiction between his findings and the Bible; he saw only confirmation. (It did not help his argument, however, that he kept so much of his explanation still to himself.) He was surprised that Baker should try to exempt ecclesiastical history from his criticism. Was it really so free from all doubt and difficulty? "Have the Composers of it in all Ages been so perfectly enlightened, so impartial, so unprejudiced, in a word so infallible, that all things are out of Dispute? Are either Clergy or Laity when they write Ecclesiastical more infallible than when they write Civil History? Tis plain the case is exactly the same." Again, however, Baker wavered; he was less skeptical than appeared. On the whole it appeared to him after all that history *was* reliable; he had meant only to show that "Historians have been ignorant in the most Antient History and often partial in the present." But Woodward's arguments were frightening; they could as easily be turned against the Jewish Canon as against secular history. "If History be insufficient and not to be relyed upon," Woodward had written, "we have no way left either to make out the Canon of the Books of Scripture, or the Divine Origin of them." Baker hoped that *he* at least had not inferred any such thing. After all, there was a fundamental difference between the sacred and the secular; the former was generally written by men of "the greatest piety and uprightness," with an integrity beyond question. (The divine and the layman could hardly be expected to see eye to eye here.) As for the Bible itself, its significance far transcended the stories of Herodotus and Livy, as did also its authority. "The Books themselves were in the Hands of all, they were read publicly and dispers'd over the face of the whole world. They had the force of Laws, and were constantly appeal'd to in all controversies of religion." Above all, "the Providence of God was concern'd in their preservation and that men should not be deceiv'd after an impartial search. What is there like this in prophane history?"[39]

What the argument needed, of course, was a genuine atheist. How tempting it would have been to take the destructive arguments of both sides and put them together! Indeed it was not to be long, although for the while almost everyone hesitated. It was turning out more difficult than had been expected to "prove the truth" of Christianity. Either one could look outside for confirmation, to science or to history, or one was forced to accept its truths as self-evident. If Dr. Woodward was unable to persuade Baker by his philosophical arguments, he was clearly doomed to failure even more certainly with those who really doubted. And if Baker was unable to convince Dr. Woodward that the argument could be won without an appeal to some external proof, then he was even further removed than his rival from the spirit of the age. It was indeed disturbing when men committed to a common object must find themselves so sharply at odds. Perhaps that is why they persisted so long in their frustrating correspondence.

Inevitably, the debate ended badly on an unpleasant note. Woodward was exasperated by his opponent's unwillingness to concede, after he had seemed so often to waver. His "evading, uncandid, indirect way of treating is what one of my Temper and Principles cannot but have a great aversion to." Baker's shifts and artifices were shocking. Woodward had suggested making the affair public; Baker had declined. Baker had offered to make amends, again and again, but never enough to satisfy the incorrigible doctor. At last he was bludgeoned into submission and offered to recant his judgments of Woodward's book as well as to commend its pious intentions. But even this was not enough; Gualter now drafted a letter for Baker to sign (to be inserted in the *Monthly Account of Books*), a submission so complete that even the gentle Baker was at last aroused. It was "neither fit for a Scholar nor a gentleman," he protested to Woodward; besides, Baker would never say publicly what he could not believe, that the Doctor's dissolution was more than a theory. Woodward insisted that he had not known about Gualter's letter; but he reiterated his intention to publish the whole correspondence, and in a parting shot reminded Baker that despite all his pretenses *he* knew who was the author of the *Reflections*.[40] It was a great many years before the two men recognized each other again.

IV.

The Early History of Mankind

[1]

At the beginning of the year 1677, Sir Robert Southwell wrote to his friend William Petty to report that he had seen the great jurist Matthew Hale just five days before his death. The judge, it appears, had written a book, part of which was now in the press, entitled *The Origination of Mankind.* In it he "confirms the Creation layd down by Moses, and soe Inpugns the business of the Atheist."[1] It was, as we have seen, a familiar employment. I do not know that Dr. Woodward ever read this book, although there is an excellent chance that he did; he must certainly have known about it from Southwell, who once described it as his "best Companion" and was responsible for a translation into German.[2]

It was a celebrated book in its time. Bishop Burnet tells us that it took seven years to write, but "only in the Evenings of the Lords Day when he was in Town, and not much oftener when he was in the Country." It was not easy for a man so much in the world to find time for so ambitious an undertaking, but Hale took his treatise (like everything else) with the utmost seriousness. The result was "one of the perfectest pieces both of Learning and Reasoning that has been Writ on that Subject"— though, to be sure, only a part of it ever appeared. According to Burnet, when it was done Hale passed it about in manuscript, concealing its authorship. One day Bishops Wilkins and Tillotson received a copy, read it through, and at once guessed who wrote it. When they saw Hale they told him that "there was nothing could be better said on these Arguments, if he could bring it into less Compass. But if he had not leisure for that . . . it was better to have it come out, though a little too large, than the World should be deprived of the good which it must needs do."[3]

The point of Hale's treatise was to show that the Mosaic description of the history of the world, "abstractively considered without relation to the Divine Inspiration of the Writer," was consistent with reason and preferable to anything else. It was a notion not unlike Dr. Woodward's afterward, as was its confidence announced at the outset that "the knowledge of Nature is useful to Mankind, to bring him to and confirm his knowledge of the Glorious God."[4] Hale was a man of exemplary piety and an experimental scientist, rather like Robert Boyle, though with much less time for nature. He had written an *Essay touching the Gravitation or Non-Gravitation of Fluid Bodies* in 1673 and another on an experiment by Torricelli in the following year, but he did not impress his contemporaries very much with either. The latter was described in the *Philosophical Transactions* as "a strange and futile attempt of one of the philosophers of the old cast to confirm Dame Nature's abhorrence of a vaccuum and to arrange the new doctrines of Mr. Boyle and others concerning the weight and spring of the air, the pressure of fluids on fluids, etc."[5] No matter. Hale was at least genuinely interested in the new science, and his ideas about the relations between experimental philosophy and religion, Nature and Moses, must certainly have pleased Boyle and the fellows of the Royal Society, whatever reservations they may have had about his particular conclusions. Moreover, his legal training was no handicap; Hale was a lawyer enthusiastically arguing a brief, defending the one true God against a dangerous but faulty prosecution.

The judge required testimony to establish the facts, and it was testimony, authoritative and indisputable, that Hale found in the Mosaic account. It was certainly superior to anything else that pretended to record the early history of mankind — more reliable than any of those doubtful pagan authors who had recently come under fire: Sanchoniathon, Berosus, or Manetho. It had been written down according to his chief modern authority, Samuel Bochart, fully 540 years before Homer and 350 years before the Trojan Wars. Neither the Chinese nor the Egyptians, whose claims had lately been advanced in Europe, could persuade him of a greater antiquity. It was the Mosaic narrative, transcribed exactly 1,656 years after Creation, that was for Hale the earliest record to be obtained of human history.

But how credible was its report of events? Even Moses had written long after the Flood, for example, many centuries after Noah and his sons had accomplished the re-peopling of the world. Hale was satisfied that Moses was a learned man, governor of his people, etc., and consequently must have had the opportunity "to furnish himself with old Monuments and Evidences of Antiquity."[6] If it was true that he had lived 800 years after the Flood, so Livy had written nearly the same number of years after the founding of Rome, and he was generally believed. Besides, the longevity of men in ancient times shrank the time; it was really not much longer than three generations. Now Abraham, Isaac, and the rest would surely have kept their traditions pure and fresh until they could be written down. And probably many things had survived until Moses' time to prove the event, perhaps the Ark itself. Thus even if one left aside completely the idea of revelation (for Hale was addressing himself to atheists), the argument seemed to him conclusive. "I say we have not greater Evidence that there was such a Man as Alfred, Edward the Confessor, or William the Conqueror."

Hale was prepared even to consider the new facts from America. "The late Discovery of the vast Continent of America," he noticed, "which appears to be as populous with Men and as well stored with cattel almost as . . . Europe, Asia and Africa, hath occasioned some difficulty and dispute."[7] How was the whole population of mankind to be traced back to its two common parents? How could one explain the different languages and customs of the Indians, the variety of new species of animals and plants? The solution of the *Praeadamite,* which had recently appeared to shock the world, that there were men before Adam, would certainly not do, for it gave more credit to the Egyptian and Babylonian fables than to Moses.[8] Hale seems not to have noticed how his argument had slipped suddenly into circularity. "Though they solve the difficulties," he admitted, "yet they cross the tenor of Mosaical History." Yet the circular argument was unnecessary, for there was independent evidence enough for such matters as the peopling of America by the descendents of Noah. Here Hale could fall back on a copious literature from Grotius to John Webb. Moreover he was willing to concede that the variation of the species was a fact, the result of differences in soils and climates, breeding practices, and so on. A land passage, he suggested, could easily have connected the continents to provide the bridge that was required to lead all life back to Noah's ark. As for the physical fact of the Flood, that too could easily be established, as Dr. Woodward was later to see, from that other kind of authoritative testimony, the evidence of the fossil shells.[9] Hale knew the controversy about their origins, but he took an independent stand based upon his own observations, which he recorded here in some detail. He thought that many of them were "but the Relicks of Fish-Shells left by the Sea, and there in length of time actually Petrified." Others, however, he was willing to allow *de novo,* anticipating the conclusions of Edward Lhwyd, "from certain seminal Ferments brought Thither, which are as it were the Seminium of their production." It was the bulk of the petrifactions, however, which were deposited by the Flood, that proved the accuracy of the Mosaic story.

Hale had much more to say on these subjects to confirm the orthodox and vanquish the skeptics, too much even for his most enthusiastic readers. He was neither the first nor (as we have seen) the last to employ these kinds of argument, although their combination here speaks well for both his virtuosity and his forensic skill. In the next generation the defense of Moses, as history as well as philosophy, grew more insistent as it grew more necessary to answer the insinuations of those like Blount or Burnet who had cast suspicion on the literal truth of his narrative. Dr. Woodward of course played his part, as concerned to secure Moses for human history as he was for natural history. But it was not an easy thing to do. To defend Moses with consistent arguments that avoided all circularity, to treat him (as Woodward had urged upon Thomas Baker) as an ordinary historian whose veracity could not be assumed but had to be demonstrated by a conviction in something else – all this was fraught with danger. It reversed the assumptions of centuries and it left Moses at risk. For what was one to do if the evidence of reason, the testimony of nature or history, departed from the Mosaic narrative?

[2]

It was a charge soon laid against Dr. Woodward, as it was pressed even more vigorously against Burnet and Whiston. Baker had hinted at it, but the Doctor's most persistent critic in this matter was in reality a great admirer, entirely sympathetic with his efforts and willing on the whole to accept his conclusions. He was John Edwards of Cambridge, by no means an easy man to satisfy, like Hale a puritan in outlook, deeply reverent of the Scriptures in their literal sense. He was indefatigable in their service, turning out treatise upon treatise, at least forty books in twenty-five years, to attack the enemy and vindicate the truth. To one admirer, he was the "Paul, the Augustine, the Bradwardine, the Calvin of his age," and when he died in 1716 Thomas Baker lamented "that we seem to be struck dumb with so great a loss."[10] He even gave John Locke fits, assaulting his *Reasonableness of Christianity*, that ambiguous fountainhead of deism, with a persistence that gave the great man no surcease. (Dr. Woodward was among the many who thought his criticisms worth reading.) It was not enough for Edwards to defend the general reliability of the Mosaic narrative; even a slight deviation was enough to provoke his displeasure. To be fair, this was also the usual view of the Church of England, as at least two Bishops made clear in 1694-95.[11] To Edwards the threat of atheism was real indeed and he found a dozen causes to explain it, not least the "Thrusting of Opinions and Theories in the world in defence of the plain Letter and Historical Part of the Bible."[12] No wonder that even Copernicus and Newton did not impress him.

Yet fundamentalist though Edwards was, he did not put himself outside reason or beyond philosophy. He was too confident on the whole that Scripture was defensible to think that Nature could be used to refute it. Like the more advanced thinkers of his time, he too believed implicitly in a "happy Mixture and Conjunction of Natural Philosophy and Religion."[13] Unfortunately, he discovered that there were some who were willing to argue

against it, who pretended that "the frame of the Primitive Earth is represented opposite to what Moses tells us," or who labeled Genesis in effect "a
Romantick Story." He had read in Whiston that the Flood could be attributed to an accidental disruption of the earth; in Burnet, that the Mosaic story
was a fiction and imaginary and that Moses was no more than an impostor.
The "mechanical philosophy" was useful within its proper limits as a description of natural bodies, but it could easily be pressed too far. It was in any
case of no avail whatever in recapturing the earliest history of the world.

There at least Moses reigned supreme and without challenge. "Scriptures
are the greatest Monuments of Antiquity that are extant in the whole World,"
Edwards proclaimed in a treatise of 1695 (the year of Dr. Woodward's
Essay), a treatise on the "Authority, Stile, and Perfection, of the Books of
the Old and New Testament."[14] There only could one find the first human
institutions and inventions. As Hale had argued, the Scriptures were indisputably the oldest and therefore the most authoritative of sources. What was the
alternative? "In Pagan Writers we have some wild Guesses at the Origins of
things and the first Inventors of Arts." But surely Orpheus and the Greek
gods and heroes were no more persuasive than the figures of Biblical history.
Edwards subscribed to the ancient view (first expounded by the Fathers) that
in their confused way the classical writers had only substituted classical
names for the true and original Hebrew inventors of things, Vulcan for
example for Tubal Cain. Edwards had little of Hale's forensic skill (though,
like most of the theologians of his day, he had a large dose of Aristotelian
logic) and less of his lucidity, but his arguments were not really very
different, at least as long-winded, occasionally circular, and decidely more
tedious. His conclusion, however, was simple. "A prying Antiquary may find
more Work and much more to his Advantage in the Writings of the Old
Testament, especially of the Five Books of Moses, than in all the Mouldy
Manuscripts and Records in the whole World besides."

When Edwards read Dr. Woodward's *Essay* he was delighted. Here at last
was no frivolous theorizer playing games with the story of Creation, but a
"Curious Observer and One who (as becometh so Learned a Head) joynes
Religion with his Philosophical Researches."[15] When Edwards determined to
compose his own argument from design, *A Demonstration of the Existence
and Providence of God from the Contemplation of the Visible Structure of
the Greater and Lesser World* (1696), he turned, like Cotton Mather afterward, to Woodward's work for support. He found the Doctor's suggestions
about springs and rivers and volcanoes all valuable (although he was more
dubious about earthquakes), and his strictures on Burnet worth repeating.[16]
As for the theory of the dissolution, which he recounted fully, he thought it
plausible, since it had been built so scrupulously on observation and matter of

fact — yet, upon reflection, quite unnecessary. After all, why had God needed
to employ so complicated a means for so simple an end? Edwards noticed
other difficulties too, like the fact that some had questioned Dr. Woodward's
use of gravity; the weightiest fossils were not always found deepest in the
earth. Even so, he was generally impressed by the argument. "I confess I
rather say this to provoke this Learned Author to make good his Hypothesis
in all Particulars in this matter, than to contradict what he hath said about
it." There were some other things to notice in Woodward's fossil arguments,
but in the main Edwards was all admiration. The Doctor had proved that the
changes in the earth were "a great and singular Work and Argument of
Providence." He had helped to show, as Edwards put it, that "Reason,
Scripture, and the Sentiments of the Wise agree . . . that what we behold in
the World is a Proof of a Deity and Providence."[17]

One would have expected Dr. Woodward to be grateful — even more so
perhaps when Edwards dedicated to him his next tract, an attack upon
Whiston's *New Theory of the Earth* (1697). "You are a Theorist as well as
he," he wrote to the Doctor, "and the Earth is your common Subject. But to
the Theoretick part you have added Experience and Observation whereby
you have cultivated this theme to a Prodigy."[18] Whiston had trifled with the
Mosaic history, whereas Woodward had upheld it. (He only hoped that the
Doctor would develop his criticism further.) But we have seen too much of
our hero to think that he could accept a compliment when it was qualified by
so many reservations. Worse still, Edwards compounded the difficulty by
writing a letter to Dr. Woodward adding a further criticism.[19] It seems that
the Doctor had departed from Moses after all, and on a crucial point. In
Genesis, according to Edwards, Moses showed that God had placed a curse of
sterility upon the earth immediately after the Fall, which brought forth with
it an abundance of thorns and thistles. (Said God unto Adam, "Cursed is the
ground for thy sake; in sorrow shalt thou eat of it all the days of thy life.
Thorns also and thistles shall it bring forth to thee; and thou shalt eat the
herb of the field." Genesis, III, 17-18.) Upon reading Dr. Woodward's *Essay*,
however, Edwards was surprised to discover there that the curse had not
actually taken place until the time of the Flood and that the antediluvian
earth had been more fertile before the Deluge than at any time afterward.
Here was an outright contradiction of Scripture, "which yet at other times
you shew yourself a very religious Asserter of." For Edwards it was an
insurmountable problem in Dr. Woodward's system; yet he thought it entirely
avoidable, since it did not seem to him essential to Dr. Woodward's theory.
He still meant, he assured his friend, to do whatever he could to "guard and
secure" the Doctor's hypothesis.

Woodward was "struck in admiration," "surprized," "amazed," and "much of a losse." [20] To suggest publicly that his work was "tinctured with disrespect to Moses" was indeed an empty paradox. "I think I may upon such an Occasion modestly say I have given the Defeat to the Opposers of those Writings by shewing the universal Attestation Nature gives to the truth and certainty of them, [such] that I have good reason to believe they can never rally again." (The Doctor's modesty was even more alarming than his braggadocio!) Nevertheless, he determined to set down his thoughts one more time, and so composed another long letter to Edwards dated May 23, 1699. [21] "I alwaies endeavor," he began, "to deliver my thoughts in the simplest and plainest manner I can." Everyone who had so far attended to his work, excepting only Edwards, agreed that his views conformed exactly to Moses. Yet despite two exchanges already, Edwards had failed to offer anything concrete from either Nature or Genesis to refute his opinions. Once again then Dr. Woodward repeated his interpretation of the Mosaic account. The curse and punishment, he wrote, consisted really of three different things: the production of thorns and thistles, which happened at once; the death of Adam, which transpired only 800 to 900 years later; and finally a "retrenchment" of the fruitfulness of the earth, which had to wait for the Flood some 1600 years afterward. (God of his goodness had suspended the execution of his curse; but finding that nothing would reclaim his fallen subjects, he had finally destroyed the earth, "giving it a Make and Constitution more agreeable to the laps'd State of Mankind and retrenching and burying a great deal of that prolifique Matter that rendered it so exhuberantly Fruitfull.") Thus the curse had begun to take effect before the deluge, with the death of Adam, but sterility awaited the Flood; the thorns and thistles were, somewhat paradoxically, actually an addition to the fertility of the earth. (They partly fulfilled the curse, however, by making labor necessary "and some sort of agriculture.") Not even Edwards could deny that God had suspended Adam's death for 900 years; why was it so hard to accept the rest? The fact was, of course, that Moses never had discussed whether the primitive earth was more or less fertile than ours. It was Dr. Woodward only, "from Reflections on the vast Number of the Remains of the Productions of the Earth that are yet preserved," who had shown the antediluvian world to be more fertile. " 'Tis evident therefore [that] Nature and the Remains of the Earth give their vote in Favour of me." The "monuments" — i.e., the fossils — were sufficient proof of this, though for the curse as a *cause* of all this Dr. Woodward had to accept the testimony of Moses. Indeed, he had not really meant to discuss that matter, since it belonged more to the province of theology than to natural science, but he had gone a little out of his way because he believed that he could offer "a seasonable interpretation of a Passage that in reality was never before understood." By so doing he could

show also "how conformable his Account is to that which we have from
observation and matter of fact; and how exactly Nature and Holy Writ agree
and mutually illustrate each other."

Despite Dr. Woodward's vehemence, Edwards (now Dr. Edwards) was not
so easily answered. Clasping his Bible to him once again, he insisted upon
reiterating his own reading of Genesis. There was one point especially that he
thought worth developing. He was quite certain that the plow had been
invented and employed in the antediluvian world, "for those became necessary for the subduing and managing of the earth which was now deprived of
its pristine fertility and easiness of production." [22] If this seemed too inferen-
tial a deduction, there was also the testimony of the pagan authors Virgil and
Ovid, writing about the Golden Age, to support it. Together, Moses and the
poets proved beyond question that the plow had been invented soon after the
Fall. On the other hand, Edwards concluded, "either the Arguments founded
on Nature and Fact are consistent with the Mosaick History or they are not.
If they are, I can alledg nothing against them; if they are not, I am sure you
think they are not paper to be made use of." Edwards at least, if few others,
was willing to allow the possibility that Nature and Scripture might disagree,
and did not doubt for an instant how he would choose. For the moment, he
was ready enough to admit that "it would be endless to solve the phenomena
of Vegetables and the Relicks of living creatures in any other way than you
have done it."

Dr. Woodward made a final effort, a letter (almost a dissertation) of about
thirty closely reasoned pages. [23] He kept a copy for himself, fairly written out
and intended ·clearly for the possibility of publication. Some of it was
reiteration, but most of it was devoted to Edwards' challenge about the plow.
How he found the time that spring amidst the press of other business to
compose a veritable history of that invention, in which he scoured and
employed all the available authorities in several languages, is hard to say. Why
he should think it worthwhile is perhaps even harder. The fact is that
Edwards had touched a tender spot. He might think that Woodward's theory
could survive irrespective of the date of the plow, but the Doctor seems to
have believed otherwise. He could not afford, like his critic, to admit a
difference between what Moses wrote and what the evidence of nature told
him; he must certainly have quailed at the thought of lightly tossing out the
papers on which he scratched out his hard-won conclusions about natural
history. Anyway, he thought he had got Edwards now, since his own classical
learning and antiquarian skill far excelled his critic's. Woodward made short
work of Virgil and Ovid; there were many difficulties of interpretation in
them, as well as a host of better authorities, Greek and Latin (all cited in the
original), that could be brought to bear on the problem. Edwards would have

to concede here anyway; and he would have to agree that Moses simply didn't say enough about the antediluvian situation to avoid relying upon the evidences of nature.

Dr. Woodward's method for recovering the history of the plow was ingenious. He tried, as systematically as possible, to compare the testimony of ancient authors on the early life of the classical world with modern descriptions of primitive communities like the American Indian, in order to develop a general picture of the way in which all simple societies had once lived without the plow. Thus he insisted that neither the ancient Britons nor the modern Indians, neither the barbarian Germans nor the natives of Guinea, were ever acquainted with that instrument. By closely examining the ancient authors, he attempted to show that each of the ancient peoples, Greeks, Romans, and Egyptians, had assigned the invention of the plow to a specific time after the Golden Age – or the antediluvian world. Although Dr. Woodward credited the Egyptians with its first application, he dwelt with particular attention on the Greeks, who attributed it to the Attican, Triptolemus. All his miscellaneous sources, but especially the Parian chronicle among the Arundel marbles, "that noble Monument of Antiquity," agreed to place the invention about the time of Erectheus or Celeus. By reviewing these authors one could recover "a good Account of the State of Mankind before the invention" and see "with what joy and satisfaction they received it." Nor was it "History and Monuments" alone that proved his point, for there was tradition too, showing "the very Field where Triptolemus first ploughed," not to mention temples, altars, and sculptures which proclaimed the fact. Upon reflection it even seemed to Dr. Woodward that, far from being shrouded in obscurity, there were few historical facts which were so well authenticated as this one. "Here are Inscriptions and Monuments Authors of the fairest Credit and most distinguished Sense . . . and a long series of Traditions in several Nations," all agreed to bear out the Doctor's notions.

Edwards had said what he would do if put to the choice. He did not hesitate on reading Dr. Woodward's letter.[24] "Sir, you must not think that some imperfect and confused relations out of a few Gentile Authors are of sufficient weight to suppress the testimony of Moses. . . . One such Text of Scripture is more prevalent with me then 500 allegations out of Pausanius, Diodorus of Sicily, etc., or from Marbles and Inscriptions. . . . So shall the pains you have taken in mustering up Authors and Monumental Records affect not the main Cause." Edwards could concede all that Woodward had done and yet still deny his conclusions. It did not matter what one did if there lacked a "due deference to the Sacred History." He did not understand that Dr. Woodward had done all that he could to demonstrate the truth of the Mosaic history, all that was possible to be done, if one were to take the

criticisms of the "atheists" seriously. One had to start somewhere outside of
Scripture in order to demonstrate its "reasonableness." For Dr. Woodward,
nature and the ancient monuments, the new science and the new history,
were the obvious sources for all the knowledge that could be employed to
corroborate faith and rout the skeptics. Apparently he kept at the subject in
his *magnum opus* and, despite his own misgivings, Holloway printed several
long extracts in which the argument with Edwards finally saw the light. [25]
But his humble claim to have routed the skeptics must have seemed more
fanciful with every passing year.

[3]

Dr. Woodward did not convince John Edwards any more than Thomas Baker,
although he seems to have overwhelmed him also into silence. To the modern
reader it must seem again that the Doctor had suborned his historical insight
to his literal belief in the Bible. But the case is really no different from
geology. What could one do, after all, about the early history of mankind,
other than to employ the "monuments" then available, the Bible included,
and compare them for their testimony? Nor was Dr. Woodward wrong to turn
to the modern primitives to try to recreate the condition of the early peoples,
a method that was destined to bear much fruit when it was combined in the
next century with an extended time-scale and a modern notion of evolution.
Of course the documents, traditions, and archaeological evidence which he
tried to put together were none of them adequate; Dr. Woodward's conclusions were always premature and (therefore) usually wrong. It is true that
suspicion had already been cast upon the Mosaic testimony, but it was a
criticism hardly developed enough either to discard the Bible altogether or to
use it very constructively to reconstitute the ancient Hebrew origins. [26]
Neither the "higher criticism" of the text nor Biblical archaeology had yet
been invented, though both had begun to stir. Perhaps the Doctor could have
extended himself a little here, but for once he lacked even the linguistic skill
(Hebrew) that was necessary. Meanwhile there was little enough to turn to
outside the Biblical account. So although the Doctor was caught up in the
prevailing bibliolatry of his time, his views were not unreasonable and his
method was essentially sound. To recover the early history of the world, he
thought that it was necessary to combine the evidence beneath the ground –
i.e., the monuments and artifacts of antiquity – with the testimony of
written history, including the Mosaic narrative. It was hard to imagine that
there might be no written testimony at all, though just when the origin of
letters had occurred was then a hotly contested matter, and no one believed
that Moses himself had witnessed the Deluge. No doubt Dr. Woodward

intended "to prove the truth of Scripture," but he insisted, sincerely, that it be treated like every other document. In the end, when all this was done (the archaeology, textual criticism, comparative history, etc.) by a paradox entirely unanticipated, the historical authority of Scripture, at least as testimony for the earliest history of mankind, was undone, and that choice which Dr. Woodward and his contemporaries had been so unwilling to make — the choice between the Bible *or* history — was forced upon the world. (It turned out to be more difficult even than the choice between *Genesis* and geology.) But who can doubt, after his exchange with Edwards, which way the Doctor would have gone, when, long after his death, the evidence accumulated and the balance tipped?

For his time, then, Dr. Woodward was a master of this subject too. Indeed he spent a lifetime at it, dividing his attention almost equally between the history of early man and the history of nature. We have noticed already that he wrote about the first peopling of the earth and about the first settling of America in manuscript works now lost but which impressed his contemporaries. (Cotton Mather was sure that Dr. Woodward had finally settled the vexed question of the "Jews in America.")[27] It is not suprising that he took an interest also in the most ancient archaeological monuments in Britain, the mysterious remains at Stonehenge and Avebury. Here close to home were hints of great promise both about the early history of man and of Britain, if only they could be understood. What were those great rocks, so evidently human in construction but without any literary record and with hardly a clue of any other kind to illuminate them? It was an intriguing puzzle both to the antiquary and to the geologist, a test of his "modern" skills. As early as 1693 both Woodward and Edward Lhwyd were deeply engaged in it.

Several theories were in contention in Woodward's time. A controversy had been launched early in the century by the great architect Inigo Jones, who had proclaimed them Roman and sketched them in such a way as to indicate that the stones must once have been a classical temple, built after the Tuscan order. When the work was published posthumously in 1655 by John Webb, it was at once attacked by Dr. Walter Charleton who preferred rather to ascribe the monument to the Danes, using analogies drawn from Scandinavian monuments. Webb replied, and the battle was joined. The question, thus opened, loosed other possibilities. "Mr. Sammes in his *Britannia* will have this structure to have been Phoenician; Mr. Jones and Mr. Webb believe it to be Roman; Mr. Aubrey thinks it British, and Dr. Charleton derives it from the Danes."[28] Only the Saxons were excluded, but not for long. No wonder then that John Ray found the many books on the subject "serve rather to unsettle than establish me in any opinion concerning it."[29]

Not so Dr. Woodward! He had been following the controversy and he had seen the monuments himself, or at least analogous remains, like the Rollright stones, in other parts of England. He fancied himself already (in 1693) an expert on Roman Antiquities in Britain; he knew the literature and he possessed examples of Roman coins, bricks, titles, and pavements, as well as jewelry and implements. He was ready at once to offer advice to Lhwyd about his contribution to the *Britannia,* especially since he feared the Welshman was likely to magnify the British achievement at the expense of the Romans.[30] The Celts, he assured Lhwyd, were only a very simple people until the Romans had conquered, a rude people with little civilization. Thus none of the early antiquities that survived, including Stonehenge, were likely to be British, notwithstanding the opinions of Aubrey.

On the other hand Stonehenge, at least, was not likely to be Roman either. In a second letter to Lhwyd, Woodward responded specifically to the Welshman's question. First, however, he enlarged on his objections to the Celts, drawing freely on ancient Latin and Greek literature to confirm his view of the simplicity of the British and the absence of all building among them, including temples. (It seems that they preferred groves of trees.) Lhwyd had defended his friend Aubrey; Woodward was sure he was mistaken. " 'Twere to be wisht that writers on these subjects would not rely so much on fancy and conjecture, but consult the Ancients." Taken together, these were unanimously opposed to the possibility. As for the Roman claim, that was more hopeless still. "There's no analogy, no resemblance in Nature." He could not think of a single monument in all Rome or Italy, France or Britain, to match it. "Here is nothing of the Elegance, Symmetry, architecture, or contrivance of a Roman Work: 'tis wholly Barbarous." Charleton was right, it must be Danish, for only in Scandinavia was there anything like it. Even here, however, Woodward was ready to correct his predecessor over the details.

It was, once again, an impressive performance for so young a man. Woodward's letters to Lhwyd show that he was already able to manipulate not only the modern antiquarian works and classical letters but the monuments themselves, with a ready and sophisticated hand. No matter that he was wrong on the main point; so (once again) was everyone else, though to be sure Aubrey and Lhwyd were nearer the mark. The significance of Stonehenge is still an elusive matter.[31] Certainly it cannot be understood without a modern conception of prehistory – i.e., of a long period preceding the classical world, with its own cultures, artifacts, and ways of life. Such a possibility was dawning; the clues had already begun to accumulate, and Woodward attended to every one. But exactly as with the fossil shells, they were still inadequately classified and too few; and the alternative view of history derived from Scripture was too solidly established to allow for anything like our modern view – a view which had to wait for the nineteenth century.[32] In another letter, written later and published in his book on fossils, Woodward considered the stone implements that were beginning then to accumulate. That they were the genuine artifacts of past cultures ("Axes, Wedges, Chizels, Heads of Arrows, Darts and Lances") he had no doubt, and he realized that they had come in time to be replaced by iron.[33] The analogy with contemporary nations, "yet barbarous," was very helpful here. The question remained, however, how to accommodate this view with Scripture and the Flood (which according to Woodward's theory must have destroyed all evidence from the antediluvian world). The Bible suggested that iron had been employed even before the Deluge, and the Doctor had to exercise his usual ingenuity to show that those skills must have been forgotten in the confusion and duress of the Flood and been learned over again at a later time. (Two successions from Stone to Iron Age were thus necessary for him.) As with geology, the Mosaic narrative was thus both an obstacle and an incentive to investigation.

At least Woodward was on the right track here again, and his method of collecting and comparing the evidence of past and present stone and iron cultures was bound to be fruitful eventually. What he failed to see, however, was that his knowledge was insufficient still for the problem. It was hard enough to establish the meaning and identity of a Roman object − and Europe had spent two centuries working at that − much less an object from any other ancient culture about which so little was known. But Woodward was never reticent in the face of his own ignorance. If he could see the weakness of the other fellow with ease, he could somehow never apply his very real critical skills to the arguments of his own invention. Lhwyd was not unfair when he wrote to a friend that Woodward was "too tenacious of his conjectures," though he was forever deriding the fault in others.

[4]

Naturally Dr. Woodward had to know also about the ancient Egyptians. In one of these early letters to Lhwyd he already shows his interest.[34] For a student of the early history of mankind it was necessary to consider their claims. Their reputation for wisdom and antiquity had been great even among the classical Greeks, and in the seventeenth century it had risen again. Where did they fit into that great succession of peoples that extended from Noah to the modern world? Until the seventeenth century, Europe had derived most of its knowledge of the matter from the Bible and the classical authors. But lately European travelers, like the Englishman John Greaves, had inspected the pyramids at first hand; and Egyptian antiquities, even including mummies, had begun to join their classical counterparts in the cabinets of the learned.[35] Hieroglyphics reasserted their ancient fascination and the Egyptian gods stirred new interest. It was a natural subject for the virtuoso, but it was especially important to Woodward for the light it threw upon the origins of human history. (In this regard Burnet had already broached the subject.) The Doctor could hardly avoid interesting himself, and an essay that survived his death shows him at his most characteristic.

The manuscript was a substantial tract apparently prepared for the press. It came to light in 1775 when it was read to the Society of Antiquities, and appeared two years later in its journal.[36] As the editor remarked, it was the natural result of Woodward's conviction that all the events of the *Pentateuch*, human as well as natural, had been there truly described. Moses was as authoritative for human history as he was for geology. If Woodward's earlier work had persuaded all sober and intelligent men of the truth of the one, the present essay was meant to furnish argument for the other. If one were to doubt the veracity of Moses, or turn him into a wily politician deliberately distorting the truth for an ignorant audience, we would be at once thrust into darkness. There was, Woodward thought, at this great distance of time, simply no other "like means of information." The Doctor was convinced that the "facts" of human history, the evidence of antiquities, like the fossils and works of natural history, could conclusively settle the point.

Woodward's essay was not, however, an attempt to reconstruct Egyptian history. It was rather an extended argument, a rebuttal of some newly fashionable notions which the Doctor thought wrong and even pernicious. He had for example been reading the recent works of Dr. John Spencer, Master of Corpus Christi College, Cambridge, which purported to show, in the similarities between Hebrew and ancient Egyptian religions, that it was the Hebrews who had borrowed most from the Egyptians. The traditional reading of the Mosaic testimony was thus undermined. It was a point that had been argued before (even to Patristic times), but no one had ever made the comparison so systematically. Even Woodward had to admire the "infinite industry" of Spencer's "collation of the Jewish and Pagan constitutions," though he thought Spencer's conclusions quite wrong.[37] With his usual confidence, however, he was ready at once to engage his learned opponent.

Before tangling with Spencer, however, Woodward prepared the way with a general review of Egyptian culture. In passing, he noticed an even greater ambition: to "set forth the first appearance of science in the world and trace her through all ages and climates."[38] Like his other large projects it was, of course, impossible of fulfillment. It required, in the Doctor's estimation, a "person of genius and real knowledge," someone "thoroughly apprized of history and the transactions of mankind in all ages," not to mention "a real understanding of the arts and of the sciences . . . a true knowledge of nature and a deep insight into things." Happily, Woodward had the right combination of abilities; unfortunately, he lacked the time. It was enough for the busy Doctor to deal with the immediate problem, the priority of Hebrew or Egyptian antiquities.

The connection between natural and human history was reaffirmed at once in Woodward's preface. Nature, he repeated there, was the same in all ages, but variations in natural circumstances (e.g., water, air, vegetation, etc.) combined with variations in cultural conditions (e.g., the character of government, war and peace, religion, philosophy, etc.) to produce the different peoples that inhabited the globe. In that lost paper on the peopling of the world which Holloway noticed, Woodward seems to have developed this notion further.[39] Natural history and geography were the original determinants of the variations among peoples and races; in aftertimes they continued to interact with the cultures they produced. (Woodward calls them here "accidental and uncertain" causes.) He at once sets the Egyptians in their geographical setting: the fruitfulness of the Nile Valley was the first condition of Egyptian leisure and hence of her thought and culture. But the following pages, rather than enlarging upon this theme, are all devoted to a debunking of the Egyptian achievement.

Woodward's arguments were an impressive compound of wide reading in both the classical sources and the recent literature, along with a firsthand knowledge of Egyptian antiquities. At the same time, his information was (as usual) insufficient for the problems he wanted to solve. He begins with hieroglyphics, for which the Egyptians had received the double tribute of priority in writing and in mystical wisdom, a claim that had been advanced in Antiquity and which had now been reaffirmed. The Doctor had seen hieroglyphics engraved in various scholarly works and read the ponderous tomes of the learned Athanasius Kircher who had pretended to decipher and translate them. Kircher, reported Sir Robert Southwell, "is a person of vast parts and of as great industry. . . . On the other side, he is reputed very credulous, apt to put in print any strange, if plausible story that is brought unto him." [40] ("Judged by scholars of today," writes a modern Egyptologist somewhat uncharitably, Kircher "would be considered an impostor. . . . As for his translations, they have nothing correct in them.")[41] For Woodward, who understood them even less than the German, hieroglyphics were defective as letters and even more hopeless as wisdom. He was certain that they merely recorded the religious customs of a superstitious people. They contained neither sense nor virtue. The Egyptians lacked even their own history; they had "no records of their nation, their kings, or the transactions amongst them, nothing but a mere loose Tradition." That Woodward could not read their hieroglyphics did not trouble him; he was convinced they were mere primitive signs and was ready to dismiss their substance altogether.[42]

Still, the Doctor was not content. Had the Egyptians been a great people, their works would remain to show it. Woodward was not impressed with the pyramids, despite their exact description by Greaves. Next to classical architecture, they were barbarous, "without any consideration of adornment or beauty." True, they had lasted remarkably well, but that was attributable more to Egyptian air than to architectural skill. The Arundel Marbles at Oxford had "suffered more in 70 or 80 years there than in perhaps 2000 in the countries from whence they were fetched."[43] As a physician, Woodward knew the effects of pollution in London, asthmatic as well as architectural. No, Egyptian architecture had nothing to redeem it; their labyrinths were merely expensive and their temples confused, barbarous, and ill-contrived. Woodward was here content with the contemptuous testimony of the ancient Greeks. His own esthetic standards were entirely and exclusively classical.

And so it was for the other arts of ancient Egypt. But now Woodward could add his own observations to his sources. He had seen ancient Egyptian images in stone and metal, "in the cabinets of persons curious in antiquities," and he possessed several himself.[44] He also saw them illustrated in various learned works. His classical prejudice was severe; he could see only that "their limbs are stiff and ill-proportioned; their bodies awkward, shapeless, and far inferior to their life." He was sure that they had aimed deliberately at the hideous, deformed, and monstrous. "They seem to have affected what was ugly and irregular, as much as the Greeks, and Romans, and others who had something of spirit and genteel fancy did what was handsome, well-proportioned, beautiful, and like nature." Of their painting he could say little, since it was certain (he thought) that nothing had survived, but it was hardly likely that it was better. Their dress and costume were rude compared to the *paludamentum* or *toga* "or the other truly noble and graceful dresses of the Greeks and Romans." Nor was Woodward impressed by their famous mummies, especially in the light of Egyptian superstitions about the transmigration of souls. Again, their longevity could be attributed more to desert air than to skill; upon removal they swiftly decayed. "I myself saw here a mummy, brought formerly out of Egypt, that after it had been for some time in our more humid air, began to corrupt and grow mouldy ... and in conclusion petrified and fell to pieces."

Finally, and more directly to the point, there was Egyptian religion, "the wildest and most fantastic that the sun ever saw ... sunk in idolatry".[45]

Woodward possessed a good number of Egyptian amulets and other superstitious toys and knew them again in the learned treatises and cabinets. He
recounts in detail the hostile criticism of Greeks, Romans, Jews, and early
Christians. Of all his sources he found Moses the most useful, his character
sketches brilliant, his history and description exact and full. It was simply
improbable that the Jews led and described by such a leader could have
borrowed anything from so debased a source. Dr. Spencer illustrated perfectly what happened when men fell in love with their own opinions. It was
clear to Woodward that Spencer, like Burnet and Whiston, had fallen victim
to the most natural of human propensities and sacrificed everything to his
own fancies and theories, though without any "real foundation in fact or
things."[46] It was the special vice of too much learning and reading. For the
"endless shoals of books at this day extant" made necessary what only a very
few possessed, "the capacity, attention, and strength of mind, sufficient to
make a fit choice of things, and mightily to digest and dispose of them." The
Doctor never failed to commend himself or to insult his opponents.

Unfortunately, Woodward knew Hebrew no better than he understood
hieroglyphics, and his arguments were not very strong in the face of Spencer's
erudition. Two hundred years later, Spencer was being credited with the
foundation of the science of comparative religion; according to a great
Biblical scholar, Robertson Smith, Spencer's was still the most important
book on the religious antiquities of the ancient Hebrews.[47] The issue,
however, was less clear in his own day: Spencer's work was widely challenged
for its orthodoxy by learned writers (several of whom were known to
Woodward). It was long before its conclusions were accepted or developed.
There was as yet no Biblical criticism sufficient to sustain it, no clear
apprehension of the development of Hebrew religion, and no understanding
of the chronological order of the Old Testament documents. It was limited,
too, by its exclusive reliance on literary evidence. Exactly as with his
geological speculations, Woodward could feel that he had argued a plausible
case in terms of the available testimony, archaeological as well as literary. In
neither case did it ever occur to him that his convictions could not properly
be supported by his "facts," that the inductive leap was too long, and that his
strictures against his opponents applied at least equally to himself. Yet the
range of his reading and the passion of his convictions, above all the compre-
hensiveness of his historical vision, are oddly impressive, and the publication
of his work a half century after his death is testimony to its continuing
persuasiveness.

V.

The Embattled Philosopher

Meanwhile the geological battle continued, even as the Doctor began to squirm within the toils of historical and theological controversy, distracting his attention and engaging more and more of the philosophical world. The war was fought out on several fronts, though the battle-lines were often confused. Inevitably, Edward Lhwyd and his friends played their part, provoked beyond endurance by what they thought the Doctor's disregard for their feelings as well as their ideas. But there were soon others in the fray as well, for Dr. Woodward's opponents were both more numerous and more harmful than the poor Welshman and his coterie. Moreover, in every case there was that strange compound of personal and philosophical rivalry, serious disagreement and childish petulance. Yet we should not think this peculiar to the quarrels of the Doctor; one has only to recall the petty squabbles of Hooke and Newton and Flamsteed (to name only a few of Dr. Woodward's friends) to see how characteristic it all was. Decorum was a literary ideal in this period, but even the wits who were publicly pledged to it rarely let it determine their behavior.

For a brief time Lhwyd stood outside the battle which went busily on. It is true that he allowed a veiled reference to appear in the new *Britannia,* where he argued against "whoever" it was who supposed that the subterranean leaves (and fossil shells) were all once real. Dr. Woodward appeared not to notice, although their mutual friend John Morton thought the remark gratuitous.[1] Meanwhile Lhwyd's friend Tancred Robinson and others kept him closely apprised and tried directly to involve him. Although Robinson at first agreed with Lhwyd that it was probably best to "neglect DW and his Scaramouch H," he nevertheless asked for suggestions that might help Arbuthnot and Wotton in the impending struggle. Harris and Woodward, he wrote, were giving out "terrible pronouncements of war, which they intended to commence as soon as the weather is fit for the opening of the campaign." He himself had been forced to print half a sheet to rebut Harris and to deny his authorship of the L.P. *Essays,* while Hooke (so Robinson wrote) was threatening some reflections, and Wotton promising a vindication of his account of Scilla.[2] At Westminster Hall the quarrel finally broke into violence. In an anticipation of the later attack by Dr. Mead, Dr. Lister this time drew his sword against Woodward, "and had it not been for another Dr. in company with him (who suddenly interposed) there had been Philosophical blood spilt and it may be the death of one or both of the brethren of the same society."[3]

So far, however, Lhwyd had not been attacked publicly, either by Woodward or his friends. He had not therefore the direct challenge to reply as had Ray, for example, who found himself continually dragged into the dispute. (In fact, Ray's criticism written out in 1703-4 awaited the posthumous publication of his *Three Physico-Theological Discourses,* 1713).[4] When Woodward wrote to Lhwyd, therefore, for his thoughts about the *Essay,* Lhwyd replied simply "that two or three Gentlemen reputed the best Philosophers here admired it very much, tho' for my Part I could not as yet subscribe to his Hypothesis . . . 'til his larger work was published." He was, however, offended by Woodward's attacks on his friends, and even more at his "disingenuousness" when he learned that the Doctor was describing him as "an invidious man" who meant to expose the *Essay.*[5] He was no doubt also annoyed that Woodward had completely forgotten to acknowledge his aid, though he had not forgotten to apologize for the oversight. (In later years Woodward came to believe that this was the sole reason for their quarrel.)[6] At the least it was an awkward situation; Lhwyd had definite and hostile opinions of his own about Woodward's theories and would eventually have to proclaim them. Thus Woodward was not mistaken in his suspicions, although Nicolson, who was on good terms with both, could still hope that the dispute might be raised above the personal level. In the end Lhwyd did send some suggestions along to Tancred Robinson for the use of Woodward's critics, and at some point even drew up some notes for a satirical work which was to be entitled "The Sympathizing Virtuosos or a Comparison Betwixt Dr. Woodward's Style in his Essay . . . and Mr. Harris in his late Remarks, with some brief Annotations." A series of parallels between the two works seem to have been intended to show that Woodward was the author of both, but fortunately Lhwyd suppressed his essay stillborn.[7]

Meanwhile his own work lagged; his friends were anxious for its publication. Apart from interruptions (his contribution to the *Britannia* was complete by 1695), Lhwyd was a painstaking worker and meant his treatise to be thorough. "The best of my Book will be figures of some things which perhaps DW his self has not seen notwithstanding his correspondence with five hundred persons (as he lately boasted to my Friend the Archdeacon of

Cardigan)." He intended essentially a "Classical enumeration" of the formed stones of Britain with figures of the rarest, *not* (he wrote to Lister), "a Rabinnical chimera about the thawing of the Globe."[8] But he could not avoid the main question, the origin of the fossil shells, and for some time he had been developing his own theory in his correspondence with John Ray. At last, after some difficulties over publication, Lhwyd's book belatedly appeared, dedicated to Lister. Inevitably it collided with Woodward's.

It was entitled *Lithophylacii Britanici Ichnographia* (London, 1699), and for the most part was innocent enough.[9] As Lhwyd had promised, it was largely descriptive, a pioneering work of its kind, and like Woodward's later catalogs valuable for the progress of the new discipline. But at the end Lhwyd appended several letters in which he too took up the question of the fossil origins. He had moved a long way from his earlier intention of avoiding the subject. In an earlier undated letter to Lister he declared his intention of making "some exceptions" to Woodward's theory, and it appears that he had been corresponding with Lister about the problems which the dissolution of the earth seemed to pose for the construction of Noah's Ark. He saw, however, that Woodward could escape one difficulty by supposing that wooden rather than metal pins had held it together. (Like other organic materials, wooden pins would have escaped dissolution.) Perhaps God had hinted to Noah not to use iron![10] Now he was prepared to add a "general discourse" to his description, to try out his own ideas, but he was not optimistic about its popular success, especially since it would be in Latin.

When it finally did appear, it was in the form of a letter to Ray (subsequently translated by Lhwyd himself). His own view was modestly proposed as a "conjecture."[11] But first he had to dispose of the theory of the dissolution, which he proceeded to do with a long enumeration of difficulties, including many original observations of his own. We have not space for them here; they concerned largely the disposition of the fossil shells in and on the earth, but also their composition and especially the existence of many without any living counterparts. Lhwyd possessed more than forty *nautili* of this kind alone. With the help of his friend Richard Richardson he was able to add similar difficulties for the vegetable remains they had both been collecting. He even repeated the notion that he had long ago reported to Woodward (which he thought not yet satisfactorily answered), that marine substances were sometimes generated in human bodies.

Yet, Lhwyd was no more prepared than Woodward to admit the possibility of the extinction of species. His own theory ascribed the generation of fossils to the "seminium or spawn of marine animals," carried by spray and rain and "impregnated with an, to the naked eye, invisible animalcula." (So too for leaves and branches.) His theory had all the possibilities of a good compromise, and indeed it was an idea that had been set out many years before by Sir Matthew Hale, though Lhwyd disclaimed knowledge of his predecessor.[12] Like its alternatives, however, it was open to many serious difficulties. Indeed, the most that could be said for it was that, in contrast to Woodward's, it was modestly argued, and that Lhwyd was prepared to consider the objections — even to enumerating them himself. Moreover, he had to admit that most of the virtuosi were being won over to his rival's theory of the flood.[13]

Clearly, then, Lhwyd's tract was no solution to the problem of the shells, nor an end to the controversy, which it simply stirred further.[14] It did not help that the ubiquitous Tancred Robinson, who saw to the publication of the work, added the letters D.W. to the margin and so made the criticism of Dr. Woodward explicit.[15] Ray admired the letter, but more on its critical side than for its new hypothesis. It was becoming clear that *all* explanation was premature. Unlike the physical sciences with their mathematical formulations, the biological and geological sciences depended for their advancement, at this stage anyway, on a more systematic and complete classification of the evidence. For the fossil shells this was just beginning. "Tis but now of late," Woodward wrote correctly in one of his manuscripts, "that the Natural History of Fossils, of Ores and Minerals has been attempted with any Application." He dated the origin of the science to Agricola just 150 years before. "Natural Studyes," he explained to Scheuchzer, were thus "only in their Infancy."[16] The great merit of the speculations of Woodward and Lhwyd was that they both realized the direction which research would have to take to answer these questions and to test their hypotheses. This was their common ground, not their specific conclusions, and in the end it was their lasting achievement; for (as we shall see) each in different ways made important contributions to the actual enumeration and description of the fossil evidence. And as Lhwyd foresaw — however eccentric Woodward's hypothesis, he had at least created new excitement for the subject and raised the study of fossil shells and geological formations to true philosophical dignity.[17] Nor was it coincidence that the disciplines of the doctor and natural philosopher were exactly analogous to those of the antiquary; that natural history was thought akin to civic history; and that fossil shells were studied like ancient coins (or shields). For the *virtuoso,* for Dr. Woodward and his contemporaries, they were all somehow part of a single undertaking out of which, eventually, the entire history of the world could be re-created.

[2]

The *Essay* made Dr. Woodward famous. From the moment it appeared, the world took notice. Nor was it only the philosophers and divines who agreed to take it seriously. Even the wits attended to it, though in their fashion they preferred to ridicule it rather than to read it. Thus the Doctor found himself soon with a European reputation — admired, criticized, satirized, all at once. For a man not yet thirty, it was a considerable achievement.

Unfortunately, the wits were cleverer than the virtuosi, read more widely and attentively, especially afterward. So Dr. Woodward suffered their caricature and (as we shall see) lost his reputation. True, he brought something of this upon himself; his arrogance and bad manners did not serve his cause. Moreover, he estranged himself from his best protection, the company and favor of his fellow naturalists, in the Royal Society and outside. Yet it will not do to let the matter rest there. Woodward had his supporters too, in London and across the seas, in his own time and for a long time afterward, and he left his mark even upon the opposition. The balance sheet is more complicated than his posthumous reputation, so we must try to do what we can to restore it. But before we attempt this any further, it remains to us still to recount what was probably the most famous, as it was surely the most frustrating, of his many quarrels, his struggle with Hans Sloane for control of the Royal Society. Moreover it is not generally known, I think, how deliberate and how nearly successful in this Dr. Woodward was.

The Royal Society was the natural arena for Dr. Woodward, who soon became one of its most active members. From 1694, he offered papers, contributed specimens, nominated new members, introduced foreigners, and participated in debate.[18] He spoke on shells and minerals, diseases and medicines, Etruscan and Roman monuments, Egyptian religion and fossil trees. He reported his correspondence from around the globe, his experiments on vegetation and on the walls of Gresham College. He described the anatomy of a dog and showed off his collections: Roman glasses from Wapping and a piece of an urn; a rhinoceros hide and a stuffed seal's skin from Newfoundland (both donated to the Society); figured stones from the Barbados; the latest antiquarian treatises from abroad, a tartar-lamb from China; and a piece of East India wood. He presented the Society with Chinese coins and a Japanese mirror, a lodestone from New England, a stuffed penguin and a shark's tail, and various shells dug up at Maidstone in Kent. He read a "Discourse upon pretious Stones," showed a large tooth found in Wirtenburg and a copper medal with "Punick" characters. And there was more, as the years passed, very much more. Woodward's interests exactly matched those of his fellow virtuosi; after a time he was elected to the council.

But not all was smooth sailing. Among the members were several with whom he fought. Besides Wotton and Tancred Robinson, for example, there was his old friend Robert Hooke. Their rival theories about fossils necessarily brought about a confrontation. In 1699 Hooke read some new lectures on the figured stones and fossil shells which provoked an argument.[19] It did not however diminish Woodward's respect for his friend,[20] and their rivalry was nothing compared to Woodward's enmity for Hans Sloane. As early as 1697 Ray heard from Tancred Robinson that Sloane had been affronted and abused by the younger man at a meeting of the Society.[21] Was it because of his intimacy with Woodward's critics? Or was it the natural rivalry of two ambitious men, both doctors, both collectors and virtuosi, both members of the Royal Society? Undoubtedly it was a little of each. Sloane was the senior, always a trifle more successful, as a physician, as a collector, as an officer of the Society. On the other hand, Woodward was the more original mind and the more accomplished scientist. Apart from the fact that Robinson and Ray were among Sloane's best and oldest friends, he was besides largely responsible for seeing to it that Lhwyd's *Lithology* was published. Moreover, although he was discreet about his geological views, he was nevertheless skeptical of Woodward's hypothesis.[22] And to make things even more difficult, he entered – along with his friend, the apothecary James Petiver – into a strenuous competition for fossils with Dr. Woodward, particularly from America where sources were few and the curiosities in great demand.[23]

By 1700 the Quarrel had become bitter. In that year there appeared a new work in the growing corpus of satire against the Royal Society: *The Transactioneer with some of his Philosophical Fancies: in Two Dialogues.* This time the butt was clearly Sloane. The author (as so often, anonymous) intended to show his leading character to be a man of great reputation but with "neither the Parts nor the Learning" to support it. His particular concern was with the ludicrous quality of the *Transactions,* the chief public record of the Society, then under the supervision of Sloane. Characteristically, it was a criticism of their style and their pedantry. The subjects were ridiculous and mean, confused and unintelligible. Most annoying to the Society were the many references to specific papers; it was easy enough to throw ridicule on them by simply recounting their contents: an arch in two stone chimneys in Northamptonshire; a foetus sixteen years in the uterus; a monstrous child in Jutland; a prodigiously large feather; a shower of butter, etc. Such information was as little useful as its style was barbarous. "Strange how readily you have learn'd to speak in the Language of the Moderns!"[24]

Such an argument must have had an ambiguous appeal to Dr. Woodward. No doubt he relished the specific allusions to Sloane, here shown comically presiding at the Temple Coffeehouse with his friend Petiver. He certainly believed, as Sir Robert Southwell put it, that "the Royal Society's Stock is so low as hardly to keep Life and Soul together."[25] But he was less likely to agree that digging up the bones of dead fish had no value; or that the language of the moderns was intrinsically bad. As Harris pointed out, the author "hath but a slight knowledge of Philosophical Matters."[26] It was true; he was

Dr. William King, one of the Christ Church wits who had collaborated to attack Bentley in the controversy over Phalaris. Like his friend Swift he was equally opposed to all forms of modern learning, whether philological or philosophical, and equally ignorant of them.[27] The Society was determined to find the culprit and to institute proceedings — a fair indication that the satire had hit its mark. Woodward was forced to write in his own defense.[28] Sloane and Petiver, he insisted, were spreading rumors that he and his friend Mr. Harris had written, or at least contributed to, the pamphlet. He thought its writer, however, "meanly qualified for the task." At the same time, he pointed out that he had often had to defend the Society from those very charges, and "in publick company too." The insinuation was clear enough; perhaps there was something to them after all.

Although the Council absolved Woodward and Harris,[29] the quarrel was not possible to reconcile and Woodward found himself increasingly at odds with Sloane and isolated from his fellow workers. To Thoresby, he wrote that Sloane was "a very tricking person." "His gambols have brought the study of nature much into ridicule."[30] It was particularly distressing to discover that the Society had voluntarily relinquished its control over the *Transactions* to the secretary, who had *carte blanche* to print what he liked — and reject the rest. Harris reminded the president that a paper by John Morton applauding Dr. Woodward's fossil notions had been deliberately suppressed by Sloane. He was even convinced that the secretary lay behind the original attacks on Dr. Woodward's *Essay*.[31] Early the following year, "it was ordered that no private Conventions shall be held at the Meetings of the Society during their Assembly nor any Injurious Language be afforded. . . . And that if any one present shall think fit to Complain of any thing or to oppose what he shall hear advanced he shall speak to the President or Vice-President in the Chair who have the only Right to regulate these Matters."[32] In 1703 there was another fracas, much like the first. Dr. Woodward insisted that a "meer Pleasantry" had been misunderstood by Sloane as a slur on his management of the Society. He had not meant to reflect on anyone, but had spoken simply "to second some Motions that were made relating to Tryals of Experiments and makeing of Observations which I take to be the true Way of preserving the Honour and pursuing the Design of the Society." Another member, however, understood the exchange rather differently. There was, he wrote, "some pleasant discourse between Rough Diamond and Woodward about the liberality some Doctors take in trying Experiments (for the good of the rest of Mankind) upon their own Patients to discover the Vertues of a new Herb or Medicine."[33] The quarrels of the members seemed about to destroy the organization.

But the Society lived on to see the bickering continue. When the naturalist

Richard Richardson came to London in 1702, he sought out the Doctor, bearing a letter from their mutual friend Ralph Thoresby. Woodward treated him civilly and showed him "the noblest collection of fossils" that he had ever seen. Richardson saw nothing of the famous temper, and found it a pity that "so ingenious a person should not have more friends."[34] He soon found out why, however. Thoresby had also introduced him to Sloane, and Richardson showed each of the rivals (separately) some admirable drawings of plants found in the North Bierley coal pits. Woodward had them for an afternoon, but Sloane was able to produce them before the Royal Society as the latest addition to his own collection. Woodward was bitter; Richardson startled; Sloane defensive. "I must do what I can to serve the Society," the latter wrote to Thoresby, "without regard to the humours of those who would discourage every thing that comes not to the support of their hypothesis."[35] In fact, Richardson had had his doubts all along about Dr. Woodward's theories; among other things, he had written to Lhwyd that there were no fossils on the mountains of Pendle and Ingleborough to support Dr. Woodward's views, a letter which appeared eventually in Lhwyd's *Lithology*. It was no wonder perhaps that their meeting went badly; in any case, Richardson was soon consigned to the ranks of the enemy, with the Doctor promising revenge. It was too bad; the Yorkshireman thought Woodward's hypothesis the best yet, if still imperfect, and was willing to help him. As he wrote wistfully to Thoresby, if Woodward's "humour was as agreeable as his way of writing, he would be the mirror of this age but his haughty temper will not down with any person that is his equal."[36] Many years later Woodward tried vainly to block Richardson's nomination to the Royal Society.

The Doctor even quarreled briefly with Thoresby, by no means an easy thing to do. In his rough and forthright way he had reminded Thoresby of the promise he had made of a gift to his fossil collection. Thoresby read the letter as a charge of breach of promise. Woodward's reply was characteristically abrasive. "I would not have you believe I am backward in communicating any thing I can spare from my own studies, to any curious gentleman," he wrote, "to you in particular. But there is no life in this sort of commerce when it makes no return. . . ."[37] The quarrel was repaired, but Thoresby's estimation of his London friend was decidedly lowered. It was not easy to sustain a friendship with Dr. Woodward and traffic with his enemies. When Thoresby sent a communication to Sloane for the *Transactions* and imprudently asked Dr. Woodward about it, he received such an "ill-natured return (because not addressed to himself) as I never had from any Gentleman." Eventually he was repudiated altogether for supporting Lhwyd in the Royal Society.[38]

Meanwhile Dr. Woodward continued to trouble that body, although the events are sometimes obscure. In 1706, "It was ordered upon complaint of some late irregular proceedings at the meetings of the Society and urgent reflections cast upon some of the members, particularly by Dr. Woodward, that the members of the Society be admonished when he is present that if any member of the Society hereafter cast reflections on the Society or any of the fellows thereof, the statute concerning ejection shall be taken into consideration. . . ."[39] Just what the Doctor was up to is not entirely clear, but the precedent was ominous. It seems to have had something to do with Woodward's Swiss friend Scheuchzer, who was then becoming engulfed in the controversy. Scheuchzer was trying urgently at this time to get help from the Royal Society in order to publish his manuscript, the *Iter Alpina,* and for that purpose seems to have appealed to both the warring parties. The result was some intense scheming and a flow of denunciations from Woodward attacking his rival (now "one of the most tricking men alive") and the various contrivances he thought were meant to take advantage of his friend. Apparently Sloane wanted to print some of Scheuchzer's work in his *Transactions,* but Woodward saw through this "design" and "underhand engaged the President and some of the most considerable Members to oppose this mutilating . . . your Works to serve their private Purposes."[40] The question then became whether to print it in England, as Sloane desired, or in Holland, as Woodward wished. This time the Doctor lost.[41]

A year or two later the rivals seemed to have patched up their quarrel, but only on the surface. A letter to Scheuchzer in 1708 shows our hero at his worst. Sloane, he writes, was a man of "very moderate Understanding. At the same time that he has ever treated me with great Regard, I would not forbear using him with Contempt; but I have of late forborn it that I might do you good Offices with him and he is very sweet upon me and very well pleased that he has a Truce and that I forbear the Sarcasms I was wont to fling upon him." To be fair, Woodward was not alone in resenting Sloane's management of the Society; even Sloane's friends seem to have entertained some doubts. "Even it it were true," wrote one, "even were it true, I say . . . that you take too much upon you, that you Govern us, and even that you bestow Places upon your Creatures . . . yet I think we ought even to contrive at all this rather than lose an old able and experienced officer who is the very life of the Society."[42] Beneath the personal squabbling there was a difference of principle. Sloane was essentially a dilettante collector with a great web of personal connections which he used to help build both his practice and his collections. He was certainly no philosopher. He used the *Transactions* largely to further his personal ends and those of his friends, confining them to miscellaneous and random descriptions of very unequal value. Dr. Woodward and others thought that they could be improved by encouraging experimental activity for more philosophical objectives.[43] How far they would actually have improved them had they secured the chance is hard to say, but the prospect was not implausible.[44] Meanwhile it did not help matters that Dr. King decided just at this time to direct still another lampoon at the *Transactions*. It was distressing, wrote the astronomer Flamstead, "how loud Men of Sense all over the Town are in their Complaints and Men of Wit in their Railery of the Society on account of Dr. Sloane's Management."[45] Clearly, something had to be done.

On top of all this, there was yet another bone of contention to divide the Society, this time over its meeting place, which continued as always to be at Gresham College. Dr. Woodward was delighted with the arrangement, which he and his colleagues found very convenient, but there were others, including Sloane, who had begun to think of an independent location. When the Mayor of London abruptly requested the Society to remove itself, it was Dr. Woodward and his friends, after some intense maneuvering, who managed to forestall the move.[46] But sentiment was growing for a change, led by Sloane and eventually by Newton. With all these grievances, therefore, it was not surprising that Dr. Woodward began to conspire against his rival.

The quarrel came to a head in 1709, provoked by his old enemy Edward Lhwyd. In the years since their first encounter the two men had dropped their faltering correspondence for a steadily increasing antipathy. On the question of fossils they still divided; Lhwyd continued to maintain the opinions of the *Lithology,* although he turned increasingly away from geology to the study of Welsh antiquities. Now the quarrel resumed with a review by Lhwyd of Scheuchzer's *Iter,* which appeared in the *Transactions* and went out of its way to take a swipe at Dr. Woodward's theory. The review had been expressly solicited by Sloane's young assistant John Thorpe and (inevitably) by Tancred Robinson.[47] Lhwyd could not (or at least would not) overlook the opportunity so conveniently set before him. Apart from the old wounds, he felt a new grievance which was caused by some satirical remarks aimed against his *Lithology* by Scheuchzer. "I have been credibly informed," he wrote to Thorpe, "that D.W. has often told his Friends that 'twas he that set Scheuchzer on ridiculing my Catalogue of English Fossils in his *Specimen Lithographiae Helveticae.*" He was relieved that Scheuchzer had elsewhere made amends by praising his work, but he was unforgiving of his old rival. When the review appeared, therefore, he made sure to refer to the theory of the dissolution as an "Opinion which has been long since sufficiently exploded in the Ingenious examination of it," and even suggested that Scheuchzer himself did not believe it.[48] Knowing our man, it was hardly surprising — least of all to Lhwyd — that when Dr. Woodward read the passage he became livid. "He expostulated with both Secretaryes about it," but as might be expected to no avail.[49] He then turned to the President (Newton), who promised to bring the question before the Council. There it was agreed, however, "that the *Transactions* were not the Act of the Society but the Secretary's who might put in what he pleased; and the Secretary declared the *Transactions* were free and that not he but the Authors only of such Papers as were printed were accountable for the same."[50] This was more than a little disingenuous, however, since Lhwyd's review had clearly been solicited with malice. Indeed, it was meant to be followed by another which would have been even more provocative. "I have received Langius's *Historia Lapidum Figuratorum Helvetiae,*" Lhwyd wrote to Richardson, "and am to give some account of it in the *Transactions* which wil undoubtedly create the Gresham Professor a further disgust, in regard Langius admits of and confirms my Hypothesis."[51]

Of course Woodward was wrong to protest and Lhwyd and his friends right to recall the motto of the Society, *Nullius in verba.* Woodward seemed to be intent on undermining that "common liberty of Philosophy" which even the French Academy of Sciences recognized.[52] Yet it was exasperating that Sloane should be allowed absolute control over the *Transactions* to whatever purpose he liked. "Afterwards Dr. W (not knowing what had passed in Councill) begins his Tale at the Meeting with all the Eloquence and Aggravating Circumstances imaginable. The Secretary heard him with a good deal of patience and told him at last that not he but the Authors only were accountable. . . . But not satisfied with this Answer, he talked of nothing but having Justice, and Satisfaction, and taking Courses and so the fray ended for the present." Thorpe, who was describing the event for the amusement of Lhwyd,[53] was not sure how the quarrel would end, "whether by the sword or by the pen. If the former, Dr. Mead has promised to be Dr. Sloane's second, and Dr. Oliver mine." (How prophetic the jest!) "But I believe he will wisely choose the latter; if so it will fall hardest on the Secretary for he seems to have the greatest Pique at him."

In fact, Dr. Woodward was not entirely beaten. It is true that Lhwyd was, later that year, successfully proposed as a member of the Society and elected over Woodward's strenuous objections.[54] But an apologetic advertisement did at length appear in the *Transactions,* ordered apparently by the president, which conceded something to Woodward's point. Sloane reaffirmed the responsibility of the authors alone for any "reflecting Expressions" that might appear, but he urged also that all future contributors leave aside anything that might give offense. It was more than a tacit admission that there was something to Dr. Woodward's wrath.[55] Lhwyd died unexpectedly within a few months, before his review of Langius could appear and in the midst of a last vituperative exchange of letters with his enemy.[56] Even so, Dr. Woodward would not forget either his "malice" or his "stupidity." "He expected I should have praised him highly in my *Natural History of the Earth,*" he explained separately both to Scheuchzer and to Thomas Hearne, "and being disappointed by what he had no cause to hope for, he published that silly theory out of meer Envy."[57]

Thorpe imagined that Woodward would respond with either the sword or the pen. The Doctor had a better idea; he soon hatched a conspiracy to undo Sloane and sieze control of the *Transactions.* "As the Members of better Sence scorn or slight him," he wrote to Scheuchzer, "so they have a Design to turn him out at the next Election; but this only betwixt you and me." Now at last the setting was right for the climactic event. Somehow Woodward was able to elect his friend John Harris as Secretary, replacing Richard Waller who had held the position since 1687. (Apparently Sloane thought briefly that Harris was on his side.)[58] Although Sloane himself was still untouched, Woodward boasted to Scheuchzer that "the greatest enemy you have is made a meer Cypher." For a moment everything hung in the balance. "I suppose you have heard Dr. Harris was chosen Secretary," Woodward wrote to Thoresby, "and that Dr. Sloane declared at the next Meeting he would lay down. He knows well enough his Management hath been long thought very meanly of and that Dr. Harris was elected for that Reason. He guesses rightly enough that the next step would be to set him aside and therefore it is that he is so hasty to lay down. His *Philosophical Transactions* have given great Scandal and whether Dr. Harris will suddenly publish any, is not yet settled. But if you please to send what at any time you design for the Society either to Dr. Harris or me, our Care shall be taken that right be done you." Dr. Woodward's cohort also went to work, even inviting the great Leibniz (already Newton's enemy) to contribute to the *Transactions.*[59]

In the uncertainty of the moment, the Royal Society was plunged now into "a great Ferment and Struggle with all things in the utmost Confusion." More than one member was alarmed and thought that it meant the end of that body. The president, Woodward complained, however good a mathematician, was unable to keep order.[60] Not long after Waller had been deposed, with Newton in the chair, Dr. Sloane entertained the Society with a translation from the *Memoirs of the Academy of Sciences* in which he maintained that the Bezoar was a gallstone and that gallstones were a cause of colic.[61] Woodward, consistent with his own theory of disease, took issue with him and (according to a sympathizer) when Sloane was unable to answer took refuge in a series of grimaces, "very strange and surprising, and such as were enough to provide any ingenuous sensible man to a warmth." Woodward said that no one with the least understanding of anatomy could maintain Sloane's opinion. Sloane returned that all medical men agreed, to which Woodward replied, "None unless the writer of the *History of Jamaica,*" and challenged him to produce another. (Apparently Sloane could only cite Dr. Mead.) The quarrel was brought before the Council and the precedent of 1706 was recalled.[62]

The questions put to the council were two: Did Sloane grimace, and did Woodward's temper precede the event or follow naturally upon it? There were witnesses on both sides, but Woodward lost his chance when he blurted out upon Sloane's denial, "Speak sense, or English, and we shall understand you! If you understood Anatomy you would know better," or something to that effect.[63] According to Halley, it was agreed "by a great Plurality of Voices" that although the grimaces were doubtful, Sloane should disavow any evil intentions and Woodward (whose words could not be denied) should beg pardon of the Society and promise not to repeat the offense.[64] When Woodward refused, it was voted that he should be removed from the Council, and Sloane's services were endorsed. The secretary was accused of packing the Council and unfairly influencing the president[65] — but although Woodward turned to the law for satisfaction, and some of his friends rallied to his support, it was without success, and he watched his insurrection collapse utterly with the defeat a little later in the year of his friend Harris and the return of Waller. In a short time, under the leadership of its president, the Society withdrew from Gresham College to a new home in Crane Court.[66] Dr. Woodward's only satisfaction was to watch Newton and Sloane begin to quarrel in their turn. Although he later renewed his correspondence with Sloane, it was perfunctory, and the Royal Society saw relatively little of him again. It is not easy, even now, to say who lost the more through that unhappy estrangement.

VI.

The Collector

[1]

Behind the endless quarrels, Dr. Woodward was earnestly at work. It is one of his admirable qualities that nothing could interrupt the seriousness of his purpose. Quite apart from his speculations, he was consumed by his collections, which were meant to supply the evidence necessary to persuade the world of their truth. And indeed it was his collections as much as his theories that brought him fame. From his first discovery in a gravel-pit in Dover Street in 1688 through the wonderful days at Sherborne until his death, Dr. Woodward never tired of the search.[1] There is something quite remarkable about his dedication, for he spared neither time nor money from a busy and frugal life to procure his specimens. Yet he was no ordinary collector; he had only contempt for those who sought the abnormalities and rarities in nature to the exclusion of the commonplace. " 'Tis not well, that Gentlemen that have not duly inform'd themselves of things most obvious and common," he wrote, thinking here of limestone and chalk, "should take upon them to write of those that are the most abstruse and difficult. This is what has laid the foundation of Amusements in Natural History and Errors without end." Nor would he justify collecting as an end in itself, as though the study of nature could be advanced simply by filling cabinets or transcribing observations. Like his master, Francis Bacon, he believed that collections of natural history were but a first step and defensible only to the extent that they were useful. They must contribute to "building a Structure of Philosophy . . . or advancing some Propositions that might turn to the Benefit and Advantage of the World."[2] Here was the inspiration for his own investigations.

That he was indefatigable no one could deny. He had traveled widely in England in his youth and sought out for himself the stones and fossils in the mines and quarries, the gravelpits, and seacoasts. Later he had employed younger men like Hutchinson to collect for him; he sent one man to the tin mines of Cornwall and another pair to the North of England. In order not to waste their efforts, he drew up some *Brief Directions for making Observations and Collections and for composing a Travelling Register of all sorts of Fossils.*[3] His northern emissaries were sent especially to Pendle and Ingelborough to settle the argument with Lhwyd and Richardson, to see whether fossils did not in fact appear on those representative mountains. When they got to Carlisle in 1700 they reported to William Nicolson, who greeted them warmly, that Richardson's account was very faulty. "They found plenty of conchitae of several kinds in both of 'em." Dr. Woodward was delighted and, though Richardson and Lhwyd remained skeptical, he eventually got the fossils to prove his point. The two young men, "The one a BA of Cambridge and the other an apothecary," landed at Hull and searched the coast to Newcastle, "examining all the Cliffs and the Shore, the Cole-pitts etc., and made weekly returns of their discoveries." At Newcastle, they met another helpful friend of Dr. Woodward's, Jabez Cay, who like Nicolson offered all encouragement. "The Dr. had obliged them to be very exact in their Searches" and he must have been very pleased with the results.[4]

Meanwhile Woodward had begun that amazing correspondence across the globe about which he boasted. He was willing to beg, borrow, trade, even steal,[5] to enrich his collections. He saw at once that his theories about fossils and the deposition of the strata, about the universal Flood and the dissolution of the earth, required evidence from all over the world to make them good. "It hath been the great defect of those that have wrote on the subject," he had complained to Lhwyd in 1692, "that their observations have been confined to one country, or some small part thereof."[6] He had sent out queries to Asia, Africa, America, and Europe, apparently as early as 1690. Lhwyd might limit himself to Britain; his ambitions were slight compared to the Doctor's. Undoubtedly something of the friction between the two men stemmed simply from their rivalry as collectors. Although they began by trading happily, it was not long before there was misunderstanding.[7] Lhwyd was not as generous in these matters as appeared; he admitted it himself. Thus he felt betrayed when an assistant told John Morton where he could find some rare stones.[8] Another time Lhwyd gave some fossils to a friend who unhappily turned them over to Dr. Woodward. "Making a vertue of necessity," Lhwyd explained to Lister, "I presented some better patterns and also rec'd some good return from him." A year or two later (in 1695) he complained that the Doctor was offering higher prices and wooing away his assistants; and ten years later, he repeated this same lament.[9] Dr. Woodward was richer, more ambitious, and outlived his rival. His collections therefore soon outstripped the Welshman's and stirred him to envy. Eventually his museum became the most important of its kind in Britain, famous at home and abroad. While the remnants of Lhwyd's collection were disappearing from view, lost to all practical purposes in the disorder of the Ashmolean Museum, Dr. Woodward's fossils were kept intact in their original cabinets to inspire generations of geologists.[10]

Their survival was no accident. As Dr. Woodward grew older he became concerned for his lifework. From 1725 he was often ill, but despite the press of business he tried to bring order to his collections and prepare for the future. The Doctor had no family whatever, at least that he recognized. His parents seem to have died before he launched his career, and he had no brothers or sisters. He never married, perhaps because he would have had to relinquish his post at Gresham College otherwise. As he drew up his will, therefore, he cast about for a public institution which might preserve and foster his work. Cambridge was the obvious choice — though he flirted for a time with Oxford; it had awarded him a degree in the past and he was always well treated there.[11] The device he hit upon was to endow a new professor whose responsibilities would be to care for his fossils while he lectured upon those subjects which had most interested him. In his will, therefore, he gave to his executors all his possessions (including his investments in the South Sea Company) to furnish an endowment.[12] He drew up the qualifications of the candidate and the procedures for election with characteristic attention to detail. The professor (like Woodward) was to be a bachelor, "least the care of wife and children should take the Lecturer too much from Study and the care of the Lecture." He should be a layman, since there was "a much greater number of Preferments for the Clergy than for Men of Learning among the Laity." His lectures should deal with "some one or other of the Subjects treated of in my natural history of the Earth, my defense of it against Doctor Camerarius, my Discourses of Vegetation, or my State of Physick." As for the fossils, it was to be the duty of the professor to take care of them, reside nearby, and attend them regularly during the prescribed hours of each term. He was "to shew the said Fossils *gratis* to all such curious and intelligent persons as shall desire a view of them for the Information and Instruction." And he was to be provided with ten pounds beyond his salary for "making observations and experiments, keeping Correspondence with Learned Men . . . and procuring additions to the collections of Fossils." Dr. Woodward was careful to set up machinery to see his will enforced. It was all to one great purpose, "the setting forth the wisdom of God in the works of nature, the advancement of usefull knowledge, and the profit and benefit of the Publick."

The terms of the will were observed, and both the museum and the Woodward Professor continue until this day at Cambridge. A hundred years after it arrived it was still thought "one of the most remarkable occurrences in the progress of descriptive geology in England," and that is a judgment which should probably be allowed to stand.[13] There is something oddly moving about entering the Sedgwick Museum (the great nineteenth-century geologist was himself a Woodward Professor) and discovering still preserved there the very cabinets described by Dr. Woodward, with their contents almost exactly as the naturalist left them. With the aid of his catalogue, one can see just how he went about his task of collection, arrangement, and classification; the labels he attached show how meticulously he transcribed the details of discovery. How annoyed he was when his correspondents failed to supply him with the necessary information! When the famous collector Dr. Charleton gave him a fossil from "somewhere" beyond the seas, Woodward could not conceal his exasperation. "Tis pity a Gentleman so very curious after things that were elegant and beautiful, should not have been as curious as to their Origin, their Uses, and their Natural History, about which he was [so] little sollicitous."[14]

Dr. Woodward had the catalog published, although he did not quite live to see it. In 1725 he prayed that "in the Hurry of my Business, I shall find Leisure, during my Life to reduce all into one common Method and Series and one Catalogue."[15] He hoped also to set beside each entry a full description (where found, what depth, how it lay, color, figure, texture, etc., along with the results of chemical analysis); he wanted also to describe the history of every specimen, its relationship to his theories, and all the references in the past literature, taken from "those Authors that have set forth Museums, the Writers of Fossils, the Natural Histories of Particular Counties," and anyone else who might have written on the subject.[16] His great library, collected as assiduously as his fossils, was meant to serve. Unfortunately, it was not to be; the published work remained a series of separate catalogs with only brief descriptions, updated from time to time but never revised or integrated into a whole.

It remains an impressive performance nevertheless. If it is not much fun reading over a long list of stones and fossils, without illustration, the catalog is yet relieved from time to time by the Doctor's pungent remarks. He never lost an opportunity to advertise his theories or to repay his opponents. Lister, Lhwyd, and Plot (all long dead) received their due, while Woodward's latest rival, Dr. Sloane, was placed characteristically amongst those writers whose fancies outran their facts and whose guesses were invariably unlucky. There was some autobiography and much reiteration of his theories. Most important, there was elaborate testimony to the astonishing range of his connections and correspondence.

A mere enumeration would be tedious, but on occasion some correspondence or other matter survives to illustrate an entry. Here we see Dr. Woodward in another context, apart from controversy: among friends or well-wishers, often separated by thousands of miles, but all agreed to contribute to the Doctor's researches. He took every opportunity to cultivate travelers in foreign lands, ship surgeons and merchants for example, who often responded by sending him fossils. The most far-flung corners of the globe were not outside his reach. One of his contributors was Henry Worsley, brother to a baronet, who had "travelled through not only Italy and other the more civilized Parts of Europe but likewise . . . Egypt, Arabia, and several other Parts of Asia where few Europeans besides have ever been." He sent the Doctor many valuable pieces from the deserts of the East.[17] One fossil was "brought back from Tartary by Padre Felix, a Capuchin friar." Others were sent by James Cunningham (or Cuninghame), a surgeon who had sailed to China, while still others were donated by the celebrated Captain Dampier, "buccaneer, pirate, and circumnavigator" whose travels had occupied him twelve years and were reported in his very popular *Voyage around the World* (1697).[18] Dr. Woodward was particularly well rewarded from the Levant, where he had several donors.

From Europe he had other sources. There were travelers again like his fellow doctor Edward Browne, President of the College of Physicians, who had visited the mining towns in Hungary and sent Dr. Woodward a number of fossil ores. There was Francis Maximilian Misson, an expatriate Huguenot, who had written a famous description of a tour through Italy, translated into English in 1695 and often reprinted. Misson came to reside in London, though when he met Dr. Woodward does not appear. In his account of Italy he had digressed to write about fossils, stimulated by the hills of sand near Certaldo "stuff'd with diverse sorts of Shells." He was however very skeptical of the Flood theory and half inclined to accept the notion that the shells were only *lusus*.[19] In Woodward's *Catalogue* we find Misson contributing several pieces (at least one from Certaldo) to the Doctor's collection, and we discover that he soon became a convert. Misson read the *Essay* and changed his mind; according to Woodward, he now "believes these Bodies Remains of the Universal Deluge."[20]

Miners and mine owners furnished another opportunity which Woodward never missed. The *Catalogue of English Fossils* shows that he covered the field, with respondents especially in Cornwall, Wales, and the North. "Almost every Mine wrought at the Time supplied him with Specimens of its valuable Ores and uncommon Productions."[21] If Mr. Harley owned mines in North Wales, then Mr. Pigg, Mr. Harley's steward, must be coaxed into help. As always, however, Woodward ranged far beyond the local scene. One of his best sources was Baron Schonberg, "Berghauptman or Superintendent of all the mines of Saxony, free Baron of the Empire and Lord Chamberlaine of the King of Poland"; another was the invaluable Mr. Weber, "a native of Hungary who has long been conversant in the Mines there, as likewise those of Saxony." Of course Woodward cultivated also the foreign naturalists, exchanging books and fossils and reporting the news.

He was particularly well served by the Germans and Swiss, who generally received his work with praise. He was right, in old age, to claim his greatest influence in that part of the world. In the fossil catalogue he records contributions by many – by Scheuchzer, of course, but also by others like the two doctors from Lubeck: Dr. Melle, who had taken up the study of fossils under Woodward's influence and addressed a learned tract to him, and Dr. Leopold, who wrote a Swiss itinerary with compliments to the Doctor. "Finding my Collections not sufficiently stored with Swedish Fossils," the Doctor recalled of Leopold, "and that I had not a satisfactory Account of the Mines there, of his own Accord, and at his own Expense he undertook a Journey thither for my Satisfaction."[22] Another contributor was J. H. Hottinger, a Swiss who dedicated to Woodward a work on crystals which he presented to him along with some stones.[23] Most important, perhaps, was Louis Bourguet, a French Protestant who was forced to take up residence at Zurich after the revocation of the Edict of Nantes. By 1711 he had struck up a correspondence with Woodward and an exchange of books and fossils which persisted for many years, despite a growing divergence of views. At the beginning, at least, he thought Dr. Woodward uniquely qualified in Europe in his field; and even when he departed from the Doctor's theory, it was with the greatest diffidence.[24] But there was scarcely a part of Western Europe from which Dr. Woodward did not receive some help.

Needless to say, America was not overlooked. By 1692 Woodward had already discovered a New England correspondent, and he soon found many more.[25] In the fossil catalogs there are American donors from Newfoundland to the Barbadoes. Most of them are now obscure, but several of them were once better known. Among the Virginians, for example, was John Banister, a capable naturalist who served Ray and Lister even better than the Doctor and who published at least one paper in the *Philosophical Transactions.* The Reverend Hugh Jones who appears there was probably the mathematician who wrote *The Present State of Virginia,* although two other ministers from the same locality with the same name make the identification uncertain. William Vernon was a botanist of some reputation who sent fossils from Maryland. But the best-known among the southerners was probably William Byrd of Virginia, who substantially enlarged the Doctor's collections. Byrd was a considerable figure, "the greatest gentleman of Virginia in his time." He was a plantation owner, politician, dilettante scientist, and fellow of the Royal Society. He spent many years in London, where he knew Dr. Woodward – perhaps through their common friend and patron, Sir Robert Southwell, certainly through their common activities in the Society and its Council. I have been able to locate only one letter from their correspondence, but in it Byrd supplied an interesting confirmation of Woodward's theory of the Deluge in the traditional stories of the Indians.[26]

The competition for American curiosities was particularly keen. For Dr. Woodward there was the possibility of proving the universality of the Flood, that "America was underwater as well as Asia, Africa, and Europe, at the Deluge."[27] For others, for Sloane or Petiver, there was the discovery of unknown plants and animals. The rivalry of the collectors was particularly intense because the number of contributors was small and the appetites of the Society insatiable. As William Byrd complained to Petiver from Virginia, "The Country where Fortune hath cast my lot is a large field for Natural inquirys and 'tis much to be lamented that we have not some people of skil and curiosity amongst us. I know no body here capable of making very great discoverys so that Nature has thrown away a vast deal of her bounty upon us to no purpose."[28] When the Reverend Jones sent a cargo of fossils from Maryland to Lhwyd while he was away, it was quickly "plundered by several Virtuosos" and only a small part was rescued for him by a friend. The competition between Woodward and the Sloane-Petiver faction was particularly intense. Thus, when Jones sent some petrifactions to Lhwyd and to Petiver, the Doctor was "mightily nettled." Never one to concede, however, he managed to secure some anyway (to the chagrin of his rivals) from his old friend Samuel Doody. Meanwhile, Petiver thought to spur Jones on by telling him that Dr. Woodward had sent out someone else with a salary from the Archbishop of Canterbury, "on purpose to collect Plants, Shells, Insects, and particularly Fossils and formed Stones or Petrifactions." "I begg you would double your diligence," he concluded, "that you may have the honour of their first discovery and he only what is left." Typically, when Banister's collection went to Woodward, Petiver was thrown into despair at ever seeing them. Yet upon reflection he was even more afraid that he might actually be allowed access — "because when I publish any thing of these parts I shall in justice be obliged to acknowledge the first sight of them from him!"[29] Thus the peevishness and bad manners were not at all on one side. Before William Vernon left for America in 1698, Dr. Woodward told him that he was unhappy that he kept company with Sloane, Robinson, Lister, and the rest. "That wou'd not doe", wrote Vernon, "I must Expect dealings accordingly." Apparently the Doctor was able to make Vernon's activities in America uncomfortable, although in the end he managed to get hold of his specimens, and Vernon seems to have accepted Woodward's theories.[30]

One omission in Dr. Woodward's catalog is curious. Among the several New Englanders to supply Dr. Woodward, there is no mention anywhere of

Cotton Mather. This is surprising because, of all his friends in America, Mather was the most devoted. The New England clergyman was much drawn to the Doctor's ideas, which he found congenial in several ways. His father, Increase Mather, had already taken an interest in natural philosophy and known several members of the Royal Society. In the 1680s he set up a small society of kindred spirits in Boston to discuss scientific matters, and it was here that Cotton Mather, fresh from the university, developed his own lifelong interest in natural history. In two early books, *The Wonderful Works of God Commemorated* and *Winter Meditations,* the young Mather already showed a familarity with the "arguments from design" drawn from natural philosophy which Boyle and Ray were expounding in England. Although as a Calvinist he was inclined always to see the immediate hand of God in the events of nature and history, he was at the same time drawn to the newer view, then becoming fashionable in England, that the ways of nature were largely uniform and that God's hand (albeit differently turned) could be seen at least as well in the regularity and pattern of natural events as in the exceptional and miraculous. If Mather was inconsistent here in holding to both views, then so too were most of his contemporaries (even Newton, who argued the point with Leibniz). Most everyone preferred not to have to choose. As one of Dr. Woodward's correspondents put it in a series of triple negatives that reveal his anxiety, "I am very well satisfied 'twas his omnipotence that first put all in motion and I as firmly believe that 'tis his providence which protects and keeps it so. But tho' we are told that 'tis Providence which protects and keeps it so . . . I can hardly persuade myself that there is nothing else but that to maintain and guide it in the order we now see it." Woodward's correspondent could not "but Think that God when he first made the world put it into such a frame . . . as to enable it afterwards to be carry'd on by itself from generation to generation ('til its parts are so worn . . . another miracle is wrought to preserve it) without forcing the first maker now and then to work a fresh wonder." [31] Unfortunately we do not have Dr. Woodward's reply or the manuscript treatise which it appears he wrote on the subject. [32] But in his effort to explain almost everything by natural laws and yet to allow for an occasional miracle like the suspension of gravity at the time of the Deluge, he may well have agreed.

Mather's mature view was expressed in a work entitled (after Boyle) *The Christian Philosopher.* It is not much more than a pastiche of quotations drawn from contemporary English works to show that "Philosophy is no Enemy, but a mighty and wondrous Incentive to Religion." [33] The Christian Philosopher's science, like his philosophy, was largely secondhand; but Mather was writing almost alone in Boston, three thousand miles away from the Royal Society, and his work is remarkable both for its range of interest, which was entirely up-to-date, and its viewpoint, which was being expressed for the first time in America. Newton was naturally the hero of this work, but so too was Dr. Woodward, who furnished Mather with his views there both about fossils and vegetation.

Indeed, it was Dr. Woodward who was Mather's chief link with the Royal Society and the scientific world of his time. And it was Dr. Woodward who was entrusted with the publication of *The Christian Philosopher,* a work very much in his own spirit, though very far from his plain and direct style. From 1712 the two men were already in regular correspondence. It was Dr. Woodward who was largely responsible for Mather's election to the Royal Society, and it was through Woodward (as well as the Society's secretary, Edward Waller) that Mather funneled that steady flow of scientific contributions which he called his *Curiosa Americana.* Some of these essays, written out in the form of letters, were excerpted and printed in the *Philosophical Transactions;* most remained in manuscript. Many covered subjects congenial to the Doctor, like the one on "fossils" of July 25, 1717, or the celebrated essays on the smallpox (1721-22), but they included papers on almost every conceivable science – in twelve years at least eighty-two papers. Inevitably there was one about the American moose, read before the Society in 1714. Dr. Woodward thought them all worth printing and had to apologize eventually for his colleagues. "The Editors," he wrote in 1721, "since Mr. Waller's death are very neglectfull and partiall by which the Society suffers not a little." It was an old complaint. But Woodward assured his New England friend that he at least did Mather justice in "making the Curious here sensible to your Diligence there."[34]

Apparently the correspondence began with a Woodward query and a copy of his work. "Your excellent *Essay,*" Mather wrote in 1712, "has obliged and even commanded the true Friends of Religion and Philosophy to serve you with as many Communications as they can." He was flattered to be asked "to supply you with such subterranean curiosities as may have been in these parts of America mett withal." It was not so easy, however, to find fossils. He turned to his young friend John Winthrop, grandson of a founder of the Royal Society, for help. In 1716 Mather lent Winthrop a copy of Woodward's book and urged him to make "as full a Collection as may be of the Fossils (the Names written on each little Bundle)" to be sent on to London. Winthrop was more than willing, and promised to add to them some of "the utensills of the Pagans." Dr. Woodward had expressed great interest in the culture of the Indians. "Perhaps I may grattifye the doctors curiosity in some of the originall instruments, ancient notions and traditions, etc., which I have lately learn't and received among them." Eventually Woodward got a box of shells, but all apparently of the living variety. He was grateful, he wrote to

Winthrop, although his real interest lay in the fossils, especially "such marine Bodyes, Remains of the Deluge, as we find in Digging, Mineing, etc. These will be of Use to the perfecting my *Nat. Hist. of the Earth* and I find much Difficulty in Procuring any from your Countrys there." To Mather he was more frank. "I much admire I hear so little of Marine Bodyes, Remains of the Deluge, dig'd up in New England. Pray be more inquisitive on all Occasions of Digging, Mineing, etc."[35]

If the net profit of a dozen years was small – a box of some twenty to thirty shells only, and a piece of limestone from Sir William Phip's famous Treasure Ship – Woodward did receive at least one other benefit from his American correspondent. Mather, it appears, was always ready to lend a sympathetic ear to the Doctor throughout his many quarrels. Not only did he support Dr. Woodward in his geological views, he sided with him in his smallpox controversy also. (Among other things, both were advocates of the strange new treatment of inoculation.)[36] When Mather appeared to waver a bit about geology, he was swiftly reprimanded. "By what you write," Dr. Woodward told him, "you seem to have forgot how near I am to supplying all the Defects of my first *Essay* in my *Observations and Reflections in Answer to Camerarius.*" For his part, Mather assured his touchy friend that he wished him forever protection against the monsters that surrounded him. He received all his communications with pleasure, and he was entirely gratified to read the account of his duel with Dr. Mead "wherein his life was in danger." [37] Although Mather thought some of Woodward's defenders as scurrilous as his enemies, he never hesitated in his own support for the Doctor. America, like many other far-flung parts of the world, was ready to do its best for the embattled philosopher.

[2]

As the fossils collected, Dr. Woodward was faced with the problem of arrangement. How was one to store the various bits of stone and ore, shell and bone, that were accumulating from all over the world? It was normal to display such "curiosities" in cabinets (exactly as the coins of antiquities which they so much resembled) whose many drawers could be opened for visitors to gaze upon and admire. Dr. Woodward had several built especially for this purpose, and they have lasted beneath their great weight remarkably well. But in what order were the specimens to be arranged?

Unlike many other collectors, Dr. Woodward was not, as we have seen, indulging an idle curiosity or even a generalized wonder at the marvels of nature. He was a philosopher attempting to turn his fossils and stones to intellectual purpose. He saw at once that the only way to discover their secrets was through the experimental investigation of their properties, accompanied by a systematic and painstaking comparison. Arrangement was thus crucial, for it must either be imposed artificially for the convenience or display of the collector or it must arise from the intrinsic affinities of the materials. In a letter to Scheuchzer in 1703, he writes that he is drawing up "an Account of all the known Minerals and Fossils for your use." He is trying to reduce them to classes "according to their Natural Relationship and Affinityes." He disapproved of Scheuchzer's "alphabetical method" except as the merest preliminary; it was necessary to discover the likenesses among the stones which alone could define true and meaningful relationships and which could serve further to eliminate those objects which did not belong, which perhaps were not even fossils. Gesner, Dr. Plot, and others, he wrote later, "have fallen into a great absurdity in giving Figures of Pyritae, Flints, etc., that are lusory accidental and of which there were never two alike. That has contributed much to render these Studyes difficult, obscure, and uncertain, to common Readers. I have some Time since, reduced all the Native Fossils into a Classical Method." The foreign fossils would follow.[38]

It was a propitious moment for the attempt, although, like many of the Doctor's efforts, it proved a little premature. His friend Harris was able to get from Dr. Woodward an outline of his ideas, and this appeared in print for the first time in his *Lexicon Technicum* (1704) in an article entitled "Fossils." The *Lexicon* was a scientific encyclopedia, a strenuous effort to gather up the latest conclusions of natural science into a single massive work, and it paid frequent respects to Dr. Woodward for his views upon such matters as the Deluge, earthquakes, metals, meteors, springs, strata, and vegetation. "If you turn over Harris's Dictionary," Lhwyd wrote in disgust to his friend Richardson, "you will stil meet with fresh Eulogiums of the *Essay* as tho' it were an Abyss of Inventions and eternal Truths always own'd by mankind."[39] The article on fossils dealt primarily with stones and minerals, then still lumped with the "formed stones" under one generic name. Harris presented the Doctor's views on geological classification in the form of a table, "extracted out of a Natural History of these Bodies . . . and founded wholly upon Experiments and Observations." To the modern historian it looks like a major step in the evolution of the science, the first attempt of its kind by an Englishman. In many ways it was like the efforts of Ray and others to bring order into the botanical and zoological worlds by classifying the different forms of life. "Several very learned Men of late Years," Woodward recalled to Newton, "have happily employed themselves, and spent much Time and Labour, in reducing all Kinds of Animals and Vegetables into Method. But Fossils . . . have been much neglected. . . ." He meant to rectify the omission.[40]

There was, however, another model for classification than the one provided by the natural sciences. As Woodward sought to discover a method for
his fossils, he was very conscious of the elaborate schemes of the antiquaries
for arranging their cabinets. At first sight they had not much in common; it
was no help for geology or paleontology to bring together fossils under
headings like *common* or *rare, mean* or *precious, greater* or *lesser use,* etc.
Something like this, he allowed, might be sensible for a history of technology
or the arts — subjects which of course very much interested the Doctor — or
for the collector assembling a cabinet of curiosities for display, but hardly for
a history of nature.[41] In general he thought that it was bad classification and
bad nomenclature that had led to those common mistakes by which the same
object could be set under different names, or different objects under the same
name. Woodward never tired of ridiculing Plot and Lhwyd on those grounds.
He thought that it was particularly difficult to distinguish among stones and
minerals, which were often mixed and very confusing compared to the "fixed
Characters of Affinity or Disagreement that Animals and that Vegetables
carry along with them."[42] But he believed that he at least knew how this
could be done. The essential forms and properties of the various objects had
to be established through chemical trial, microscopic investigation, etc., above
all through the simple comparison of large numbers of apparently similar
objects. Once this was done it could not be hard to detect the mistakes in
identification that bestrewed the works of Lister, Lhwyd, and the rest.
Weeding out such errors was very like detecting the forgeries and mistakes in
a series of ancient coins or antiquities, except of course that the principles of
arrangement and points of comparison were necessarily different. Undoubtedly Dr. Woodward was the master connoisseur in his time of the stones
beneath the earth.

He tried to prove it in a last ambitious work written toward the end of his
life, entitled *Fossils of all Kinds digested into a Method* (1728). Here he
developed the scheme first announced in the *Lexicon Technicum* by trying to
classify all the ores, metals, earths, and petrifactions then in his possession.
This required especially "chymical analysis" and "subjecting them to Experiments and particularly of the Fire," and if the Doctor was aware that his
efforts were not always successful, he was confident (no doubt rightly) that
they were at least better than what had passed before.[43] As always, he was
delighted to show how readily his predecessors had erred. In the absence of
sufficient comparison, how easy it was to make a mistake! Dr. Merret, for
example, had written an ambitious work of zoological, botanical, and geological classification entitled *Pinax Rerum Naturalium Britannicarum* (1666).
There he had compared the teeth of the wolf-fish (known as *buonitae*) with
the magical stones grown in the heads of toads and worn in rings (known as

busonii), and had triumphantly shown them to be the same. "By this Method he imagin'd he had made a Discovery of a Counterfeit and Imposture of the Lapidaries in selling these Teeth for the true Toad-Stones." What he had failed to see, of course, was "that there are naturally no such Stones in the Heads of Toads, that these are really all of them Teeth of the Wolf-Fish, tho' thus found in the Earth. . . ."[44] The Doctor, needless to say, was right; his confidence was based on a relative grasp of the materials which outstripped that of most of his predecessors, although it naturally depended upon them. As a modern, Dr. Woodward understood the cumulative nature of his science, though he sometimes forgot his debts.

He was himself also frequently led astray, sometimes by his theory — which, despite his protests, necessarily affected his investigations — and sometimes by lack of sufficient evidence. Thus he insisted upon denying that either the fossil corals or the belemnites were once alive, even after the work of the despised Buttner appeared in 1714 to settle the question. As Nicolson pointed out to Lhwyd: "Dr. Woodward says, the outward crust of the true natural Corilline is of the same kind of substance with marble and limestone and therefore his hypothesis having dissolved all these at the Deluge, he cannot admit that the other was able to ride it out."[45] In *Fossils of all Kinds* there is a letter to an unknown correspondent "Of the Origin, Nature, and Constitution of the Belemnites." These fossils, Woodward insisted, were "native," although Lhwyd even before Buttner had claimed them as "extraneous." Woodward's correspondent thought them likely to be the horns or teeth of once-living animals, (although we now classify them in fact among the molluscs). Woodward had no trouble disposing of this idea, partly by comparing them with other fossils and living remains, but also by invoking his theory. As he had shown in the *Essay*: horns, bones, and teeth, "being lighter than the common Sea-Shells subsided last and consequently being lodg'd near the Surface of the Earth, and there exposed to the Weather and external Injuries, are generally perished and destroyed, few of them remaining at this day." Since belemnites were frequent, obvious, and everywhere, even to the depths of the earth, they could not be such remains. Woodward seems not to have noticed the circularity of his argument, for he agreed with his correspondent at the outset of the letter that it did not matter to his hypothesis either way, whether the belemnites were or were not "as I have asserted a meer stone."[46]

Actually it did matter, and Woodward was wrong. The fact that belemnites were entirely of a "talky" constitution (i.e., of the same specific gravity, texture, and constitution as talc) betrayed the Doctor, for although he realized that they could have become entirely petrified, he thought it impossible that all should *always* be found thus. He believed that some at least would have been unchanged or still alive, like the buried trees he knew so well. Like Ray, he could not conceive of a whole genus, with countless thousands of relics, being lost completely out of the world. But perhaps we should not press the belemnites or the corals too far; Dr. Woodward was more often right than wrong in his identifications, and he was neither the first nor the last to be caught employing a theory to determine a fact, as well as the reverse.

[3]

It was thus not only his speculations but even more his collections and their purposes which drew to Dr. Woodward the respect and admiration (albeit often grudging) of his contemporary naturalists. He was recognized in time as the leading expert of his day on fossils, surpassing all his rivals in knowledge and understanding of them. One might remain suspicious of his theories, but it was hard to argue with the Doctor over his identifications, wrong though they often were, and harder still not to be impressed with the scope and system of his collections. It is not surprising therefore that he inspired others with his learning and enthusiasm and instructed many by his method and example. Indeed, Dr. Woodward was an excellent teacher, as long as his students remained properly deferential.

Of course the Doctor always had younger disciples, some of them as we have seen trained deliberately to assist him. But he had older "pupils" besides, some of them very distinguished and influential men. There was Sir George Wheler, for example, who had won fame for himself in 1675-76 with an unprecedented tour of Greece and the East which he recorded in a very popular book of travels. Wheler was an antiquary and botanist of distinction, but he came to fossils rather late in life, perhaps after a visit to Dr. Woodward's collection about 1700.[47] "I'm glad to find you pay any Regard to Fossils," Woodward wrote to him that year, "and are in earnest to make a Collection of them. 'Tis indeed very worthy of a Gentleman of your sense and Curiosity." Wheler sent the Doctor some fossils and received a promise of duplicates, as many (in Wheler's words) "as he could find it in his heart to spare." When at last the Doctor returned a box, he had to apologize for sending no more; a "great Number of my Friends and Persons of Generosity desire Samples from me."[48] He was not exaggerating. But mutual interest brought a mutual return, and Dr. Woodward's catalog is full of acknowledgments to his older friend. Sometime later he drew up for him and for the Bishop of Man (two of his "most generous Benefactors") a set of special directions and queries relating to the "Oeconomy of the great Abyss, and Damps in Mines, Fogs and Mists on great Mountains and Meteors."[49]

Sir Robert Southwell was another older disciple. He was very much a man of the world, diplomat, clerk of the privy council, eventually principal secretary of state for Ireland. He was also an original member of the Royal Society and eventually its president, an intimate of Boyle, Petty, Pepys, and the rest. It was Southwell who had helped to win the Gresham College post for Woodward.[50] And as we have seen, it was Southwell among others who had come under the spell of the Doctor's ideas. Dr. Woodward's theory was very appealing to a man who had long speculated about the same problems from something of the same perspective. Twenty years before, Southwell had shared his ideas with William Petty who, however, found some serious objections to his "Theory of the Deluge." (Petty preferred to think, like so many others, that the Mosaic account was "a Scripture Mistery which to explain is to destroy.")[51] Apparently Southwell soon shelved his own ideas for those of Matthew Hale, another friend who had given even more thought to the subject and whose work he did much to encourage. If there was always something of the amateur about him, Southwell was neither ignorant nor unsophisticated on the subject; he frequently attended the sessions of the Royal Society and he kept the respect of some of its most famous members. His conversion by Woodward was therefore a real achievement, another sign of the Doctor's influence.

But the relationship was not one-sided. Southwell responded to Woodward's tutelage by furnishing him with fossils for his collection and by introducing him to others, especially William Cole of Bristol, who could also help. Southwell owned a great house near Bristol and had known and encouraged Cole for a long while before he wrote to him in 1694 about some topazes. "Mr. Woodward is a great judge of these Things, admires them much and hath oblig'd me to furnish his collection with two or three." He was about to publish "some admirable Reflections upon his View of the Inward Parts of the Earth, and undertakes to illustrate thereby the manner of the Deluge and other profound thoughts which will certainly entertain you." Southwell thought it particularly commendable that Woodward was "exactly orthodox in all."[52] Cole had himself been collecting natural curiosities for almost half a century, and while old and a trifle cantankerous could prove to be a great help to Dr. Woodward. The topazes were slow in coming, but at Southwell's urging Cole read Woodward's book soon after it appeared and was impressed. Dr. Woodward had written (so Cole wrote to his friend) "more to purpose and with more seriousness" on that subject than anyone yet. This was praise indeed from a man who had his own independent view of the fossil origins. (He believed that the fossils were once real but had changed their forms while beneath the earth.) Woodward, Cole assured Southwell, need not fear his critics, for "no one could write to any purpose against his

Essay but such as have not only laboured under ground more than he hath done but allso were forward in the principles of divine and Naturall Philosophy." There were few enough of those, though Cole had heard already of five or six who intended to write against the work.[53]

Cole and Woodward were soon directly in correspondence, Cole promising to write "with that freedom which is allowed between true freinds." He remembered when he had lived on the Isle of Wight, some twenty years before, finding fossil leaves in the clay along with some hazelnuts. The country people thought that they had been left by Noah's flood, "than which vulgar traditionall opinion I never had a better."[54] Unfortunately, he no longer had any in his collection. Woodward was extremely interested; he thought to send one Thomas Cole at once to Cowes to see if he could obtain some of the nuts. They were especially intriguing for his theory about the exact date of the Flood, which he had set in the spring before they were ripe, and here was possible evidence on the point. Eventually, he accumulated a great number to support his notions. Cole's own effort to provide some for him were stymied, though he did make some other discoveries which he promised to send on to London.[55]

In short, Dr. Woodward seems never to have lacked for supporters at home or abroad, among the young or among the most respected naturalists of the time. In his fossil catalog he invokes a circle of botanists in London, all of whom appear to have been friendly and who certainly sided with him in the controversy over the origin of the fossil vegetables. "Before I drew up the following Account, I took along with my own, the Judgment and Opinion of four Persons that have been very conversant with Vegetable Bodies, and are very eminent for their Skill and knowledge in Botany. These were Dr. Plukenet, Mr. Stonestreet, Mr. Buddle and Mr. Doody and we carefully view'd and examin'd every individual Body herein mentioned."[56] None of them is much remembered today, but they were all accomplished naturalists or collectors, well-known in their time. Leonard Plukenet was probably the most celebrated, at one time supervisor of the Royal Gardens at Hampton Court, the "Queen's Botanist," and a prolific author. He seems to have approved Dr. Woodward's theory at once and, discovering his own reasons for quarreling with Ray and Sloane, no doubt relished the new alliance.[57] Adam Buddle was a specialist in "mosses," and his collections, which came eventually to Sloane and the British Museum, have been praised for their accuracy, diligence, and knowledge. Samuel Doody was the superintendent of the Apothecaries' Garden in Chelsea, and was particularly renowned for his knowledge of London flora. But the most extraordinary of the quartet was the most obscure, lost now even to the pages of the *Dictionary of National Biography*. Parson Stonestreet was a London clergyman who built one of the really immense collections of curiosities, natural and human, in London. According to Nicolson, "Mr. Stonestreet's shells are wondrous fine. So are his fossils and his coins. . . ." But best of all was the man, learned, modest, good-natured, and religious. Dr. Woodward visited him often and especially admired his "Sagacity in searching into natural things and success in methodizing them"[58] We shall meet him again as an antiquary.

Finally I must not pass by John Morton, the Northamptonshire historian, though he was neither the first nor the last naturalist of consequence to find the Doctor's work important. Morton was a country parson with a passion for antiquities and natural history. "My acquaintance with Mr. Ray," he recalled, "initiated me early in the study of plants; from the reading of Dr. Lister's books, I became an inquirer after fossil shells; and my correspondence with Dr. Woodward, Dr. Sloane, and Mr. Lhwyd, has supported my curiosity."[59] He managed somehow to avoid taking sides, although solicited by all parties. When the quarrel between Woodward and Lhwyd came to a head in 1696, he undertook to mediate between his two friends, and it is interesting to see that of the two it was Lhwyd who proved to be the more stubborn and recalcitrant.[60] Morton had from the first adopted Dr. Woodward's theories, for which he was rewarded in 1700 by being nominated by his mentor for the Royal Society. But he kept Lhwyd's friendship and entered also into a fruitful correspondence with Sloane, even while continuing to support the Doctor's theories with his many fossil discoveries.[61] At length he published his ideas in his justly praised *Natural History of Northamptonshire* (1712). "The Author," noticed William Nicolson, "guides himself wholly by Dr. Plot's Method and chiefly by Dr. Woodward's Hypothesis." "I have had for many years," Morton wrote in his preface, "a Correspondence with other Persons, the most eminent in Natural Learning, and particularly with Dr. Woodward . . . As in these Papers there are various Subjects that are treated of in his *Natural History of the Earth*, I have been frequently led to compare what he hath there deliver'd with Nature and things which that Work every where answers with great Truth and Exactness. On which Account I have had Occasion to quote it very often."[62] Morton was a meticulous observer and collector, one of Woodward's most helpful contributors, respected by everyone and in the end one of his most persistent and capable disciples.[63] His advocacy is an excellent reason for admitting the persuasiveness of Doctor Woodward's ideas.

[4]

But before we consign Dr. Woodward to his most effective critics, the wits, we must make a last effort to place him in the science of his day. It is right to see how influential were his notions, how highly regarded his collections, and how jealous his rivals. There can be no question that he looms large in the natural history and philosophy of his own time, and larger than has been suspected. But how does he measure up to the truly great men of his day, to Locke and Newton, for example, or to Leibniz? Needless to say, they all knew his work; if nothing else, Dr. Woodward was unavoidable. And though we know less about their relationship than we should like, less about what they thought of him than he of them, there is enough anyway to be suggestive. Unfortunately, I have not been able to discover much to his relationship with Locke, nothing more than that Locke gave a few fossils to Woodward, Woodward some medical advice to Locke, and that the Doctor admired his controversial manners.[64] But Woodward's connections with his other philosophical contemporaries were more elaborate. Typically, they were complicated both by intellectual and personal recrimination. Dr. Woodward had a genuine respect for both Newton and Leibniz which was (for him) unusual, coupled with a jealousy and suspicion which was much more characteristic. More importantly he differed with both of them on fundamental matters about the nature and method of natural philosophy. Woodward believed himself to be a thorough Baconian (whatever his practice), a pure observer and experimentalist, in an age which reason and mathematics had come to dominate.

It is unlikely that Woodward had much to do with Newton before the latter became president of the Royal Society in 1703, except through his work. I have tried to indicate above how Woodward employed the idea of gravity for his theory about the dissolution of the earth. It is probable, however, that he did not really understand Newton's views very well. Certainly he had little mathematics himself, and he saw no application for it in any of his work. As for Newton, he appears to have sympathized first with the theories of Burnet and later with those of his disciple William Whiston rather than with Dr. Woodward, although he did not lack interest in the Doctor's work.[65] Indeed, at some point he gave to Woodward one of the most prized possessions in his collection, "a Stone . . . that was brought over from Antwerp into England by Fr. Merc. Van Helmont, as his father's Ludus." Later still, he gave him another of these "waxen veins" or, as they were better known, *Ludus Helmontii,* as well as some other stones. Woodward returned the compliment with a copy of his *Essay.*[66]

After 1703, as an active member and officer of the Royal Society, Woodward came often into contact with the great man; in 1704 he even claimed him as his "particular Friend" in a letter to Scheuchzer. At that time and for some years afterward he seems to have hoped to influence Newton against Sloane — not without real possibility, since Newton had no love for his rival and seems to have been jealous of Sloane's position and independence.[67] Newton did not have much use for natural history, certainly not as an end in itself, and he must have preferred Woodward to Sloane on philosophical grounds. Nor did he have much respect for the antiquarian activities of his contemporaries, at least insofar as they remained prizes in the cabinets of collectors. Although he was consumed (in private) with an interest in the past, it was centered on chronology and Scripture. (One interest he shared with Dr. Woodward, however, was in the settlement of Americans after the Flood).[68] "Sir Isaac Newton," someone told Spence, "though he scarce ever spoke ill of any man, could scarce avoid it towards your virtuoso collectors and antiquaries. Speaking of Lord Pembroke once, he said, 'Let him have but a stone doll and he's satisfied. I can't imagine those gentlemen but as enemies to classical studies; all their pursuits are below nature.' "[69] It is likely however, that he thought better of Dr. Woodward, who was never a mere collector. Indeed, to one very competent observer who knew them both (the famous scholar Richard Bentley), they were a complementary pair whose activities deserved almost equal praise.

> Who nature's Treasures would explore,
> Her Misteries and Arcana know
> Must high as lofty Newton soarr,
> Must stoop as delving Woodward low.[70]

For Woodward, Newton remained still "a man of sense" in 1708,[71] but with the crisis of 1710 he began to have his doubts. Upon his expulsion from the Council Newton is said to have commented that Woodward might be an excellent natural philosopher but he could scarcely qualify as a moral one. For Woodward that was the last straw. He saw now that Newton was an old man "and very humorous," that he aimed at power, and that between Newton and Sloane the Society was likely to be ruined. In 1711 he pointed out to Scheuchzer that no *Transactions* had been published for nearly two years, and he was soon able to add, with wry satisfaction, that Newton and Sloane "are become bitter Enemies to each other." Through their management, he wrote, "The Royal Society is brought to a very low Ebb and indeed to the utmost Contempt."[72]

It is no wonder then that Woodward took sides (at least in private) against Newton in his quarrels with the astronomer Flamsteed and with Leibniz. His letters to Scheuchzer furnish an interesting commentary on those discreditable incidents, with the Doctor aroused against the insolence and tyranny of the new dictator of the Society. "Indeed Sir Is. Newton makes it his constant Indeavour to decry and depress every man that pretends to Science beside himself." That alone, Woodward insisted, explained his ill use of Flamsteed and Leibniz. "I may tell you it privately as a Friend, he is a Man of most arrogant and invidious Disposition and very apt to lessen all learned Men and to use them ill."[73] This is, incidentally, not far from the sentiments of Newton's latest biographer, nor indeed inconsistent with the encomiums of his admirer Stukeley.[74] There was but one way to treat the great scientist in those years, and that was (in Woodward's phrase) to "condescend to trumpet up his Fame." This surprisingly he was willing to do, however, and (as we have seen) he dedicated one of his pieces in 1714 to Newton — whose request, so he claimed, had both inspired and guided the whole venture. To the end he remained absorbed in the works of the great man, the optics equally with the physics.[75]

Curiously, when Sloane was forced to resign in 1713, leaving Newton in complete control of the Society, Woodward's sympathies repaired momentarily to his old rival. With Sloane gone, he saw that Newton was free "to transact all as he pleases." No matter that Sloane had once done the same. "There's a good Affection and Understanding," he told the surprised Scheuchzer in 1716, "betwixt Dr. Sloane and me." As for his collections, "I know no man so curious in such things, or of a more generous Disposition." It was too good to last; in a few months' time, Woodward reverted to his old posture and Sloane was, once again, "very deep and designing," entirely given up to his own affairs.[76]

With Leibniz it was a little different, perhaps because their acquaintance (like Woodward with Scheuchzer) was entirely by letter. Woodward read a piece on fossils by the great philosopher in 1709, but was not impressed. "I admire that he has no further Insight into those things, at this Time of day."[77] Leibniz was very interested in the problems of the Creation and Flood and had ideas of his own, many of them inspired by his supervision of the Hartz mines.[78] It was not long before the two men began a correspondence during which they discussed their theories of the earth and other matters both natural and antiquarian. In one letter Leibniz asked Woodward's views about the possibility of antediluvian metals. The Doctor replied that he believed that they had existed, but he had not written about them for want of evidence. "For I had determined neither to advance nor to assert anything which was not either manifest from observation of the thing itself or declared by the authority of Holy Scripture." Woodward could not accept Leibniz' views, which he thought resembled Descartes' in the liberty he was willing to take with both Scripture and the geological facts. Thus he was unable to supply Leibniz (who had asked) with any indication of burning in his large collection of fossils. (The great philosopher was here exploring the idea of the igneous nature of many rock formations, an idea of great potential promise.) Leibniz, of course, defended his views and the method of Descartes, and suggested that it *was* possible to deal more freely with Scripture than

Woodward had allowed — a view which, as we have seen, he shared with Newton. There was obviously no settling a dispute which rested on so fundamental a ground; the empiricist and the rationalist (if I may be allowed to simplify) could not easily be reconciled.

Still the two philosophers continued to correspond anyway, and to share at least a common hostility to their mutual enemy. "No man," Woodward assured him in 1714, paid "a greater Deference to your Learning and Judgment than I do." It was a most extraordinary tribute for someone who did not share his views. At the same time he sent Leibniz by messenger the full story of his enemies in the Royal Society, "the Bottom of the Affair of Sir Is. Newton," and offered his services. "I shall be very forward to do you Right in that and all other Respects." Unhappily, I cannot find that anything ever came of the offer, or that their correspondence continued beyond this last English letter that passed between them in 1714.[79]

If the relationship of Dr. Woodward with Newton and Leibniz thus reflects no particular honor upon him, it nevertheless helps to define his place in his time. Both of his great contemporaries (like everyone else) took notice of him and treated his work with respect. Both had to overlook his cantankerous personality — and he theirs — to do so. Neither was really taken by his theories, but both of them deferred to his superior knowledge as a naturalist, and at least neither found his speculative notions contemptible. Like each of them, Dr. Woodward was a philosopher with a large view which embraced both nature and history; and like each of them (but perhaps more like Leibniz than Newton) he left his system in fragments, only partly visible to his contemporaries.[80] Unlike them, posterity has not found it worthwhile to resurrect and assess that system — perhaps rightly so. For Dr. Woodward was not the equal of the greatest thinkers of his age; he was only one among the second-best.

VII.

The Virtuoso Satirized

[1]

Next to collecting fossils, Woodward had told Lhwyd, he loved antiquities best. The combination was irresistible and was shared by many of his contemporaries. What could be more natural than that the inquirer into the earth should find human remains as fascinating as the others? The artifacts that lay preserved beneath the ground were as mysterious and challenging as the fossils, and offered as we have seen a similar possibility. To divine their meaning, they also had to be collected and classified; to discover their significance, they too had to be placed correctly in the past. Indeed, the one appeared the natural extension of the other; put together properly, the "curiosities" of the collectors could be made to recount the whole story of the world from its first beginning. Thus geology, paleontology, archaeology, history, slipped imperceptibly one into another; for Dr. Woodward they appeared aspects of a single great field of discovery. To assemble the evidence and find the key to their arrangement was his lifework.

Of the two pursuits that Dr. Woodward undertook, the quest for antiquities was the older and the more commonplace. It also had a different origin; it was essentially a byproduct of the revival of antiquity that had begun with the Italian Renaissance. By the sixteenth century it was widespread throughout Europe. Its initial impulse was the restoration of classical Latin and Greek letters through the critical examination of texts — and so it was from the first tied to philology. The Middle Ages had retained an interest in ancient letters but it was largely indifferent to the precise original (i.e., historical) meaning of the ancient works. The Renaissance classicist wished to imitate their virtues exactly. But to do this he had necessarily to restore the whole lost world of classical antiquity. How else could one understand the works of Cicero in ignorance of the historical circumstances of his life? What could one make of his letters and orations without knowing the political and social institutions, the religion and customs, etc., of ancient Rome? The antiquarian impulse was the result of this desire, and was fed and sustained by it throughout the Renaissance and beyond. It turned humanists from the classical literary texts to the other remains of antiquity, to the documents, coins, inscriptions, and monuments that could illuminate them. In the end, the commentaries on authors escaped the confines of that limited genre and took on independent life. Separate works on ancient theaters, weapons, calendar, costume, and so on, began to appear, while philological treatises and handbooks of antiquities were produced to meet the growing demands of the classical curriculum. Interest soon began to extend beyond ancient Greece and Rome to the Egyptians, Celts, Saxons, and so on.

When John Woodward was a child, the standard grammar-school education consisted entirely and exclusively of the reading and imitation of classical authors, i.e., the teaching of Latin and Greek grammar. For a century and a half the curriculum had hardly varied, and several generations of Englishmen had been nourished on that restricted diet. The informing idea behind it had been to educate the governing classes to their appropriate responsibilities; and indeed, the ruling classes in England and throughout Europe had been won easily to the idea that it was the ancient literary texts, in fact, that best trained the political man. No one figure therefore so dominated the ordinary thought of the period than Cicero, the Roman citizen and statesman, orator and philosopher, the author of the most influential of educational treatises, the exemplar of Latin and the exponent of Greek. In the original and translation, in quotation and paraphrase, his works and his ideas may well have been the most popular throughout the early modern period.

But if the schools were restricted to grammar, they inadvertently opened windows on that wider world of antiquity without which the authors remained unintelligible. If Cicero was in the curriculum, so also were the Latin and Greek historians and, even more significantly, the new philological manuals and antiquarian compilations that accompanied them.[1] An example that may well have been known to the young Woodward, since it was widely used as a school text, was Thomas Godwin's *Romanae Historiae Anthologia* subtitled *An English Exposition of the Roman Antiquities wherein Roman and English Offices are parallel'd and divers obscure Phrases explained*. Its four chapters dealt with the Roman city, religion, state, and military art, in 277 pages. There were more ambitious books of the kind, like those published by Woodward's contemporaries Basil Kennett on Rome and John Potter on Greece, and still grander treatises in Latin. In time, Woodward came to know and to read them all.[2] Here indeed was an innovation, for the ancients themselves had had relatively little concern, as they had not the same need, for antiquities. (They were after all, not reviving antiquity but creating it.) Paradoxically, therefore, the antiquary was a "Modern," though his subject matter was ancient — and even more disconcertingly, the exponents of imitation (the "Ancients") began to quarrel with the antiquaries! One of the basic differences between Temple and Wotton in the "Battle of the Books" was whether "philology" ought to be admitted to or banished from among the arts and sciences.[3] It was difficult, perhaps impossible, to reconcile the demands of imitation, with its concentration on literary style and expression (on "eloquence"), with scholarship and its accumulation of detail. The problem was never resolved.

The interest in antiquities advanced by slow degrees in England in the sixteenth century. By 1634, however, Peacham's *Complete Gentleman* could introduce a whole new chapter on the subject. According to this standard manual, the gentleman must now add to his other attainments the collection and discrimination of classical remains. He was to become a connoisseur and antiquary, adding to his "polite" literary skills a smattering at least of the new classical scholarship. Peacham dwells especially on the pleasures furnished by the antique statues, inscriptions, and coins, and notices that the gardens of gentlemen in France, Spain, and Italy (like those of ancient Rome) were being beautified by them. There was nothing, he thought, "more delightfull, more worthy observation, than these Copies and Memorials of men and matters of elder times whose lively presence is able to persuade a man, that he now seeth two thousand yeares ago."[4] For the gentleman steeped in the classics, it was natural to experience a special thrill at the sight of the ancient figures. Aesthetic delight was mingled therefore with educational and practical advantage in the new taste for antiquities. Peacham was able to describe the exact moment when an Englishman appeared to rival the Italian virtuosi. It was, he thought, to Thomas Howard, Earl of Arundel, that the nation owed its first glimpse of Greek and Roman statues (the famous collection known eventually as the *Marmora Arundeliana* and settled finally at Oxford), and to Charles I its first royal patronage.

But if statuary was beyond the means of the ordinary gentleman, inscriptions and coins were easy enough to obtain and had indeed been admired and collected by Englishmen from the beginning of the revival of learning, from the days of Thomas More and Erasmus. Peacham is insistent upon their utility. "Those old memorials tend to the illustration of Historie, and of the antiquitie of divers matters, places, and cities, which otherwise would be obscure, if not altogether unknowne unto us." He gives an example: on the reverse of a coin of the Emperor Nerva there is a team of horses set loose and an inscription *Vehiculatione per Italiam remissa,* "Whereby wee learne (which no Historian remembers) that the Romane Emperours did command all the carriages of the Countrey every where; that Nerva did resist that burden and acquitted them of it; and that this grievance was so heavy, that Copies were stamped in remembrance of the Emperors goodnesse that eased them of it."

Whether everyone found this news important may seem a little doubtful, but Peacham was persuaded that it was a matter of general concern. More dramatic, however, were the figures on the coins. "Would you see a patterne of the *Rogus* or funerall pile burnt at the canonization of the Romane Emperors? would you see how the Angurs Hat, and *Lituus* were made? Would you see the time and undoubted modells of their Temples, Alters, Deities, Columnes, Gates, Arches, Aqueducts, Bridges, Sacrifices, Vessels, *Sellae Curules,* Ensignes and Standards, Navell and murall Crownes, Amphytheaters, Circi, Bathes, Chariots, Trophies, Ancilia, and a thousand things more?"[5] One had only to look at the old coins to find them vividly and accurately represented. Through them the whole world of ancient Rome could be recovered and experienced. No wonder the English gentlemen, despite the misgivings of some, hastened to the new pursuits and began to adorn their cabinets with the antique remains. The Italians had lost their monopoly; the English virtuoso had come into his own.[6]

Yet even as he did, he became the object of fun and abuse. The very success of the antiquary provoked criticism and concern, especially perhaps in those who felt the danger of the new interests to the polite attainments of the gentleman. The quest for ancient things, especially where it seemed to become an end in itself or where it required too much erudition, seemed incompatible with the other activities of the man of affairs. The collector or the scholar did not necessarily sit well with the wit and man of the world; inevitably, they became the object of satire. Already in John Earle's *Micro-Cosmographie* (1628) the mock "character" of the antiquary was complete.[7] And there it remained for a century and more, as vigorous in Woodward's day as upon its invention. For Earle, the antiquary was "a man strangely thrifty of times Past."

Hee is one that hath that unnaturall disease to bee enamour'd of old age, and wrinkles, and loves old things (as Dutchmen doe Cheese) the better for being mouldy and worme-eaten. . . . A great admirer he is of the rust of old Monuments and reads, onely those Characters, where time hath eaten out the letters. . . . His estate consists much in skekels, and Roman Coynes, and he hath more Pictures of Caesar than James or Elizabeth. . . . He loves no Library but where there are more Spiders volumes than Authors and lookes with great admiration on the Antique worke of Cob-webs. . . . He would give all the Bookes in his Study (which are rarities all) for one of the old Romane binding, or six lines of Tully in his owne hand. . . .

Allowing for exaggeration, it was not an unrecognizable portrait of Dr. Woodward or his antiquarian contemporaries. What it failed to take into account, of course, was the ultimate seriousness and the real significance of these new activities. For even at their most frivolous (and collectors then as now could merely follow a fashion), the virtuosi were opening up new pathways to the past, an historiography unknown to the advocates of imitation and eloquence. Like the men of the new science, their achievement was in the end more fruitful of consequence than anyone yet suspected.

[2]

The Royal Society had not meant to include antiquities in its *Transactions*. But it was impossible to keep them out when almost every member was interested.[8] From the first, then, the minutes of the proceedings as well as the *Philosophical Transactions* recorded the antiquarian recreations of the fellows. At times historical and philosophical interests could intersect precisely, as when Halley's mathematical skills were applied to a problem of ancient chronology or geography.[9] More frequently they simply illustrate the independent, though parallel, activities of the virtuoso, describing and measuring the newfound world of space and time and reveling in the freshly observed fact for its own sake.

Whatever the impulse, there was nothing more common by Dr. Woodward's day than the collector who mingled both natural and human remains in his cabinet. This was true from the time of the first large assemblage of natural curiosities in England, the collection of the Tradescants (cataloged in 1656), which soon came to embrace antiquities when it was transferred into the Ashmolean Museum. The repositories of the Royal Society and Gresham College likewise combined both interests, while the antiquaries began increasingly to notice the curiosities of nature. And if there were some (like the naturalist Richard Richardson) who declined to pursue them together, it was a bit self-consciously and defensively. For the physicians, especially, the combination was somehow irresistible.

Of course a collector needed money. Yet it is surprising how much even poor men like Humfrey Wanley or John Bagford (both friends of our Doctor) could accumulate. The great aristocrats naturally formed the greatest cabinets and the greatest libraries; it had become almost mandatory to display them by Woodward's day. The competition for old books and manuscripts between Somers and Harley, Sunderland and Pembroke, was notorious. But the most prominent physicians held their own, helped no doubt by their great wealth and connections. "One busies himself with paintings, antiquities or prints," a French visitor noticed,[10] "the next with natural curiosities in general or with particular departments of them. . . ." Nor did it interfere seriously with their practice. "The apparent inattention with which the English practitioners exercise their calling," he noticed, "is sometimes of incalculable value to the patient. Nature, it is suggested, frequently takes advantage of their negligence to exert all her own efforts in effecting a cure." It was obviously better than blistering or bleeding anyway. No one in Woodward's time outdid his two great contemporaries and rivals, Doctors Sloane and Mead. And Dr. Woodward, himself, was not far behind.

When Ralph Thoresby, the Leeds antiquary, visited London in 1701, he dutifully made the rounds of the great collections. He himself on somewhat limited means had for years and with single-minded intensity been collecting curiosities, both natural and historical. Like his friend and fellow antiquary in Carlisle, William Nicolson, he divided his time between the two, though like his friend he was probably more interested in history than in botany or geology. A letter as early as 1692 is typical in its exchange of views: the inscription on a recently discovered coin of Canute, a parcel of stone specimens (including a copper-piece, and others of lead and sulphur), followed by a request for information about Roman and British antiquities for their respective northern counties. Each was working on a local history. When at last they met in 1694 they retired at once to Nicolson's museum, "where he showed me" (Thoresby kept an admirable *Diary*) "his delicate collection of natural curiosities (and very kindly bestowed several of them upon me) some coins and medals . . . many choice authors in print, but above all . . . his own most excellent manuscripts."[11] Thoresby's chief interests were Roman antiquities and English coins; to those he added autographs. Nicolson was more concerned with Anglo-Saxon antiquities. But neither felt any limits to their historical or natural interests. In this they were representative virtuosi.

In London, Thoresby first visited the Earl of Pembroke's "incomparable museum of medals," where he was entertained for several days. On one occasion, after being guided about the collections by the knowledgeable Earl, he was taken to dinner with Dr. Wake (later Archbishop of Canterbury), Dr. John Locke, Sir Robert Southwell, and Dr. Woodward. He was astounded at the extent and variety of Pembroke's museum: even its series of forgeries were remarkable! Besides many rare medals that Thoresby had never seen, Pembroke showed him "a strange variety of counterfeits, in some the metal genuine, but inscription false; in others, one part of the metal right, the other side soldered to it wrong; with a medal of the two famous Paduan brothers, whose counterfeits are not only hard to be distinguished from the originals, but preferred to bad ones, though genuine." Thoresby took the autograph of the collector as well as copious notes on the collection and a few gifts for his own museum. He was quite overwhelmed by the experience.[12]

While in London he visited the collection of the Royal Society, to which he had earlier been elected, and met its most distinguished virtuoso, "the famous Dr. Evelyn." The diarist was now an old man; with his good friend Mr. Pepys he had helped to bridge the generations from Arundel to Pembroke. Dr. Woodward knew him, though not apparently very well. He was the very model of the wide-ranging savant, bibliophile, and connoisseur, and it was no accident that he had received a dedication to the life of the greatest of the continental virtuosi, Nicolas Peiresc, some years before. From those men, from Peiresc and Evelyn, the English translator had drawn a lesson: "that knowledge . . . must be the principal accomplishment of the Gentleman and that the compleatly knowing man, must be Janus-like, double-fac'd, to take cognizance of time past, and to understand the world from its cradle, as far as any Monuments of Antiquity can give Sight, as well as of the late-past, or present times, wherein he lives."[13] It was the aspiration of the age.

In London in 1701 there were many private collections to look at. Parson Stonestreet, Dr. Woodward's friend, had a good collection of Roman coins "and a most surprising one of shells," several thousand from all over the world. Above all, Thoresby was impressed by the museum of Dr. Sloane. "He has a noble library, two large rooms, well stocked with valuable manuscripts and printed authors, an admirable collection of dried plants from Jamaica, the natural history of which place he has in hand. . . . He gave me the printed catalogue and some Indian seeds: he has other curiosities without number, and above value. . . ." Nicolson told him that it exceeded the collection of many foreign princes. A year later Sloane was bequeathed the collection of William Charleton. Thoresby had earlier seen it in Charleton's chambers at the Temple. He thought it "perhaps the most noble collection of natural and artificial curiosities, of ancient and modern coins and medals, that any private person in the world enjoys." There was the greatest variety of insects and animals, corals, shells, and petrifactions, that Thoresby had ever seen. He spent most of his time with the coins, however, British and Saxon, Greek and Roman. "He has also a costly collection of medals of eminent persons in church and state, domestic and foreign reformers."[14] Now, all belonged to Sloane. And yet Sloane had only just begun to accumulate the vast collections that subsequently became the foundation of the British Museum.

Inevitably Thoresby visited with Dr. Woodward, and he found in his home also a fine and representative museum. It was "most curious in natural curiosities, fossils, gems, minerals, ores, shells, stones, etc., of which he made me a noble present." It was the beginning of an exchange that brought to Thoresby a startling number of new accessions.[15] But Dr. Woodward's museum was not lacking in books, either, and "a curious collection of Roman antiquities, not only of urns, but gems, signets, rings, keys, *stylus Scriptoriis, res turpiculae,* ivory pins, brass *fibulae,* etc." The Doctor had obviously more than the ordinary credentials of an antiquary. Thoresby, echoing his friend Richardson, thought it only a pity that "so ingenious a person should not have more friends."[16]

All this Thoresby reported to Nicolson. But his energetic friend had soon seen everything for himself. In his years at Carlisle Nicolson had advanced his career from Archdeacon to Bishop, and politics now brought him often to London and the House of Lords. When there, he never missed an opportunity to visit with his fellow virtuosi. He too kept a diary; he too recorded with relish the details of his visits, writing often to Thoresby. And despite his continuing skepticism of Woodward's hypothesis, he managed skillfully to retain his friendship and to see the Doctor often. In 1702 he was shown the fossils, noticed Woodward's busy practice, and remarked on his progress toward the great work. He too saw Pembroke's cabinet of medals, "the best I ever observ'd," with its rarities and counterfeits.[17] He visited the Tower and the libraries of the Bishop of Norwich and the King to look at rare manuscripts. He dined with Stonestreet and examined the parson's museum, "such a Collection of Natural Curiosities as I had never seen before."[18] He returned often to see Woodward.

Their friendship ripened. On Nicolson's visit to London in 1704, the two men dined together and Woodward showed his visitor a reputed Celtic coin. He was very angry at the Archbishop of York for renouncing once and for all what Nicolson continued to describe (in his Diary anyway) as "the Hasty-pudding Doctrine of his Theory of the Earth."[19] He showed Nicolson his method of arranging his fossils, also a "Northern monument" with inscription, and other curiosities both natural and human. On another visit to Gresham College, Woodward would hardly let the Bishop view the College collection but hastened him to his own, "as he thinks it, richer Museum."[20] Still, Nicolson was impressed by the Doctor's immense collections, and found agreeable conversation there with Mr. Harris, Mr. Hutchinson, and others in Woodward's entourage. Among the vegetable remains that the Bishop viewed there were some mosses which Woodward insisted were an "infallible Argument that the deluge happened in May." Woodward was obsessed by the subject. When it got dark, "just as we came to the first Drawer of his *Cornu Ammonis,*" they withdrew to dinner. Afterward they returned to the Library and were shown several of Woodward's manuscripts, especially his *Catalogue of the Rarities of the Museum,* "with discourses on the chief of 'em in a good Method." (This was the work published, still incomplete, twenty-five years later.) There was also a *History of Metals* in two parts, a work that was ready for the press. (It was recopied again in 1724, but never published.)[21] Most interesting was his *History of America.* Here Woodward tried to prove that at a very early time America was populated by Europeans who had brought with them their culture and implements. (He argued from the similarity of their remains.) While discoursing, Woodward enlarged on a favorite theme, running down the Egyptians "as mistaken Masters of antient Learning." (As we have seen, Woodward left a manuscript on this subject, but the American History seems to have perished.) Nicolson was impressed, but Woodward had been hurt by the fossil controversy. He told the Bishop that he was inclined to withhold his new labors since he had been "so slenderly rewarded by the Publick" for his earlier effort. Nicolson and Harris urged him to reconsider.

Woodward's touchiness was growing. Only a few days after this encounter Nicolson spent another evening with the Doctor,[22] only to discover that he was now running down his old friend Harris, "whom he represents now as a forward scribbler not to be imitated." It was a strange remark (unfortunately without explanation), since Harris was Woodward's chief exponent in these troubled days. Whatever the cause of their quarrel, it was apparently soon patched up.

For Nicolson there were other (friendly) visits through the years. At the beginning of 1706 he was shown more of the Doctor's manuscript works, several of which were completely ready for the press.[23] Foremost among them were his two catalogues of fossils. His own collection now included 1,760 English specimens alone! Nicolson saw works on the weather (*On Foggs and Vapours*); on the natural history of metals and minerals; on the practice of assaying metals, ancient and modern; on the learning of the Egyptians; and *Of the Origins of all Nations from Noah and his Sons.* Woodward was delighted with his friend; in 1705 he proposed him successfully for the Royal Society. In return Nicolson confined his skepticism largely to his *Diary.*[24]

Others were less tactful and less impressed. Controversy seems to have made Woodward more cautious about exhibiting his collections, less communicative and certainly more cantankerous. Foreign visitors repeated Thoresby's and Nicolson's tours, but were left sometimes with a different impression. The cultivated traveler was expected to visit the public and private cabinets of his hosts. When Dr. Lister returned from a tour of France, he scrupulously recorded his impressions of all the collections of the virtuosi that he had seen there. Unfortunately his work was satirized almost at once in *A Journey to London in the Year 1698 after the Ingenious Method of Martin Lyster,* a parody by William King, the author two years later of *The Transactioneer,* in which a Frenchman was shown to visit all the famous (though here quite mythical) collections of rarities, pictures, and statues to be found in the city. There was infinite amusement to be wrought in describing the enthusiasms of the virtuosi for mussels, webbed toes, tadpoles, and butterflies, not to say rare coins, old cats, and a wonderfully ancient Scottish stone with the names of all who had fallen at Chevy Chase.[25] More reliable was *The Relation of a Journey into England and Holland in the Years 1706 and 1707* by Christopher Erndtel, a friend of J. J. Scheuchzer's. Inevitably the traveler visited London, called at Gresham College, and met Dr. Woodward. He was enormously impressed with the collections of minerals and fossil shells which he was shown, and with the library, though less with Woodward's Latin, which he found hard to understand. Even so, it was "with much Difficulty and straining of the Voice that he shows his Curiosities, which when you see, you must take care you Touch not with the tip of your Finger, neither look into his Books except he hold 'em to you in his own Hands."[26]

When a German collector, Zacharias Conrad von Uffenbach, came to London in 1710, he too made the rounds and reported his observations. He was especially struck by the contrast between the two rival virtuosi, Sloane and Woodward.[27] The former had addressed his visitor amiably in French and personally escorted him through his growing museum. Uffenbach was impressed by the quantity and diversity, but also by the high quality of the specimens. His host told him that the Venetian ambassador had offered him fifteen hundred pounds for the collection, but that he had been refused. He was shown a great variety of stuffed animals and strange fishes, a large accumulation of ores and formed stones, and a handsome collection of insects. Sloane's shells were very choice. There were Indian costumes and weapons, antlers as large as those at Windsor; there were four hundred varieties of agates and many other costly stones and materials. They drank coffee and looked at the books and coins; the hours fled. Uffenbach was grateful for the time of so busy a man.

Earlier he had met Dr. Woodward; it was a very different experience. After calling upon him five times in vain, he was at last admitted. There he waited for fifteen minutes until an apprentice appeared to ask his name, another fifteen minutes until he returned. Woodward was still in bed, having retired late the evening before; it would be a half hour more till he arose. Uffenbach retired to a nearby coffeehouse. When he was called and told to hurry over, he was again kept waiting. At last the Doctor appeared. "He stood there stiffly in his silk dressing gown and in a vastly forced manner with eyes rolling he asked as who we were and where we came from. But when we requested to see his collection he made excuses. . . ." After further indignities Uffenbach and his companion departed, to return (upon invitation) the next day.

Once again they were kept waiting, though Woodward afterwards regretted they had not arrived earlier. Uffenbach was by now very angry. "This is the discourteous little ceremony that the affected and pedantic mountebank makes a habit of going through with all strangers who wait on him." Nevertheless, he was impressed by the collection. Besides the natural curiosities, there were all sorts of ancient urns and vases. Woodward showed them his manuscripts, the future works he had so often promised. He wanted to show them he had not been "idle." The German visitors could not immediately recall the meaning of the word in English, and thought that the Doctor had said he was not "eitel" (vain). "Since he was making such a boast of his own works we could scarce restrain our laughter." One object especially interested the visitors, a Muscovy vegetable sheep, "one of the greatest curiosities that we saw here, or indeed, in the whole of our travels." [28] Unfortunately, Woodward showed everything with such extravagance that again they nearly burst into laughter. "He requires everyone to hang onto his words like an oracle, assenting to and extolling everything. One has to listen *ad nauseam* to his opinions *de diluvio et generatione antediluvia et lapidum postdiluvia.* He recites whole pages of his writings, accompanying them with continuous encomiums. The most ridiculous thing of all is that he never ceases looking at himself in the mirrors, of which several hang in each room. . . ."

[3]

If Uffenbach's description has the ring of satire, it was in a long tradition. We have seen the antiquary early transformed into a caricature. It was not long before the natural philosopher followed. They differed only (wrote Samuel Butler) "as things from words," for the one "uses the same affectation in his operations and experiments as the other does in language." [29] From the first, the members of the Royal Society came in for their share. Sir Nicholas Gimcrack, the lead in Shadwell's *Virtuoso,* reprinted in 1704, set the type. [30] He was followed by a string of others, till the natural philosopher and the antiquary – and even more the two together – became familiar comic figures to every literate man. As though to demonstrate the truth behind their caricature, Woodward somehow came more and more to resemble the type. Was it life imitating art, or were his critics using against him a familiar literary weapon? No doubt there was a little of each in their descriptions. One day in 1702 the naturalist Samuel Dale was invited by Woodward to see his collections. He was disturbed by the Doctor's slights at John Ray, his friend and neighbor, but avoided a quarrel. Unfortunately he whispered something to a companion later about fossil shells, only to be reprimanded abruptly and told that "it was not manners to speak while another was speaking." He thought he had never met such "an ungenteel Banquet after a seemingly courteous invitation to a Rara show as at that Gentleman's Lodgings." [31] Woodward's airs confounded everyone. Even Thoresby, who much admired the Doctor's "noble museum," had to admit that he never really enjoyed it because of Woodward's peculiar and haughty demeanor. (He contrasted it again with Sloane's easy and obliging conversation.) When the Scottish virtuoso Sir John Clerk came to visit Woodward in his old age (1727), he thought The Doctor himself, though he admired his work in natural history, the greatest curiosity of his extraordinary collection, "a vain, foolish and affected Man." [32]

Yet we should not let Woodward's personal idiosyncrasies obscure the fact that there was behind the satire also an intellectual issue. The fact is that many of his critics found the whole business of examining nature irrelevant, however it was conducted. They discovered the same weakness in the natural philosophers that they had in the philologists and antiquaries, a concern with matters obscure and impractical. This essentially had been Temple's argument against Wotton, or Swift and King against the Royal Society. What difference did the advancement of that learning mean (assuming one could speak of progress there) to the man of the world? Inevitably, they drew the contrast (it is *The Tatler* speaking in 1710), between the active men of affairs and those who can "discover the sex of a cockle or describe the generation of a mite" but are "so little versed in the world that they scarce know a horse from an ox." The "speculative genius" was too far removed from practical reality and usually wrong. (Did not their eternal disagreements prove it?) When "the minute Examiner of Nature's Works —" it is Shaftesbury now in 1714 — "proceeds with Zeal in the Contemplation of the Insect-Life, the Conveniencys, Habitation and Oeconomy of a Race of Shellfish; when he has directed a Cabinet in due form and made it the real Pattern of his Mind . . . he then indeed becomes the Subject of a sufficient Raillery . . . the Jest of common Conversations."[33]

It had been Woodward's fate from the start. No sooner was his first work abroad then he was identified with the stock figure of the virtuoso. His "character" was drawn early in *An Essay in Defence of the Female Sex* (1696), first generally in the commonplace fashion of the seventeenth century, then more exactly.[34] The "virtuoso," it began, was one who had "abandoned the Acquaintance and Society of Man for that of Insects, Worms, Grubbs, Maggots, Fleas. . . . He is ravish'd at finding an uncommon shell or an odd shap'd stone. . . . He trafficks to all places and has Correspondents in ev'ry part of the World. . . . He visits Mines, Colepits and Quarries frequently, not for . . . gain but for the sake of the fossile Shells and Teeth that are sometimes found there." With the accretion of each detail, the portrait of Woodward was gradually drawn. "He is a smatterer of Botany. . . . He is the embalmer of diseas'd Vermin, and dresses his Mummyes with as much care, as the Ancient Egyptians did their Kings. His cash consists much in old Coins, and he thinks the Face of Alexander is one of 'em worth more than all his Conquests." At last the Doctor was unmistakable. "His Collection of Garden Snails, Cockle Shells and Vermine compleated . . . he sets up for a Philosopher and nothing less than Universal Nature will serve for a Subject. . . . Hence forward he struts and swells, and despises all those little insignificant Fellows, that can make no better use of those noble incontestable Evidences of the Universal Deluge, Scallop and Oyster Shells, than to stew Oysters or melt brimstone for Matches." At that point, the sketch continues, he gives the world an *Essay*, defends Moses, and shakes the World to Atoms, pumping out even its center and clearing it of all imaginary loopholes in order to "make his Hypothesis hold Water." Finally, the portrait concludes, "he is a Passionate Admirer of his own Works without a Rival and superciliously contemms all Answers, yet the least Objection throws him into Vapours." The stock character and the real Doctor had become one, and the ironic portrait in this obscure but popular work was alarmingly close to the real one.

It must have been thoroughly exasperating for Dr. Woodward. Indeed, at the very time that his philosophical activity was being derided, his medical career was also coming under satirical attack. Thus the miscellaneous writer and wit (the author of the famous verses "I do not like thee, Dr. Fell") Tom Brown chose the Doctor for the centerpiece of his essay on "Physic." He recounts there a visit to the College of Physicians and a "surprising" incident. A young boy had accidentally swallowed a knife and the doctors were busy at work. The surgeons claimed the patient first, and one of them would have thrust a crane's bill down the boy's throat to pluck it up again had he not been restrained. Next the physicians consulted together and offered two possible remedies. The first was that the patient should swallow as much *aquae fortis* as would dissolve the knife. The second was Doctor Woodward's suggestion, "the more philosophical and therefore better approved . . . to apply a loadstone to his arse, and so draw it out by magnetic attraction." Woodward was no quack, Brown assured his readers. But he would not let the Doctor alone; he mocked his geological theories also in an essay on "The Philosophical or Virtuosi Country."[35]

Of course it was all very unfair. Say what you will about Dr. Woodward's activities, they were not idly intended. Even William Wotton, who had his reservations about the Doctor's theories, saw that he deserved better. In the second edition of his *Reflections on Ancient and Modern Learning* (1705) he devoted an entire chapter to defending the new science of geology.[36] It was true, he conceded, that the knowledge of fossils had so far lagged behind the other sciences — the subterranean world was not easily accessible, and no chemical analysis was as yet ready — still the "Learned and Ingenious Men" (including expressly Dr. Woodward and William Whiston) who had thought it worthwhile to investigate the subject had been able to invent hypotheses which, he thought, would some day (with further evidence) "be a means of solving some of the greatest Difficulties in the Mosaical History." These men deserved better than the ridicule of the *Essay in Defense of the Female Sex*, "as if what they had done had tended no more to the Advancement of valuable Knowledge than if they had gather'd Pebbles upon the Shore to throw away again." The wits might laugh, but "if immediate Usefulness had been the sole Motive of Men's Enquiries" there would never have been any advancement of learning.

It was no use, however; the Doctor's trials had only just begun. Amidst all the vituperation that began to pour down upon his head, the most effective and the most amusing satire still awaited him. In January 1717, on the eve of a smallpox war, a play was produced in Drury Lane with the Doctor portrayed in the leading role. The author of the entertainment, entitled blithely *Three Hours after Marriage,* was John Gay, assisted by his friends Dr. Arbuthnot and Alexander Pope. It ran for seven tumultuous nights, a successful run but troubled by partisan catcalls and near-riot. The doctor's friends rallied to his cause, but the play was printed. It was a popular success ("For five Days together, the Talk of the Town") and called forth two *Keys.* Anyone who had not caught the resemblance of its characters to real London figures was not to be left in doubt. [37]

The plot is classical farce. Dr. Fossile, the main character, has married a much younger and not very reliable woman who is sought after by the rivals Plotwell and Underplot, just three hours after the ceremony. Their machinations give the movement to the play. Fossile is a physician, an antiquary, and a natural philosopher. His collections form the setting of the climactic scene when the two rogues disguise themselves in order to enter the house, the one as an alligator, the other a mummy. Even the willing lady is taken aback by their sudden appearance, and turns to the other specimens around her. "Nay, I don't know but I may have Twenty Lovers in the Collection. You Snakes, Sharks, Monkeys, and Mantegers, speak, and put in your Claim before it is too late." Fortunately, they hear Fossile approaching with his friends, Drs. Nautilus and Possum, and resume their places in the collection before they are discovered. Nautilus is particularly ecstatic at Fossile's good fortune. "To have a Mummy, an Allegator, and a Wife all in one Day, is too great Happiness for a Mortal Man!" The ensuing Dialogue between the virtuosi is a model of foolish and pedantic learning interrupted however by new involutions of plot that bring the play to a swift-moving close. Suffice it to say that the impostors are discovered but the Doctor is undone and loses his wife — not however without some sense of relief. He is a ridiculous but harmless enough character. In the end the type was more important than the person. [38]

[4]

It would be wrong and misleading to leave the Doctor in defeat or ridicule. To the antiquaries, as to the naturalists, he was a formidable figure, with more than the knowledge of an amateur and with a collection of objects that few could rival. His correspondence was as staggering here in its volume and variety as it was over scientific matters, and the range of his acquaintance almost as broad. He was as likely to entertain the Royal Society with his antiquities as with his natural curiosities: Roman glasses and a piece of an urn from Wapping, some ancient coins found in the fens in Lincolnshire, a copper medal with Punic characters, and so on. His range was impossibly wide. "Besides Greek and Roman," he wrote to Humfrey Wanley, "I have here by me a great number of Gothic and Saxon Antiquities."[39] He was often helpful to others: bartering or giving from his collections, reporting on antiquarian treatises published abroad, introducing scholars to each other, supporting their work. He tried to keep abreast of every new development, philological or antiquarian, assisting in new editions of the classics like Barnes' *Homer* or Wasse's *Sallust* and pursuing the quest for classical monuments and inscriptions overseas.[40] He boasted to Thomas Hearne of his collection of rare English tracts and ancient Roman inscriptions.[41] When Dr. Harwood showed him a model of an ancient Roman hypocaust found at Wroxeter, it was Woodward, characteristically, who persuaded him to publish an account in the *Transactions*.[42] He was accepted by the connoisseurs as one of them — the Earl of Pembroke, Lord Coleraine, Sir Andrew Fountaine, the Duke of Chandos — and he freely offered advice. Always he displayed that same "confidence" that we have come to expect of him. John Strype thought that his personal collection of antiquities was worth engraving entire, like the museum of a great prince. Woodward himself began a catalog which he hoped to publish together with his fossils, but death here as elsewhere interrupted his activities. Even so, the auction catalog which did appear afterward was "fraught with great Curiosities" and seemed to Thomas Hearne, who certainly knew about such things, to "exceed most of the Auction Catalogues I ever saw or heard of."[43]

In one respect only was Dr. Woodward's collection disappointing. Somehow in that vast assemblage of *Antiquitates monumenta*, statues, inscriptions, vases, gems, urns, *imagines Egyptias et Romanas*, amulets, etc., there were no coins or medals. Without doubt these were the objects most prized by collectors, the most popular anyway and most thoroughly studied. Yet Dr. Woodward seems to have preserved only a few. Thomas Hearne was not alone in his astonishment, but there was after all a simple explanation. It was not that Dr. Woodward did not understand them or know their value; it was that he saw that, since nearly everyone else owned at least a few and desired more, they could furnish him with the most useful single item for barter. Dr. Woodward's neighbor Mr. Miller told Thomas Baker that "he did collect some Medals yet never kept them himself, but presented them commonly to Forreigners and others and had Fossils or other things in exchange for them."[44]   Indeed, he knew a great deal about the subject, and so knowledgable a collector as his friend and patient Henry, Lord Coleraine, applauded his skill in numismatics — as well as medicine.[45]   He may also have thought that for the purposes of use, i.e., their value to the historian, it was as good to possess the many treatises on the subject with their copious illustrations as the objects themselves. Even the greatest collector could not hope to possess them all, while the treatises on ancient medals were the oldest and most systematic antiquarian exercises in existence. Anyway, Dr. Woodward was after bigger game. Among the many thousands of lots that passed before the bidders in that memorable auction, undoubtedly the most important single item, deserving indeed a full Latin dissertation as an appendix to the catalog, and the one that drew the most attention, was a *Clypeo sive Votivo, sive Equestri Utpote Quantives pretii Monumenta.*[46]   No one present had to be told that it was the famous Dr. Woodward's shield.

Part Two
DR. WOODWARD'S SHIELD

VIII.

Roman London

[1]

The Great Fire of London (1666) proved an unexpected boon to the London *virtuosi.* The old city had to be rebuilt, new foundations laid, and large areas excavated; as a result all sorts of antiquarian finds were made. Especially exciting were the bits of Roman London that suddenly came to light.[1] Roman Britain held a particular fascination for Englishmen of Dr. Woodward's generation. It was, in the first place, the intersection of their two chief historical concerns, classical antiquity and the national past — their education emphasized the one, their patriotic feeling the other. It was the beginning of English history, the first point at which the past came sharply into focus. Before that there were only Celts, the painted savages of Caesar; and their indifference to written history had left a large empty gap. There were a few Welshmen like Woodward's rival, Edward Lhwyd, who protested, but it was in vain. Classical prejudice triumphed, and the earliest Britons were for the most part ignored.

What then had Roman Britain been like? While the Latin writers had sometimes mentioned that remote province, description was slight and references few. The literary sources were a little like Moses on the antediluvian world; they described the outlines of the tale, but left most of the facts and lots of the story in doubt. For the historian who wished to explore that forgotten world, the problems were just like those of the geologist. The classical writings, like Scripture, had to be read and interpreted; classical philology was the special exegetical equipment that was required. Beyond that there were new sources of information, especially Roman coins and inscriptions that were everywhere coming to light. Like the fossil shells, they must be systematically assembled, classified, and studied; they too could resurrect a forgotten world otherwise irretrievable. Naturally Dr. Woodward was as interested in the one as he was in the other.

The rebuilding of London was, of course, accomplished under the genius

133

of Christopher Wren. Here was the very epitome of the Restoration virtuoso:
"That rare and early prodigy of universal science," his friend Evelyn had
called him. He was an accomplished classicist, an astronomer and mathematician praised by Newton, an experimental scientist of amazing range, and a
chronologist and historian; he was only belatedly an architect.[2] He was a
self-avowed "Modern"[3] and a founder-member of the Royal Society, one of
its main props throughout its early years; this is probably where Dr. Woodward met him. His wide-ranging intellect remained alert to the end of his
ninety years. Naturally he took an interest in the new discoveries and even
wrote some interesting observations about them in some "scatter'd Papers"
edited long afterward by his son. According to Woodward, he intended a
treatise on the subject but it was never completed.

However fragmentary, his work shows the possibilities of the new
antiquarian discoveries. In one paper, Wren recalled the rebuilding of the
church of St. Mary le Bow Cheapside where were discovered "Walls, and
Windows also, and the Pavement of a Temple, or Church of Roman
Workmanship, entirely buried under the Level of the present Street." In
digging deeper still, he had struck a "Roman Causeway of rough Stone, close
and well rammed, with the Roman Brick and Rubbish at the Bottom, for a
Foundation, and all finely cemented."[4] He measured it exactly; his son
noticed that it was just the thickness of the Via Appia. Indeed, wherever
Wren dug there were evidences of Roman London. And he was convinced
that he could trace the boundaries of the ancient city from them. The
causeway, for example, was a highway that ran along the northern boundary
of the colony. The city lay between Cheapside and the Thames, Tower Hill
and Ludgate, with Watling Street as its principal avenue. Beyond the
causeway lay a great fen, discovered when sinking a new foundation for the
church of St. Lawrence near the Guildhall. On the West side was the
Praetorian camp, walled in to Ludgate. There he found a stone with an
inscription and the figure of a Roman soldier which he gave to the
Archbishop of Canterbury; it later joined the Arundel Marbles at Oxford.[5]

But it was at the new St. Paul's that the most plentiful discoveries were
made. The northeast corner toward Cheapside turned up a remarkable pit
"where all the Pot-Earth had been robb'd by the Potters of old time."[6] Here
were discovered great numbers of urns and vessels, pottery of every
description, much of it broken. The most remarkable finds were "Roman
Urns, Lamps, Lacrymatories and Fragments of sacrificing Vessels etc. . . .
generally well wrought and embossed with various Figures and Devices." To
one he took a special fancy. It was a fragment of a vessel in the shape of a
basin with figures on it. These he thought represented Charon with oar in
hand receiving a naked ghost. The conjecture was not implausible, but the
modern historian tells us it was merely two fisherman.[7] It was no easy task to

identify a Roman image where no inscription was supplied. Most curious of all was a large Roman urn of glass with a handle, which Wren bestowed upon the Royal Society Museum.[8]

Wren's interest was real if not profound. What his conjectures might have become in a large treatise is hard to say. But the brief remarks he did make are suggestive at least of a critical and independent historical intelligence. He thought pre-Roman history impossible; "the Accounts before that, are too fabulous." But he described the Britons in Caesar's time with some sophistication. On one point he was particularly concerned; he did not believe the tradition that an ancient temple to Diana had once stood beneath St. Paul's Cathedral. He had simply discovered no archaeological evidence to support the theory.[9] On the other hand, he did believe that an early Christian church had once rested there. While he discounted the "monkish tales" of the apostolic mission to Britain by Joseph of Arimathea along with other "legendary fictions," he approved the "authentick Testimony of a Christian Church planted here by the Apostles themselves and in particular very probably by St. Paul." He thought that the first cathedral had likely been destroyed during the persecutions of Diocletian but been restored under Constantine on the pattern of the Roman basilica in the Vatican. In any case, the discovery of nine wells beneath the cathedral showed clearly that the early church had been smaller than the medieval one.[10]

Wren's use of archaeology to resurrect Roman London was suggestive, if brief and incomplete. The obstacles were serious. "Public indifference and unscientific excavation," runs a complaint of 1909, "have rendered most of the spoil from the city worthless for archaeological purposes."[11] The problem has always existed. The search for Roman antiquities in the seventeenth century, as in our own, was entirely haphazard and accidental. "In digging Fleet Street," we are told, "in the Year 1670 between the Fleet Prison and Holborn-bridge, at the Depth of fifteen Feet, divers Roman Utensils were discovered; and a little deeper a great Quantity of Roman Coins of Silver, Copper, Brass, and all other sorts of Metal . . . and at Holborn Bridge were dug up two of their Brazen Lares or Household Gods, about four Inches in Length . . ." Such finds would have been very valuable but for the fact that there was no description of the coins (so important in dating the site), no information about the whereabouts of the two icons, and of course no subsequent digging.[12] Even Wren was too busy on the whole with the new St. Paul's to investigate systematically the old.

[2]

Indeed, modern archaeology lay well in the future. What could be done to illuminate Roman London without it is best seen in the work of Wren's contemporary, Edward Stillingfleet. Where the former was the very type of the secular virtuoso, the latter was the epitome of the scholar-cleric. Like Wren, Stillingfleet combined great erudition with the active life, becoming eventually Bishop of Worcester after playing a central role in the religious and political controversy of his time. He wielded a prolific pen in the service of his church, and he found in history a favorite weapon. His biographer remembered his library as "the choisest perhaps of any private Person of his time." Here the great philologist Richard Bentley mastered the classics and rose to eminence; here Stillingfleet assembled the erudition that is everywhere apparent in his works.[13]

He was an austere man not given to frivolity. "In no Time or Place," his biographer recalled, "was he idle or triflingly employ'd." He was sharp as well as learned, "not all Quotations and Authorities," though the margins of his folios are heavy with his erudition. Only his wife could match his temper and his learning.[14] In each of his historical works he begins with a survey of the authorities and an attempt to discriminate their value as testimony. He was a master of textual criticism, a philologist after the fashion of the new learning. And his tract on Roman London displays both his learning and his critical acumen at their best. It was a treatise "of no great Length, but of much Reading and exquisite Judgment wherin are interspersed an unusual Variety of excellent, ingenious and useful, critical, etymological and topographical Observations . . . diverting as well as instructing."[15] The work was published posthumously (in 1704), but derived undoubtedly from Stillingfleet's London days and perhaps even from the same excavations that had stimulated Wren.[16]

If so, it is at first glance disappointing. There is not a single reference to any of the discoveries that had moved Wren to his conclusions. The *Discourse of the True Antiquity of London* is exclusively a labor of book-learning. But here it is remarkable; it demonstrated the skills and possibilities of humanist philology at their most able. Stillingfleet had combed all the possible literary sources and critically compared their accounts. His main argument was a rebuttal of Geoffrey of Monmouth and his followers, in favor of the existence of a Celtic London. The twelfth-century chronicler had assigned London to the British King Lud centuries before the Roman conquest, and vividly described its many walls and towers. This was incidentally the same work in which Geoffrey had attributed the founding of Britain to the Trojan, Brutus, and recounted the exploits of pre-Roman British kings in full and exact detail. It was not the first time that Stillingfleet had rebutted the story as a monkish invention, nor was he by any means the first to be critical of it. But there is something impressive anyway in his massive assault on it here. It leads him far from London, to a consideration of the other twenty-seven British cities of which Geoffrey wrote, each here identified as Roman. In reviewing the early sources, Stillingfleet (like Wren, but now with a fuller argument) can find no testimony earlier or more authoritative than Caesar's, and "the truth is Geoffrey's Relation and Caesar's can never stand together; not only as to London but as to the whole Expedition."[17]

The authority of his argument derives from its completeness. Not only does Stillingfleet review all the ancient sources directly — Caesar, Tacitus, Dio Cassius, Xiphilinus, Ammianus Marcellinus, Strabo, etc. — but he knows the modern commentators and scholars also: Sigonius on ancient law, Spanheim on Roman coins, the *Marmora Oxoniensa* (a printed version of the inscriptions on the Arundel marbles), Weaver's *Funeral Monuments*, and so on. Moreover, he is not content to be simply destructive. Having disposed of Geoffrey and Celtic London, as well as later writers like Sheringham and Bochart, Stillingfleet goes on to describe the state and condition of London under the Romans, setting the city into its place within the larger Empire and describing by analogy its government, trade, etc.

The work was thus, as Stillingfleet's biographer believed, an epitome of the best in seventeenth-century scholarship. Its weakness lay in its unconcern with the archaeological evidence which was (with all its difficulties) piling up before the Bishop's very eyes. It was not that he was altogether indifferent to the value of coins and inscriptions or even monumental remains; he refers to all three. It was that he preferred to encounter them in his library than at the sites.[18] And as long as he thus confined himself, his information was restricted and his critical faculties inhibited. Within these limits his *Discourse* on ancient London was most impressive, the best perhaps that could then be done. Its strength lay in the application of "modern" philological techniques to the texts. Here Stillingfleet showed the rich possibilities of his classical training when applied to solving an historical problem.

An instance is his treatment of the twenty-eight cities. Stillingfleet was aware that there were several manuscript sources that differed significantly one from another. The learned Archbishop Ussher had transcribed a copy from two manuscripts in the Cotton library and compared it with nine others. Stillingfleet at once understood the value of the variant readings that were supplied there, and he was eager to consider them in the tricky business of establishing the etymologies of the different place-names. He was ready also to employ the other miscellaneous evidence of the *Antonine Itineraries* and of published inscriptions, coins, and the like. Here he was building, of course, upon the labors of his many predecessors, especially on William Camden's *Britannia*. And if he was not entirely successful in his answers, it was the new means employed that was more significant than his conclusions. Here indeed (but together with the absent archaeology) lay the future of Romano-British historiography.

In general, then, the recovery of Roman London appeared to the later seventeenth century either as an archaeological and antiquarian problem *or* as a philological one. For Wren, the literary evidence was an aid to what he saw unearthed; for Stillingfleet, the antiquarian evidence was a device to elucidate a text. On the whole the Bishop preferred the word to the thing. "I have one great Argument," he writes, "to prove that Canterbury was built by the Romans, not from Roman Coins found almost in all parts of the City, as Mr. Somner affirms; but from Caesar's Account of his 2nd Landing."[19] By interpreting the Latin words here — for Caesar was not at all clear on this point — and by employing the argument of his silence, he proves that no town existed there in Caesar's day. To others, however, it was becoming obvious that a stronger argument could be made by wielding the nonliterary evidence together with the language of the texts, by fusing antiquities and philology, archaeology and textual criticism.

Toward the end of his work, Stillingfleet turned to the question of the Temple of Diana which he found told in Camden.[20] Like Wren, he was not at all persuaded. He challenged each of the alleged evidences for it: the *Camerae Dianae* nearby; some ox-heads taken up in Edward I's time; and a ceremony in which a stag's-head was traditionally paraded in St. Paul's. The *Camerae Dianae* he shows were tenements built later in the Middle Ages. With characteristic learning he then proves that neither the Britons nor the Saxons worshiped Diana at all, and that the Romans employed stags or sheep but not oxen in their sacrifices to her. The margins are heavy with citations, one of which alone contains references to fifteen classical and modern treatises. But Stillingfleet would not leave Roman London without any temple at all. He conjectures that at the site of St. Paul's there must once have stood a Roman capitol. Most provinces had them; the situation of the place on a hill favored it; and the sacrifice of oxen was now appropriate (here Virgil, Servius, Festus, the Trajan Column in Rome, and John of Salisbury are unequally arrayed). This is as much as his essentially literary deduction can do, and he concludes his tract with a learned account of the building of the first St. Paul's by King Ethelbert.

[3]

If the obvious thing to do was to combine the approaches of Wren and Stillingfleet, the most immediately pressing business was to keep track of the archaeological evidence that was now swiftly accumulating. It was one thing to chase after a magnificent statue or a gold coin; it was quite another systematically to collect broken pottery or to examine the brickwork of an exposed wall. Fortunately there were others besides Wren who found the whole thing fascinating and did what they could: the obscure apothecary

John Conyers, the shoemaker and book dealer John Bagford, and, most indefatigably of all, Dr. Woodward.

The best record of their work was John Strype's new edition of Stow's *Survey of London*. The *Survey* had first been compiled by the Tudor antiquary John Stow at the very end of the sixteenth century and passed through many editions since.[21] It was a work remarkable for its industry and charm, a ward by ward account of the antiquities of the city. Stow was a self-educated man, a London tailor by trade, quite lacking in classical attainment but with a special (and unusual) interest in the Middle Ages. The *Survey* reflects his extraordinary grasp of the chronicle materials and the miscellaneous records of medieval London, many of which were in his possession. It also shows the author busily examining the sites themselves and recording the inscriptions and other monumental remains. But the *Survey* was notoriously deficient in its account of early London. Stow clung to the medieval convictions of his sources about the origins of London, and the *Survey* describes only a very few Roman remains. Even for these he was probably indebted to Camden. Apparently Strype had been approached as early as 1694 with the idea of editing and enlarging Stow,[22] and by 1702 he was preparing the work for the press.[23] But it was only many years later, in 1720, that the new *Survey* finally appeared. Among Strype's additions was an appendix entitled "Of divers Roman and other Antique Curiosities found in London before and after the great Fire." It is still a useful record for workers in the field, though inevitably incomplete and sometimes vague. As Strype concluded apologetically, it was all he could do "by diligent Enquiry of my Friends."[24]

Among these friends, apparently, was Christopher Wren; another was Dr. Woodward, to whom we shall return.[25] Still a third was Bagford: "He was a Man of very surprising Genius," wrote Thomas Hearne, "and had his education (for he was first a Shoemaker, and afterwards for some time a Bookseller) been equal to his natural Genius, he would have proved a much greater Man than he was."[26] He had taught himself to be an antiquary and his two great interests were the history of printing and the antiquities of London. Humfrey Wanley, the greatest bibliographer and paleographer of his time, described him as "a Person (tho' not Master of the learned Languages) very well skill'd in the different sorts of Ink, Illumination, Binding, Hands, Parchment, Papers, or almost any sort of Workmanship not to mention Books, or amongst them those relating to our English History."[27] Bagford collected for both his interests; he was also an agent for many others: Robert Harley, Dr. John Moore (Bishop of Ely), Dr. Sloane, and Dr. Woodward. Woodward's first excusions into Roman antiquities seem to have been his purchases from Bagford in the 1690s.

Unfortunately Bagford was never able to complete his projects. He intended an ambitious history of printing for which he assiduously assembled materials. Indications of other interests and projects remained among his papers at his death (1716), most of which went to Harley (by way of Bagford's friend and Harley's librarian, Wanley) and rest now in the British Museum.[28] Their awkward hand and bad spelling did not keep them from being extremely useful to later students. His only published piece was a letter to Hearne which Hearne printed in his edition of John Leland's *Collectanea*. It is the best remaining testimony of his interest in Roman London, and a useful document to set beside those of Wren and Stillingfleet.[29]

The letter (dated Feb. 1, 1715) attempts two tasks: a survey of antiquarian discoveries known to Bagford and an account of the Roman experience in London. He confined himself entirely to the archaeological evidence. "I shall not insist upon what hath been noted by former Authors," he writes, "but will only relate what hath been discovered within my own Memory." The enumeration and description of recent antiquarian finds (including his own) is therefore the most valuable part of the letter, although Bagford's speculations are interesting.[30] He begins by tracing the Roman progress from Dover to London through the evidence of Roman stations at various places in Kent and Surrey. In one instance he recalls the discovery of a number of coins, some "much esteem'd by the worthy Mr. Charleton" and many shown to him by his friend Mr. John Channop. In another, he describes the antiquities unearthed at Clapham in digging for gravel. "They are still in being and have been view'd by Mr. John Kemp who as he is as great Judge in those Affairs, so he ownes that some of them are extraordinary." In London itself he remembered a pavement dug up in Bush Lane, evidence for him that this was where the Romans had first encamped. A piece of it could still be seen with "several other valuable Remains of Roman Antiquities" in the Museum of the Royal Society. They were "very good Hints" indeed for the antiquarian recorder of Roman London. Bagford had a keen appreciation of the value of inscriptions. "As new Monuments are discovered," he wrote, they "very often give a greater Light to these early times than any written Books now remaining." The Arundel Marbles were a perfect example; he had viewed them in Oxford with Hearne the previous year. The Parian chronicle, inscribed there, was as valuable for the Greek as the *Fasti Capitolini* were for the Roman history. Perhaps something as useful would still come to light for ancient Britain.[31]

Unlike Wren and Stillingfleet, Bagford was persuaded of the existence of the Temple of Diana. His remarks are brief; the quantities of sacrificial vessels discovered in digging at St. Paul's are evidence that a pottery was there to furnish the people conveniently with the wares necessary for their worship. The boar's tusks are sufficient for him; he seems not to have read Stillingfleet.

Characteristic is his lament that "if further search had been made on the South-Side, and nearer Doctors Commons, there might probably have been found many other Roman Antiquities, which would have given several curious, learned and judicious Men each greater light into these Matters than 'tis possible for them now to obtain without the help of such Assistance." The task was certainly beyond the means of an ordinary book-dealer. [32]

Most interesting perhaps of Bagford's observations were his personal reminiscences. He had himself seen many of the antiquities discovered in Goodman's Fields in 1678-9, "and had some of them formerly in my own Collections." Apparently he had watched with care the "subterraneous parts of London" as they were gradually exposed after the fire. In this respect he was rather like his industrious and honest old friend John Conyers, "an Apothecary formerly living in Fleet Street who made it his chief Business to make curious Observations and to collect such Antiquities as were daily found in and about London." Conyers had gone to considerable expense to obtain specimens for his collection. Bagford thought the most remarkable was the body of an elephant unearthed in Conyer's presence while digging for gravel not far from Battlebridge. How had it got there? Bagford offered an ingenious explanation. Since a flint spear had been discovered also nearby, like those used by the ancient Britons, he conjectured that the elephant had been brought by the Romans under Claudius and had been killed in some fight by a Briton. He knew that others (unidentified) preferred to attribute it to the Universal Deluge. (No one, of course, thought of claiming it as a mammoth or dinosaur.) But he preferred his own conjecture, "for a liberty of guessing may be indulged to me, as well as to others that maintain different Hypotheses." The flint weapon was now among Kemp's curiosities and was drawn by Bagford. [33]

The book-dealer's industry was sounder than his criticism. He dismissed King Lud but otherwise retained Geoffrey of Monmouth. Thus a Roman Temple near the Steelyard could be vindicated on the chronicler's authority, along with the story of a Roman captain named Gallus who was cast into a brook by the Britons, thereby giving it his name Gall-brook, later Wall-brook. Bagford thought that all the Roman names had likely survived through the Middle Ages: Cornhill, Grace Street, Watling Street, even Old Fish Street, were probably of equal antiquity. His lack of training in the classics was a serious handicap to his conjectures, as indeed to his whole reconstruction of Roman London. Most ingenious was his effort to restore the ancient Roman recreations through their survival in present London activities. Vineyards were his first example, but Bagford thought that prizefighting, bull- and bear-baiting, may-games, and morris-dancing might all be traced to their ultimate source in the Roman past. In the end, it is amusing to find Bagford dismissing his tailor-predecessor, John Stow, as having "little skill in Affairs of this Nature."[34]   Still, in this one respect anyway, as a recorder of London archaeology and a collector of classical antiquities, Bagford might well feel a measure of superiority.

[4]

When Strype printed the new Stow, Bagford was dead and Conyers almost forgotten; now the apothecary is almost past retrieval. Conyers was very well known to Dr. Woodward — the source, indeed, of many things in the Doctor's collection, including above all his famous shield. He had put together a wonderful accumulation of miscellaneous objects which it was proposed at one time should be made public for its use "to the Divine, the Naturalist, Physician, Antiquary, Historian, or indeed any person of Curiosity."[35] But even less than Bagford did he commit his thoughts to paper. Only one manuscript seems to have survived to give a clue to his activities. Bagford recommended it to those interested in Roman antiquities; it would afford, he thought, "abundance of useful Diversion to such as are studious in those Inquiries."

Conyers was certainly one of the first in the field. From his apothecary's shop he was able to follow the reconstruction of St. Paul's with care and obtain many of the antiquities as they were unearthed. But he busied himself elsewhere also: in the gravel pits around town; at Goodman's fields, where he found many Roman utensils; and in the book-sellers' shops, where he bought rare books and manuscripts. Most exciting were the finds at the east end of St. Paul's. He describes the earthenware and inscriptions, brass coins, and glass pots discovered there (1675). In his notebook he drew the form of a Roman kiln filled with coarse vessels, lamps, etc., dug up there twenty-six feet beneath the surface a year or two later.[36] The kiln and many of the urns came eventually to Dr. Woodward, the manuscript by an odd chance to his rival Dr. Sloane. It was a pity, Woodward thought, that Conyers had had so little encouragement to enlarge his thoughts; with only limited resources he had done what he could.[37]

The Doctor was more ambitious. When Strype came to describe his antiquarian collections, he was quite overcome. They were, he was convinced, the most impressive he had seen. The list of Woodward's antiquities was endless: an ancient marble bust of Jupiter, a marble head with Phrygian tiara, a Grecian bas-relief, a variety of amulets, gems, cameos, and intaglios (Egyptian, Greek, and Roman); ancient medals of all varieties; Roman weights, urns, lachrymatories, and other things, "procured from Alexandria, Constantinople, Rome, etc." There was a Roman altar from the Picts Wall in

Northumbria with a valuable inscription; there were ancient weapons as well as urns and other objects from the remote parts of the kingdom. Finally, "he hath a vast Variety of ancient Instruments, Utensils, *Vasa,* and the like, that have been discovered in the several Places in and about this City." In particular there were religious vessels of different kinds, and twenty sepulchral urns. "As to their Forms, they are universally very elegant and handsom. And indeed, the Doctor, the Possesser of them, well observeth, that the Remains of these Works of the Romans shew them to have been a People of an exact Genius, good Fancy, and curious Contrivance." [38]

Woodward's cabinet of antiquities was thus a fitting companion to his cabinets of fossils; Strype was impressed particularly by the excellent condition of the pieces. Besides the major works, he noticed also some coins (British, Saxon, Danish, and Norman, as well as Roman), innumerable beads, tiles, bits of pavement, etc., "so that Dr. Woodward's Museum is a Treasury of all Sorts of Commodities and Utensils, sacred and prophane, of ancient Heathen Rome." One of Woodward's proudest possessions, purchased for fifteen pounds, was an urn that had been mentioned by Horace himself; another, added later, was a head of the comic writer Terence on an emerald gem, "indubitably ancient and the more valuable because there's no other Monument remaining of that illustrious Genius." [39]

But the Doctor was never simply a collector. He might admire the workmanship in some of his collections, but his real interest lay in their utility. His great intention, he assured Strype, in putting them together ("and that with so great Diligence, Trouble and Expence") was "in order to clear and give Light to those ancient Writers who mention and Treat of them, viz. the Greeks and Romans, which he hath read and studied with great Exactness." [40] For Woodward, philology and the study of antiquities were closely related; the Roman object might illuminate the text, and vice versa. "Another of his Ends herein," continues Strype, "was to illustrate the History and Antiquities of their great and noble City; out of the Ruins of which these things were received, upon the Occasion of that great Digging . . . after the late great Fire." And indeed, many of the medals and coins, even the inscriptions on the vases, were meant to contribute to that purpose. Woodward hoped to determine from the evidence of their recovery (where, for example, the funeral urns were discovered) the boundaries of the ancient city and to show, from the sacrificial vessels unearthed, just where the temples were located; even to solve the old problem of Diana. Strype only wished that the whole collection could be engraved and that Woodward would find the time to write the observations and reflections which they so richly deserved. He might more justly have wondered that the Doctor had found any time at all for so complicated a pursuit in the midst of a life devoted to so many other things.

[5]

Yet Woodward had in fact found the time once for an antiquarian tract, and Strype quotes approvingly from it. As long before as 1707, the Doctor had drawn up some observations on Roman London and sent them along to Strype for use in the *Survey*.[41] A few years later he retrieved them, and prefixed a letter to his colleague in the Royal Society, Christopher Wren, which was printed in 1713. He was apologetic for its brevity. " 'Twas wrote," he explained to Hearne, "during a few Days that I was under an Indisposition." Studies of "greater Application" had to be deferred; meanwhile some exciting new antiquarian discoveries had just been made. His letter was a diversion, to pass away the time while recuperating; it was also written to show a friend (who was normally clever enough but deficient in this one matter) that antiquities were not useless. Writing in 1711, he is not satisfied with his performance; he has only been able to add some "Authorities and Passages of the Antients" since. He offers it anyway, tentatively, as a memorial for further inquiry and consideration. It should in no case be read dogmatically.[42]

Under the circumstances it was a creditable performance. His letter was published as *An Account of Some Roman Urns and other Antiquities Lately Digg'd up near Bishops-gate With Brief Reflections upon the Antient and Present State of London.* It appeared both separately and as part of Thomas Hearne's edition of Leland's *Itineraries.*[43] Though brief and incomplete, it took account of all the recent literature and discovery. Woodward had read Stillingfleet and knew Wren's opinions; he had bought from and conversed with Conyers and Bagford. Behind these, he knew what Stow, Camden, Weever, and Burton on the *Antonine Itineraries* had had to offer. His classical attainments were genuine and he could cite the ancient authors fluently. (When a German visitor met the members of the Royal Society, he found only Woodward and one or two others able to converse in Latin.)[44] Most admirable, however, were his observations on the antiquities themselves, precisely and objectively reported. Throughout, as we should expect, the Doctor exercised an independent and original judgment.

His first purpose was to describe the recent discoveries at Bishopsgate.[45] In pulling down some houses in Camomile Street and laying new foundations, some exciting finds had been made. Four feet underground, beneath some houses still standing, there was an ancient tesselated pavement at least sixty feet long and ten feet wide. Below the pavement was a stratum of clay in which were discovered several urns of different forms, "but all of very handsome Make and Contrivance." Woodward thought them all Roman work. There were other earthen vessels as well: a *Simpulum,* a *Patera* of very fine material, a bluish glass viôl — all these seem to have found their way into Woodward's collection, though he was sorry to add that they had been broken by careless workmen. A coin with a head of Antoninus Pius and a partly discernible inscription was also discovered. Most interesting to Woodward was the city wall which was here exposed. His description is remarkable for its detail and exactness. He discriminates between the Roman work, which he defines by its materials and workmanship, and the later additions of the Middle Ages. His measurements are precise to a fraction of an inch. In the letter to Hearne, Woodward recommended the student to a part of the wall that was still visible, a tower still intact, "the most considerable Remain of Roman Workmanship yet extant in any Part of England that I know of." In 1876 further excavations confirmed the Doctor's account of the wall, its accuracy and its correct attribution to the Romans. Later students were not always so scrupulous or correct. [46]

Woodward's discussion of the Roman bricks was typical of the growing precision of antiquarian studies. He described those that had just been unearthed and compared them with the measurements given by Pliny. From this he conjectured a measurement for the Roman foot. The problem of course, was just what constituted a Roman "foot" or indeed any Roman measurement at all, and it had puzzled generations of students from the beginning of the revival of antiquity. The Renaissance humanist needed a modern equivalent in order to convert the Roman distances discovered in his literary sources to contemporary ones. It was but one of a whole class of similar problems, exactly like that most absorbing of antiquarian pursuits, the conversion of Roman money into English or modern European denominations. From the early sixteenth century, great ingenuity had been lavished upon the latter problem, which naturally absorbed every collector and numismatist as well as philologist. The very first antiquarian monograph in England was probably Cuthburt Tunstall's *De Arte Supputandi* (1522), the last part of which dealt with the meaning of the ancient currency. Thereafter, learned treatises on the subject were legion. But whether currency or distance, weight or length, the problem was the same: how to determine a quantity described in classical literature from the archaeological evidence. The convergence of philological and antiquarian interests is at once apparent. No antiquary doubted, as Woodward put it, that it was "a Thing of great Use." And he reminded his readers of the indefatigable Mr. Greaves who traveled throughout the Mediterranean measuring the monuments and collecting materials. Woodward very much admired the resulting *Discourse of the Roman Foot and Denarius* (1647) (whose subtitle continues, "From Whence as from true Principles, the Measures and Weights used by the Ancients may be deduced"). [47] Here he offered a modest contribution of his own in the measurement of the Roman brick.

In fact, the precision of Woodward's descriptions and his care in taking direct observations, rather than repeating hearsay, are everywhere apparent in his antiquarian work, as they were in his natural science. He remembered some remains of a Roman highway near Shooters Hill; it was just like those he had seen in Oxfordshire, Gloucestershire, and other parts of England. But to be sure, since he was recalling observations made some years before, he got Mr. Hutchinson (his collaborator in fossils) "to ride thither this Morning and take a Review of this Way." [48] The new descriptions could then be precisely reported.

But again the observations were only a means to an end. As in his natural science, Woodward was anxious to draw out conjectures from the evidence, to attempt to find a historical meaning in the discrete facts. By combining the evidence of the recent excavations with the hints in classical literature, he thought that the boundaries and organization of Roman London could be ascertained. He was willing thus to combine the approaches of Stillingfleet, Wren, and Bagford, of philology with archaeology, to the historical problem that had concerned them. But Woodward is surprisingly tentative here, much more so than with his fossils, aware apparently of the limitations of his evidence. His letter is merely suggestive of how to approach the problem. By its method "some Advances may be made toward ascertaining the old Boundaries," he wrote, "and by a carefull collection of all, towards adjusting the form, and settling the Extent of the whole; whenever One, who has Leisure, and due Information, shall engage in the Undertaking."

For the old London wall, the literary evidence was clearly deficient. It is true that Burton, in his commentary on the *Antonine Itineraries,* had attributed it quite precisely to Suetonius Paullinus, sent as governor by Nero. [49] But his source, Tacitus, seemed to Woodward (as it had to Stillingfleet) to read rather as though London hardly existed then, much less been able to defend itself with fortifications. In any case it was not the present wall, because the urns that the Doctor had found inside it were proof that London had in early times not extended nearly so far, since Roman Law forebade burying within the city. It was the archaeological evidence, especially the burial places disclosed in the incinerary urns, that offered (he thought) the best clue as to the Roman boundaries. Unfortunately the problem was complicated because it was difficult to determine exactly when Christian burial practices (which eschewed burning) replaced the Roman ones. However, since learned men generally agreed that burning fell into disuse toward the end of the Antonines, and the coin of Antoninus Pius found among the Bishopsgate urns seemed to confirm it, it was probable that the date of the burials here, and hence the Wall, was quite late.

Woodward's argument rested on a combination of literary evidence and antiquarian scholarship, on the text of Macrobius (the *Saturnalia*) and the criticism of Scaliger, Mabillon, and especially Octavio Ferrari's *Dissertationes de veterum lucernis sepulchrabibus.* These furnished the context for his archaeological evidence. Even so he was aware of the "want of Records and Testimonies," so he remained cautious. Still the modern scholar is agreed; the use of incinerary urns for mapping and dating Roman London is an important device; furthermore, the wall was late just as Woodward suspected. [50]

Inevitably, the Doctor had to confront Geoffrey of Monmouth and the early Britons. His discussion is parallel to Stillingfleet's (whose work he knew and approved), but seems to rest immediately on the authorities themselves. There is no reason to doubt that he went back over the ancient works directly — especially since his citations do not exactly correspond with the learned clergyman's. (We have seen his early correspondence on the subject with Lhwyd.) Woodward was not likely to accept anything simply on authority, and Geoffrey of Monmouth is dismissed because he cannot be reconciled with Caesar and Tacitus. Herodian, Pomponius Mela, Xiphilinus, Strabo, and other ancient writers are again adduced to show the primitive condition of the Celts in Britain. Some recent writers who had tried to elevate the Druids are dismissed along with Geoffrey: "These Gentlemen may be allow'd to indulge their Imaginations as far as they please; but if we rightly reflect upon what we find on Record concerning the Notions and Practice of the Druids, 'twill not carry our Ideas to any great Height." Now it is Claudian, Caesar, Diodorus Siculus, Pliny, and some others who furnish the "record" to dispel the myth. "This is the main," he concludes confidently, "of what Antiquity hath transmitted down to us." [51] In a similar fashion and in close detail, he argues against the possibility of the "Remains of Temples and other Noble Structures," the walls and cities, or indeed "any Appearance of Art" alleged for the ancient Britons. "Nor will it be thought strange that our Progenitors should be, in those early Times thus rude and uncivilized," he finds, "when 'tis known that several other great Nations were likewise so till lately." Had not all peoples been originally so, even the Greeks and the Romans?

Woodward's ideas of Roman London thus fit neatly into his larger views of world history, though he only hints at them here. The *Natural History* is several times cited, once to show that human bones could easily be preserved in the earth, since the remains of the Deluge had survived. Woodward reaffirms his view that in the beginning, in the ages immediately after the Deluge, mankind was primitive; "This the Histories and Accounts of the Assyrians, the Egyptians, the Chinese, and all others agree in." [52] Eventually the Greeks taught the Romans, and the Romans the Britons, until at last Woodward's contemporaries, "the Britons at this day," could be judged inferior to no earlier people. Thus the doctor held to a "modern" progressive view of history; primitive man could be seen to arise from the flood to assume gradually a civilized status in the world. (Had we Woodward's tract on the subject, we would not have wanted for detail.) If it was not clear whether modern Britain would surpass the ancients, there was at least no doubt in Woodward's mind that his own London was a fit rival to ancient Rome. And he would surely have claimed, with his fellow member of the Royal Society, William Wotton, that in the disciplines in which he was engaged anyway – in natural philosophy and antiquities – the ancient world had had nothing comparable to offer. Yet his brief essay on the urns was only a promise of what might be done.

[6]

Woodward did not overlook the Temple of Diana. Unfortunately, he reached it only in his last fading paragraphs. As the burial places could help to locate the old walls, so the sacrificial vessels, he believed, could discover the sites of the ancient temples. Here again the concrete evidence of the excavations was crucial. The tusks of boars and the horns of oxen and stags, the representation of the deer and even of Diana herself, were proof that the ancient tradition was correct and that the Temple of Diana once lay beneath St. Paul's. And the proof was tangible in Woodward's own collection, as Strype among others saw. Most dramatic was the discovery of a "small Image of that Goddess that was found not far off." If Stillingfleet had known these things, he would never, Woodward was sure (and Strype after him), have questioned the tradition. [53]

The discussion in the essay on the Roman urns was insufficient, however, on this point. Woodward knew it; unfortunately the press of other business, as always, interrupted his labors. Still he continued after his fashion, collecting materials and scribbling away on manuscripts which he hoped one day to publish. [54] One of these, a brief fragment, survived his death and was known and employed by several antiquaries afterward. [55] In it he recapitulates his argument from the evidence of the urns and offers a more complete discussion of the small image, now identified as "an incunculus of Diana, made of brass, and two inches and an half in height." According to Woodward, it had been discovered about forty years earlier, a little to the southwest of St. Paul's; it was further proof, and of a most convincing kind, that the goddess had indeed once been worshiped there.

Most of the argument was devoted to defending systematically the authenticity of the little statue. Had someone questioned it? For two centuries, Woodward reminded his readers, "learned men have been very solicitous in their inquiries after the remains of antient works, inscriptions, basso-relievos, statues, incunculi, medals, intaglios, and the like." No doubt he looked around with satisfaction as he wrote. "General books have been wrote to fix a standard and settle the rules of judging these things, to distinguish the genuine from the counterfeit, and those that are truly antient from those of a later date." Was this what the skeptic had suggested? "This small image, though it has the good fortune to be well preserved, and very entire, yet has marks enough of time upon it to put its real antiquity quite out of all question." What they were, the Doctor does not say; he hastens rather to call upon the witness of "the best judges of Italy of France and of other nations, as well as England." All were agreed that it was an authentic Roman piece. If it were not of the very best manner, it was "good and artfully enough done," especially for a provincial piece executed at a great distance from the capitol.

Woodward's principal argument for its antiquity was iconographic. "What greatly confirms the opinion that the figure is antient is, that the habit and insignia are the same with those exhibited by the other representations of this Goddess yet extant." There follows an extensive review, in very exact detail, of a series of Dianas known to Woodward – in the gallery of Versailles, the palace of the Farnese, the Brandenberg collection, the cabinet of M. Smetius, etc. There was the Diana on an ancient lamp in the custody of Pietro Santo Bartoli in Rome, "so nearly after the manner of the small image, that they could not well have been more alike." And of course there were the innumerable Dianas of the coins – of the Ephesians and the Mytilenaeans, the Delians, Cretans, and Raphiensians, not to mention those of the family of the Postumi at Rome, of Augustinus, Gordianus Pius, and Gallienus. In each case Woodward compares the details, the bow in the left hand, the quiver at the back, the habit girded up above the knees, to establish the resemblance. And finally, to clinch his case, he turns to the literary evidence, to Callimachus and Claudian in particular, to confirm even the slightest details, especially Diana's hair, "plaited, wreathed and gathered into two knots, the one upon the crown, the other behind the head." Taken all together, the statues, gems, coins, and lamps, along with the literary descriptions, there was no mistaking the authenticity of Woodward's Diana.

Or was there? In the absence of the criticism which he seems to have been rebutting (and of the object itself), it is a little hard to be sure. [56] What the reply does demonstrate pretty well is that the small representation was indeed Diana. Woodward's essay is an excellent brief example of the Renaissance iconographic discipline at work. His comparison of the visual and literary evidence was very well done; he was not, like Wren, content with a casual ascription. However, to show that the image was a Diana was not (by itself) to prove its authenticity as an ancient Roman Diana. Nor in the end was its existence sufficient to prove that a temple of Diana had once stood beneath

St. Paul's. Such inferences were simply too large for the evidence; once again Dr. Woodward's imagination had outrun his science. (One suspects as much for Horace's urn, as also the head of Terence.)[57] But perhaps there is too little to go on here; the manuscript appears to have been incomplete and has now disappeared. If we want to judge the Doctor's skills, we had better look elsewhere — to his famous shield, for example. Here was a problem exactly analogous to the Diana, more complicated but much more fully documented. Indeed, Dr. Woodward's shield was really quite the best object in his collection, perhaps in any European collection, to test the skills of the new antiquarian science. Undoubtedly that was one of the secrets of its fascination to a whole generation of learned men; it remains in any case its chief interest to us.

IX.

Dr. Woodward's Shield

I t was in 1693 apparently that Dr. Woodward found his shield. He had purchased it, he recalled, from the daughter of the apothecary, John Conyers, shortly after his death. It seems to have come to him with some other Roman objects from that surprising collection. The shield, however, had not been unearthed at St. Paul's; Conyers, himself, had bought it from an ironmonger in the city, sometime late in Charles II's reign. Where it had been before that was obscure; in any case, Conyers and the ironmonger were now both dead. [1]

Woodward's enthusiasm was boundless, though it took him a few years to get around to his new acquisition. As soon as he found the time, however, he canvassed the learned world for its opinions about his new treasure. He had it drawn, engraved, and copied in plaster. He pored through learned tomes to discover its meaning and proclaim its importance. He scribbled away unceasingly in his notebooks, jotting down whatever he could find to illustrate his object. And he scrupulously filed the replies that began to flow in from his many inquiries. [2] In the end, here as elsewhere, he was too busy to consolidate his work and he tried instead to provoke others to the task. The correspondence mounted; several Latin treatises were projected and eventually one was published.

It was natural for Woodward to turn to his learned friends abroad. His correspondence with foreign savants was well established by 1700, and he took a special interest in exchanging information and specimens with them and in squiring them around the precincts of the Royal Society when they came to visit. The first notice of the shield that I have been able to discover appears in a letter to the Danish antiquary Otto Sperling, although Dr. Woodward did show an "old Roman Target" (otherwise unidentified) to the Royal Society the year before. [3] Woodward knew Sperling through his works and through his brother, who for a time took up residence in England. Although Sperling had little interest in natural philosophy, he was willing to send the

151

Doctor some geological information, to exchange antiquarian views, and especially to purchase Scandinavian books for his English friend's amusement. (Woodward's interests extended, as we have seen, to Danish antiquities.) The Doctor responded by nominating Sperling for membership in the Royal Society, and he was elected in November 1700.

Woodward remembered reading one of Sperling's works on coins as early as 1690. Now his friend Robert Molesworth, the English ambassador to Denmark, brought him news of more recent antiquarian enterprises, including a projected dissertation on ancient shields (*de Scutis Veterum*). Here was a fortunate chance! The Doctor was proud to announce that he had, among his assorted oddities, an iron shield with figures embossed upon it. Its antiquity was undoubted, its variety and excellence beyond cavil. The shield portrayed the story of the attack of Brennus upon the Roman capitol. If Sperling thought it useful, he would be only too happy to send him a careful rendering of it. Sperling was delighted; he hoped Woodward would describe both faces of the shield, as well as its size and weight. He was interested too in other ancient arms that belonged to the Doctor, since he was presently writing about various weapons depicted on the coins of some ancient peoples.[4]

The correspondence prospered; Sperling's brother acted as agent. The promised drawing was soon dispatched, though not as perfect as Woodward had hoped and only approximately like the original. The Doctor sent other sketches of arrows and spurs drawn from his collection, some discovered in the Fleet-Ditch, others from the Orcades and Virginia. Sperling responded with more questions and some doubts. He was concerned as to whether Woodward's find was a combative or a votive shield. Did it have a handle? Was it convex or concave? His knowledge of the arms on ancient coins led him to cite many examples there of shields engraved with histories and figures. But he was not convinced that this was the Gallic Brennus attacking Rome; it seemed to him more likely to be that Brennus who, according to Plutarch, had laid waste to ancient Greece.

At first Woodward was too busy to supply the "very full and particular Account" that he had promised. He did send some further description: weight forty-one ounces; diameter fourteen inches; concave interior, convex exterior. He told Sperling's brother that he would eventually convince the antiquary that the figure depicted on the shield was not the Brennus who had attacked Apollo's Temple at Delphi, but another and older Brennus who had assaulted Rome. Here indeed was a difficulty; two years later, however, it was still not settled. Apparently some letters went astray; Sperling's brother appears to have been unreliable. Otto Sperling in any case remained unconvinced about Brennus; he could not recall that Brennus had ever been named as leader of the Gauls in their attack on Rome.

It was not until the summer of 1703 that Woodward found time for an adequate reply. He at last dispatched a long letter to Copenhagen (in English) describing the shield in detail and defending his opinions about its iconography.[5] He had not been idle in the interim. The letter began with thanks and compliments to Sperling, as well as news from Ralph Thoresby about some ancient arms found in the north of England. The Doctor was apologetic for the bad drawing of his shield. But he was adamant about his original opinion: the shield portrayed the earlier Brennus. His proofs were elaborately and formally argued, impressive in their mastery of classical learning and the new and budding science of iconography.

It was really not hard to dispose of Sperling's opinion. The shield could hardly depict his candidate, the Brennus who had destroyed the Temple of Apollo at Delphi, since the scene on the shield was very different. "In the upper part of it is shewn, not one single Temple only but a whole City, with a Variety of Buildings." That this was Rome one could hardly doubt, for in the upper righthand corner of the shield, just as the ancient historians had suggested, was the Capitol, which alone had escaped the destruction of the barbarians. (The rest of the buildings were in flames.) Moreover, at the center of the shield one could see the Gauls clearly demanding their tribute, exactly as Plutarch and Livy had recounted. Nothing of this had appeared in the accounts of Brennus attacking Delphi. Furthermore, Strabo (quoted in Greek and exactly cited in a footnote) had specifically referred to the Greek conqueror as "the other or later Brennus." And finally, a host of ancient writers had quite distinctly named Brennus, despite Sperling's contention, as the leader in the sack of Rome. The coincidence of names did not trouble Woodward; he was inclined to agree with Camden that Brennus was the generic Celtic title for a prince rather than a proper name.

Woodward's argument was characteristically vigorous and thorough, his account of the ancient sources particularly impressive. He had apparently run down every mention of Brennus that he could find in classical literature, tracking the Delphic conqueror in Greek and Latin through Diodorus Siculus, Plutarch, Pausanias, Polybius, Florus, Justin, Valerius Maximus, Callimachus, and Propertius; and the Roman conqueror (the older and more appropriate Brennus) in a host of early and later writers, all of whom attested specifically to his name. He noted each citation precisely, quoting freely in both classical languages. Thus, besides Livy and Plutarch, he was able to find an earlier Brennus in Zonaras, Festus, Servius, Orosius, Suidas, and, most interestingly, in Silius Italicus. In that poet Dr. Woodward discovered a passage describing what seemed to be the very shield in his collection. "The Description that he gives of that Shield agrees so exactly to this in my Repository, that one would be tempted almost to believe 'twere the very same...."[6] Thus if one put the various accounts together, both those truly ancient with those later (but all agreeing exactly), it was impossible to doubt his conclusion about the story on the shield.

Sperling agreed. The coincidence of detail was far too striking to admit of argument.[7] Moreover, the story on the shield was one of the most memorable in ancient history; in Woodward's words, "one of the most considerable and extraordinary in all the whole Roman Story." Both Plutarch and Livy had told it in very similar versions, and they were among the best-known and most authoritative of classical historians.[8] Essentially, it went like this: The Gallic invaders led by Brennus had descended swiftly on Italy and threatened Rome. After the failure of negotiations, the murder of the Senate, and a famous incident in which some geese sounded the alarm to warn the Romans, the Gauls assaulted the city itself, burning most of it to the ground. However, they foolishly allowed the remaining Romans to recover sufficiently and to fortify themselves in the Capitol. There a siege was laid, and both sides began to weaken. In the countryside the Roman, Camillus, rallied his forces even while the beleaguered city was beginning to submit. The scene was thus laid for the climactic moment represented on the shield. As Plutarch reports it (in an account somewhat fuller than Livy's); "Sulpicius, Tribune of the Romans, came to parle with Brennus, where it was agreed, that the Romans laying down a thousand weight of Gold, the Gauls upon receipt of it should immediately quit the City and its Territories." When, however, the agreement was concluded and the gold produced, the victors proved suddenly deceitful; "the Gauls used false dealing in the weights, first privily, afterwards openly, pulling back the balance and violently turning it." The Romans naturally complained, but "Brennus in a scoffing and insulting manner pull'd off his Sword and Belt, and threw them into the Scales; and when Sulpicius asked, What they meant, what should I mean (says he) but *woe to the conquered?* which afterwards became a proverbial Saying." It was indeed a dramatic moment – the arrogant taunt, in Livy's telling, gaining from its alliteration, *Intoleranda Romanis vox, Vae victis.* But this was the turning-point. As the Romans divided about how to respond to the affront, "Camillus was at the Gates and having learned what had passed, he commanded the body of his Forces to follow." He snatched the gold from the scales and repudiated the agreement, explaining that "it was customary with the Romans to deliver their Country with Iron not Gold." Both historians agreed that this riposte was followed at once by the solid defeat and withdrawal of the invaders. Camillus was a new Romulus; Rome had restored her honor and dignity.

It was an especially vivid tale in either Greek or Latin. And now one could see it represented strikingly on Dr. Woodward's shield, "in a manner very full and distinct." The Doctor thought it quite wonderful that so much of the story could be depicted on so small a space; "The Capitol entirely unde-molished, standing on the Mons Capitolinus, the rest of the City, part in Flames and Ruins and not much of it whole and standing. The Scales with the Money in one and the Sword in the other with the Belt . . . the Dead Carcasses lying upon the Ground, which those who have recorded this Transaction say were become very offensive and also a great motive for the Gauls to accept the Money . . . the Roman Army marching up to the Relief of the Place, with Camillus at the head of them on the left side of the Shield and that of the Gauls flying and in Retreat on the Right." Moreover, the details of the setting were equally precise. "The Buildings were done in such Manner that the very Order and Architecture of them appears, the Clouds above, and the Trees and Herbage without the City . . . the Vase with the Money in it, the Scales, the Arms and Habits are all finely designed." The arms and accoutrements of the two sides he thought would be particularly interesting to Sperling: the helmets, swords, lances, and *thoraces.* Thus the work not only related the story told by Plutarch and Livy, it also illustrated and illuminated that history. "By this means," Woodward was confident, "we have here in View a very noble Prospect of Antiquity and that affords us a vast deal of usefull information." That it was also beautifully executed was an added bonus. The workmanship was "Masterly and exquisite." The Doctor thought that he had never seen an object with such admirable variety, so "just and natural" and so finely designed. In the end, it was hard to say whether its historical or its aesthetic appeal moved him more.

[2]

The Doctor had indeed made a discovery. The shield was superior to any fossil — much superior to even the rarest ancient medal. Still, it was like them in its effect: it furnished a direct and vivid link with the past. Moreover, the particular past that it evoked was never so much admired as in Dr. Wood-ward's day. The Augustan Age was no misnomer; never perhaps had the reputation of Roman history and the Roman historians stood so high. Livy's translator spoke for his time when he wrote that "amongst all prophane Histories, none can for Greatness of Action, prudence of Counsels, and Heroick Examples of all sorts of Vertue compare with that of the Common-wealth of Rome."[9] Indeed, the only possible rival was Imperial Rome; on that point even the moderns tended to agree.

There is no trouble, therefore, in understanding the enthusiasm of Dr. Woodward for his new shield. It is a little harder perhaps to share his confidence in Livy and Plutarch. From the modern vantage point the story of Brennus and Camillus looks very suspicious. It is true that Woodward was able to find allusions to it in a wide range of classical literature. But it is also true that none of his sources, not even Livy and Plutarch, wrote within several hundred years of the event. The earliest historian to discuss the episode was Polybius, who describes the Romans as simply and ignominiously paying the ransom and who omits all mention of either Brennus or Camillus. Anachronisms and inconsistencies in the later writers serve also to weaken their authority (at least for us), while it is plain that Roman pride was sufficient to embroider the tale in the Romans' favor. While no one today doubts the Celtic sack of Rome (here archaeology confirms the literary accounts), there is also no one who accepts the story of the Roman historians in its detail. Dr. Woodward's shield portrays a thoroughly fictional scene. [10]

Why then did Woodward accept the story and defend it with so much erudition? The answer lies in the peculiar authority that the classical historians exercised over Renaissance Europe and later centuries. The humanists were agreed that the classical authors furnished the standards of style and form. In 1700 it was still hard to imagine how a modern history could be written which could surpass the classical works. If it was to be attempted – and some of the moderns at least were willing to try – it would have to be with a later subject matter. Who would dare to compete with the ancients on their own ground? Consequently, almost no fresh narrative was written for the ancient period and it was not likely to occur to anyone to rewrite the story of Camillus, when it had been told so effectively by Livy and Plutarch. [11] Reverence for style enforced the authority of the subject matter. Thus Livy was generally praised "both for Copiousness and Elegance [and] for Accuracy and unblemish'd Fidelity." [12] It was hard to doubt the veracity of a writer whose story was so effectively told and whose stylistic authority was so great.

There was thus an overwhelming predisposition to believe. The superior air of the nineteenth-century writer was not altogether unfair. "In the first two centuries after the invention of printing, the history of Rome, for the regal and republican periods, was principally studied in Livy or the classical compendium of Florus and Eutropius, and in Plutarch's Lives. . . . The entire early history of Rome was in general treated as entitled to implicit belief: all ancient authors were put upon the same footing and regarded as equally credible; all parts of an author's work were moreover supposed to rest on the same basis. Not only was Livy's authority as high as Thucydides or Tacitus, but his account of the kings was considered as credible as that of the wars with Hannibal." [13] Certainly Dr. Woodward's indiscriminate parade of authorities for the elder Brennus seems to bear this out. "It is now incomprehensible to us," wrote Niebuhr, "how even very ingenious writers, men far above us, took details of ancient history for examples without feeling any doubt as to their credibility." [14]

Yet it was more complicated than the nineteenth century was willing to allow. In fact doubts had been expressed occasionally in the earlier period about the classical historians, and Livy's first decade had been singled out especially for suspicion. The philologists had not been slow to discover inconsistencies in the ancient narratives, and the antiquaries had been willing to correct points of detail with their coins and inscriptions. It was the Moderns, of course, who led the way — nor were they oblivious to the fact that some of the classical writers had written "as smartly against their Predecessors as we can write against ours, and laid as many Accusations of Partiality and Error to their Charge, as even we pretend to lay to ours." [15] The details were therefore not above challenge, and it is not surprising that Sperling suspected that Livy and the others had confused one Brennus with another in their accounts of the Gallic incursions. But sporadic questions of detail were not the same as systematic criticism. The discovery of occasional errors did not usually lead to systematic doubt.

Yet even to this there were exceptions. A scrupulous reading of Livy of itself suggested a problem. In his preface, the Roman historian had noticed expressly that the Celtic sack of Rome had destroyed the written records of the early period, leaving the city bereft of authority for its early history. But if Livy himself had lacked the sources, what was one to make of his long account of the Roman tale from Romulus to Camillus? From the time that Philip Cluverius (1580-1623) announced the problem and cast suspicion on the history, antiquaries began gradually to notice the difficulty and to attempt to meet it. [16] In Dr. Woodward's day the debate was reopened by the Dutchman Jacob Perizonius, and it gathered increasing strength in the eighteenth century. It was fed, no doubt, by an increasing general skepticism, by the appearance of a "historical pyrrhonism" that was beginning to cast doubt on the reliability of all historical sources. The French Jesuit Hardouin announced paradoxically that the classical authors were the invention of medieval monks, since they were found only in medieval manuscripts; and La Mothe le Vayer wrote an essay entitled *Du peu de certitude qu'il y a dans l'histoire* (1668). [17] Yet the vast number of scholars and the even larger number of literate laymen were unaffected. The criticisms of Livy could be met by new arguments, as for example by Perizonius, who noticed that Livy had not said that *all* the sources had perished and who invoked an oral tradition to supply the evidence for the early Roman history. The philosophical skeptics were even less successful; here the coins and inscriptions, the monuments of the antiquaries, furnished an irrefutable confirmation of the classical literary works. Hardouin was rebutted; it became fashionable to decry his "paradoxes," and the authority of the ancient historians was left largely intact. [18]

Thus while it was possible to be skeptical altogether, or at least critical in part, of the early Roman history, the overwhelming predisposition remained to belief. What was lacking still was a clear alternative: in the absence of a systematic archaeology, dependence on the literary texts remained overwhelming. It is still a temptation today, despite the characteristic disclaimer of the modern historian who, for example, argues that "For a study of early Roman history, the archaeological material must form the basis, because it constitutes the only primary source of evidence." [19] Perhaps so, but there are few who have despaired altogether of employing the literary texts; and archaeology has remained, with all its achievements, insufficient for the task of reconstruction. The example of the eighteenth-century writer Nathaniel Hooke is illuminating. He was apparently the first Englishman to attempt to rewrite the Roman story in a thorough and scholarly fashion. [20] He was aware too of the general criticisms of Livy, now full-blown, first as a result of the controversies in the French Academy of Inscriptions (1729) but especially because of the work of Louis de Beaufort, *Sur l'incertitude des cinq premiers siecles de l'historie romaine* (1738). Yet Hooke is only half-persuaded. He is willing to concede that the story of Camillus (for example) is "romantic in the air of it" and that Polybius' earlier account suggests that it was "a mere fiction of Roman vanity." [21] But if that and some other points of detail in Livy seem faulty, he thinks it hardly enough to reject the whole history. And furthermore, when he reaches Camillus, in his own account, he tells the story as he finds it, resigning his skepticism to a footnote. [22] What else was there to do? For Dr. Woodward, writing fifty years earlier, the alternatives were even less clear, and it is doubtful that he was much aware of the skeptical argument. Moreover, through all his vast correspondence with the antiquaries of his day, there seems to have been no real doubt expressed that the story on the shield represented anything but an actual and famous historical event.

[3]

Dr. Woodward had solved the basic iconographic problem of the shield by assembling the literary evidence to match the details of the picture, and he had thus answered Sperling's objections. But he was not satisfied. His researches left a series of questions unanswered. Why, he asked Sperling, were both armies accoutered in the same arms and dress? [23] Had the Gauls acquired their costume from their vanquished foe? "The *Thoraces* in particular appear at first view to be Roman and such as occur commonly in the

Basso-Relievos, Statues, and other Remains of the Antiquities of that great People." (He noted here too the description of Diodorus Siculus.) Not all the persons on the shield were easily identified. Who was it amidst the Roman Army with a shroud or veil flying loose in the air above him, and who in the smooth arms standing at the left of the scales? Who was it holding the balance, and who resting on his hands just behind? Perhaps most importantly, what was the age of the shield? What was it used for? Who made it? "Everything upon it shews it very antient, particularly that the Horses are without Bridles and the Riders without Stirrips or Spurs, which you know was the first most antient Manner." The best and most knowledgable workmen he could find agreed that it carried with it "the most indubitable Marks of its being truly ancient," though it had quite remarkably escaped the ravages of time. Still, how ancient was it? Woodward hoped that Sperling would freely offer his conjectures.

Sperling could be helpful here; as an accomplished numismatist with a special interest in arms, his opinion was welcome. But Woodward had begun to think of others at home and abroad who might also be able to assist him. It was clear, however, that before they could venture their opinions, the shield would have to be copied more exactly than in the rough draft sent to Sperling. Everything depended on the precision with which the details were rendered. The first job therefore must be to have it copied exactly and engraved. His instructions to the artist emphasized the necessity for exactness.[24] He insisted that the figures be reversed on the plate so that the representation would correspond precisely with the original. Above all, he urged the engraver to use the shield itself as his model, not the drawing, which was imperfect; he was able to point out three errors himself. The result was a full-size rendering of the shield which the Doctor now began to send to his friends in England and to his correspondents abroad.[25] Dr. Woodward's shield was thus publicly announced to the world.

The Doctor was very pleased with the response. The shield was treated at once as a major discovery. By now Woodward was himself a celebrated figure among the virtuosi, and the representation on the shield declared it to be an object of highest importance. It was a great prize for the collector, a great stimulus to the scholar. It could only be compared with the large classical statues and reliefs that were so much sought after by the wealthy and written about by the learned. Indeed, a few years afterward, Woodward had plaster casts made and sent to the King himself, to the Princess, to some of the great aristocratic collectors (including the Earl of Pembroke), to the universities, the French king's library, and the library of the university at Leyden.[26] Meanwhile, Thomas Hearne employed it for his Livy (1708); the Dutch classicist Drakenborch used the shield to illustrate his new version of Silius

Italicus (1717); and it appeared also (not altogether appropriately) in that greatest of all the illustrated classics, Samuel Clarke's sumptuous edition of Caesar (London, 1712). Two hundred years later it was still thought useful enough to illustrate the article on Brennus in Harper's *Dictionary of Classical Literature and Antiquities!*[27]

All the while the replies poured in as antiquaries throughout Europe offered their opinions about the engraved shield. Especially useful to Woodward was his friend Petrus Valckenaer, the Dutch ambassador to the Swiss cantons and himself an antiquary and geologist. Through Valckenaer, Woodward's inquiry was distributed widely to a long list of classical scholars and antiquaries drawn up for him by the Doctor. "I have rolled every Print round a Stick," he wrote to Woodward, "that the Beauty of it might not be spoil'd with folds."[28] In this way the greatest of the Dutch scholars, then the greatest in Europe, were solicited and responded. Among them was Perizonius himself, the celebrated professor at Leyden. His *Animadversiones historicae* (Amsterdam, 1685) had brought him early fame as a repository of philological erudition; it was resurrected long afterward as an anticipation of Niebuhr's ballad theory of the history of early Rome. Perizonius was an indefatigable worker and his erudition was universally respected; at the same time he was a judicious critic, praised even by the skeptical Bayle, though much concerned to answer the historical pyrrhonists then rearing their heads in France. He took the middle ground in the controversy over the early Roman history, anxious to submit the story to rigorous criticism but in order to preserve it where possible. He believed therefore that there was an element of truth throughout the history of the kings from Romulus, though it could not all be accepted literally. In short, the Leyden antiquary was, in the words of an early biographer, "l'un des plus savants philologues, et des critiques les plus judicieux dont s'honore la Hollande."[29]

Valckenaer reported having conferred with Perizonius about the shield.[30] The scholar had admired it and offered his opinion (the same we are told as M. Vries, M. LaFaille, and other nameless savants) that the object was a "votive" shield made for the family of Camillus to commemorate the great event. He did not believe that it was contemporary with the sack of Rome, but rather the production of a later time made under the first emperors when the arts of ancient Rome had received their highest perfection. The existence of a well-known Furius Camillus under Claudian helped to strengthen the conjecture. It was possible, he thought, that the shield had been made by Greeks who were by then living and working in Rome. The buildings, dress, and arms all suggested the later date.

Perizonius' opinion was echoed by others. Valckenaer was able to report that Antoine van Dale, another distinguished Dutch antiquary, held much the same view.[31] As Woodward noted with satisfaction, Van Dale had also found that the shield was votive.[32] He too thought that it must be later than the event, when the arts of Rome had become perfected under Greek influence. Besides, in the time of Camillus there were no baths, theaters, or amphitheaters in stone, as clearly depicted on the shield. And finally, the dress and arms of the Gauls and Romans should not have been identical (as on the shield), since they differed greatly in those early years. All these anachronisms, together with the classical shape of the shield, showed it to be beyond question a later Roman production. Valckenaer's letter closed with a characteristic report of some elephant's teeth dug up in Tartary.

Valckenaer also talked with Jacob Gronovius. If Perizonius was the most judicious of the Dutch antiquaries, Gronovius was the most prolific and quarrelsome — a great man despite an "ill temper" that "render'd him odious to most Learned men."[33] The son of an even greater and very much better-tempered scholar, he had traveled everywhere in Europe (including England) before settling down at the University of Leyden, where he enlisted against the most eminent of his contemporaries — Perizonius and Bentley, LeClerc and Kuster — and almost everyone else. Meanwhile critical and antiquarian works poured from his pen in a torrent that only ceased with his death in 1716. Innumerable were the editions with commentary of the classical Latin and Greek authors that he compiled (Nicéron lists forty-six large works) — one of which alone, the *Thesaurus antiquitatem Graecorum* (a collection of antiquarian treatises that he edited), filled a dozen stately volumes.[34] Gronovius promised Valckenaer a full account of the shield. In due time a long and formal Latin letter made its way across the Channel, to the delight of the Doctor. In more detail now and with exact citations, Gronovius placed the shield in later Roman times, calling attention especially to the testimony of Pliny and to the analogous shield of Scipio illustrated by Jacob Spon.[35] He identified several of the figures besides Camillus and Brennus — for example, the Tribune, Sulpicius, and the cavalry commander L. Varenius; he expounded on the gold-pieces that passed for the tribute money, the military costumes, Capitoline architecture, etc. The details of the shield allowed the scholar ample opportunity to dilate almost indefinitely on his favorite subject.

And still they came. Jacques LeClerc, the friend and biographer of Locke, universal scholar, theologian, philosopher, philologist, historian, journalist, and critic, wrote first to Valckenaer, then directly to England upon receiving the engraving.[36] He too connected it with the buckler published by Spon, thought that it belonged to the imperial period, and admired its surpassing artistry. It was no ordinary shield, he believed, for they were not usually so decorated; it must be a later commemoration of the great event. In any case, the artist had given free reign to his fancy, and the details should not be literally understood. Hadrianus Relandus, another Dutch classicist and Hebraist, professor at Utrecht, wrote two letters to Woodward to much the same effect.[37] The greatest numismatist of his time, Ezechial Spanheim, promised a Latin treatise, and the erudite Gisbert Cuper sent three long letters which he expected Woodward to publish. Inevitably J. J. Scheuchzer received a print and responded with a learned note. He was struck most by the absence of bridles on the horses, but was happy to relay a passage (pointed out to him by Cuper) from Livy showing that the Romans had sometimes fought that way. In a second letter he was able to report that the shield had even penetrated into France and arrived in the hands of the celebrated Abbé Bignon. And there were others still.[38] The continental correspondence was indeed impressive testimony to the interest and importance of Doctor Woodward's possession. By 1707 the engraving had made the shield known everywhere. And discussion had revealed a broad unanimity of sentiment; Dr. Woodward's buckler was a beautiful example of classical Roman workmanship at its pinnacle under the early Caesars. It was an accurate and useful depiction of a great event and it was a provocative exercise for antiquarian scholarship. To date it, author it, and to explicate its least details, furnished a series of problems that promised both the delight and the utility that the best of classical scholarship was said to afford. For its explication all the new disciplines could be invoked, philology and archaeology, numismatics, epigraphy, iconography — in truth, the shield was a virtuoso's dream.

[4]

The antiquaries were agreed in most of their conclusions. It could hardly be otherwise, since they followed a common method. In the case of Dr. Woodward's shield they simply inverted their customary procedure of finding the nonliterary evidence to support and illuminate a classical text. Thus, where they were used to citing the coins and inscriptions to explicate a passage in Livy or Plutarch, they now simply reversed the process and sought in the literary texts a passage to illuminate their new discovery. It was a practice familiar from the ancient coins but equally appropriate to any classical object. The shield was like a great medal or relief; the first problem was to define and classify it, the next to identify the image on its face. The first required knowing something about ancient shields; the second was iconographic and required identifying the picture from classical history or poetry. But once these problems were resolved, the shield, like any authentic ancient object, could be used in its turn to illuminate the literature. The reasoning was thus circular — or perhaps spiral — as gradually each new piece found its place in a growing mosaic that was meant eventually to portray the whole of classical civilization.

In the case of Dr. Woodward's shield, there was an obvious comparison that served both to identify the object and to furnish an iconographic model. It was the famous shield of Dr. Spon, cited again and again in the Woodward correspondence.[39] Here was an object strikingly similar to the Doctor's, discovered only a few years before when it had been published and illustrated. Spon was a Frenchman, a medical doctor like Woodward, who had made an unusual voyage to the East in the company of Dr. Woodward's friend George Wheler. In 1674 he set out for Italy, where he met his English companion; the two traveled all the way to Constantinople, passing through Greece and the principal cities of Asia Minor. Spon lingered for half a year in Greece, making observations and examining antiquities; on his return he brought back over three thousand inscriptions as well as many manuscripts. (Wheler brought fewer antiquities, but hundreds of rare plants.) The *Voyage d'Italie* (1678) brought Spon immediate fame. Not long afterward (1685) he died prematurely at the age of thirty-eight.[40]

Spon had written about the local antiquities of Lyons as early as 1673. The voyage to the East enlarged his range, and soon after his return he published his magnum opus. It consisted of thirty-one different dissertations on ancient works, the first of which was a marvelous antique shield found in the river Rhone.[41] In each case, Spon identified the object and tried to explain the picture, using his skill in coins and inscriptions and his wide reading in the classical authors. His compilation thus marked a milestone in the new science of iconography. Many of the objects were illustrated. The work was very successful and appeared in a Latin version two years later. The shield, which furnished the frontispiece as well as the first essay, was at once famous. Gilbert Burnet took special notice of it when he passed through Lyons in 1685.[42]

Spon had discovered his shield in the cabinet of Octavio Mey at Lyons. It had been found in the Rhone near Avignon by some fisherman who had failed to see that it was made of silver (since it was covered with lime) and consequently sold it. It was a striking work, twenty-one pounds and more than two feet in diameter; in Burnet's words, "the noblest piece that has been handed down from old times to ours." Spon was convinced that it was a *clypeus votivus*, which he defined as a shield representing a memorable action of antiquity, designed to hang in a temple of the gods. He distinguished it at once from a combative shield. Its weight and value alone made it an improbable weapon. Moreover, the usual shape of the fighting shield, as shown on coins and reliefs, was long and oval. A *clypeus votivus* was, Spon thought – like many of the ancient medals, inscriptions, bas-reliefs, and statues –

meant to commemorate someone's exploits and enhance a reputation. Already in Homer one could read of the ornamented shields of Achilles and Ajax. Often they bore only portraits, but sometimes they described whole histories. The coins were particularly helpful here, for they frequently depicted the votive shields. Spon was able to describe and illustrate seven examples from medals to make his case. Since the votive shields were usually made of precious metal, they were themselves (like the other gold or silver objects of antiquity) excessively rare. Spon had never seen another, although he claimed to be familiar with most of the great cabinets in Europe.[43]

It remained to explain the figures that were vividly portrayed on the face of the shield. Mey had correctly identified them, Spon thought, with still another famous story from Livy, this time about Scipio Africanus in Spain.[44] Following his victory there (so the story went), a young and beautiful maiden was brought to Scipio, who was known to be fond of women. The great warrior learned, however, that she was betrothed to a young Spanish prince named Allucius. He summoned the couple to him and, to everyone's surprise, announced that he would give up his own claims in favor of their love, whereupon the relatives of the maiden entreated Scipio to accept a gift from them. Scipio agreed and ordered it to be set at his feet, adding it unexpectedly to the dowry for Allucius as a gift of his own. Needless to say, the youth returned home a new friend to Rome, filled with the praises of Scipio. Lord Chesterfield could think of no better example to instruct his son in virtue and generosity.[45] And Spon recalls that Aulus Gellius could find only one action to compare it with: the somewhat similar feat of self-denial by Alexander the Great.

Spon recognized Scipio as the figure seated on the throne in the middle of the shield, half cover'd with a Mantle onely, his Beard close cut after the manner of the Romans at that time, holding a Speare in his hand."[46] This allowed him to discourse learnedly of ancient beards, to fix the time of the event, and to separate the different characters on the shield into Romans and Spaniards. Four of the relatives of the bride, he suggested, surround Scipio (distinguished by their beards) and beg him to accept as a gift that which they had brought as a ransom. The young woman "stands modestly by with her Spouse, who gives her his Hand and Faith, and embraces her with the other Hand." (Spon here added some remarks on their dress.) Two officers of Scipio were also identified by name, citing Livy. A nude figure on the ground suggested to Spon a Spanish prisoner, adoring Scipio's virtuous act. Spon learnedly described the arch in the background, identifying its ornament as a Triton and a Nereid, symbols of Scipio's victory. The arms at the great man's feet were further tokens of his victory, each identified by the antiquary. Spon commented too on the swords carried by the soldiers and on the table to one side, with its vase and two loaves of bread. These he thought were meant to recall the pagan marriage ceremony. Thus every detail in the crowded scene was described and identified by the antiquary as far as possible. It was an exemplary performance, unique for its time, as far as I can see, in its description of a classical shield. In the dissertations that followed, Spon turned to a variety of other classical objects to apply his technique both to the traditional coins and inscriptions and also to statues, reliefs, and mosaic pavements. His work thus furnished a compendium of iconographic models.

There remained a problem or two about the shield. Spon was concerned to date the object as exactly as he could. The event of its face he assigned confidently to the year 543 AUC (210 B.C.). The shield, he was equally certain, must have been contemporary with the event. Indeed, what purpose could it have had (as a votive offering) afterward? Its iconography appeared authentic, and even the location where it was discovered added plausibility to the date. The Rhone at Avignon lay exactly on the route that Scipio had taken in returning to Rome from Spain. No doubt, in crossing the river, some accident had caused it to be lost there. The shield was thus exactly 1,892 years old.

Finally, Spon was aware that there were some who questioned the traditional story. The poet Naevius, for example, had cast aspersions on Scipio's chastity, while Valerius Anteas had also written heartily against him. But Spon thought that the testimony of the other writers was sufficient refutation. One could even allow Scipio the temptations of youth; when the episode on the shield took place, he was twenty-seven or twenty-eight, old enough to withstand them. But best of all was the evidence of the shield itself. As a sacred object intended for a temple, it could hardly lie, especially with Romans and Spaniards still alive and able to attest to its fidelity. Thus Spon found that the historians explained the shield and the shield confirmed the historians. However circular the argument might appear, it only served to show the importance of the shield and all other such monuments for the reconstruction of the past.

[5]

In advertising his possession, Dr. Woodward did not overlook his English friends. On January 8, 1707, he presented the Royal Society with a copy of the shield, for which he was thanked.[47] John Strype naturally received one and applauded the publication of "so venerable a Piece of Antiquity." To Cambridge he sent a print for the Greek scholar Dr. John Covel, who shared his interest in fossils as well as antiquities.[48] To Oxford he sent another for his friend Dr. Hudson, the keeper of the Bodleian and also a famous Greek scholar. Woodward heard that Hudson intended to "say Something of the Design, the Age, the Persons and Things, exhibited by the Shield, which will be a very great Gratification to men of Learning and Curiosity."[49] Nothing came of it, however. The connoisseurs – the Earl of Pembroke, Sir Andrew Fontaine, John Kemp, and the visiting Italian Abate Benedetti – were all sent copies.[50] Far away in Florence, the English envoy Henry Newton also received one for which he returned an ornate Latin poem. As far as I can see it is of neither aesthetic nor antiquarian interest, but it was well received in Rome and must have given Dr. Woodward great satisfaction.[51] And of course the Yorkshire antiquary Ralph Thoresby received an engraving also.[52]

As we have seen, Thoresby was a kind of Yorkshire counterpart to his London friend. Their interests certainly were very much alike. As collectors they exchanged objects and information; both were active members of the Royal Society, and both took an interest in natural curiosities as well as antiquities. Neither was a great scholar, but both could understand and criticize the most abstruse labors of their contemporaries. Dr. Woodward, it is true, aspired to philosophy, while Thoresby was content by and large to remain a collector; and of course temperamentally they were very different, with Thoresby's piety, generosity, humility, and even his growing poverty, standing in thorough contrast to the irascible Doctor.[53] But their common enthusiasms helped to bridge the gap and keep them friends for years, despite the frictions we have seen developing. Although Thoresby's antiquarian concerns (like Dr. Woodward's) were various, he too took a special interest in the Roman antiquities of his neighborhood. His father before him had been a collector and had taken particular account of Roman objects. One Roman altar that had passed down from father to son had been noticed by Dr. Lister as early as 1682 in the *Transactions* of the Royal Society. Fourteen years later the younger Thoresby reminded the Doctor of that contribution and described a second altar that had meanwhile come into his possession, along with a discovery near Leeds of a Roman pottery. Lister unexpectedly passed the letter along to the Royal Society, where it was quickly printed. It was the first of a long series of similar reports to the Society by Thoresby, many devoted to Roman antiquities, others dealing with a variety of natural phenomena in Yorkshire.[54] It was his neighbor, the learned Dr. Gale, who proposed him for membership in 1697 after reading the Society a description of some Roman urns, bricks, and coins that had newly entered the Yorkshire-man's collection. Thoresby's descriptions were always precise, sometimes illustrated, and usually explained and interpreted. Although he complained of a faltering Latin, he was not unwilling to try his hand at the emendation of a classical inscription, and in 1704 he even began to improve his Greek. His conjectures, however, were always proposed modestly and usually submitted first to one or another of his innumerable antiquary friends.

In the catalog of Thoresby's museum there was a whole section devoted to Roman antiquities, quite apart from a very full description of Thoresby's coins. It is interesting to see there, among the Roman vessels, the mouth of a *Praefericulum* "given me by Dr. Woodward," as well as "the Handle (half a Foot long) of a large Vessel, found at St. Paul's [with] thick white Glass from the same Place." He remembered, too, seeing molds for these vessels "in the Learned and Ingenious Dr. Woodward's Museum at Gresham College."[55] Most of his antiquities, however, were assembled in the North. There were, for example, sepulchral monuments from a newly discovered Roman town, a discovery that led to a contribution in the *Transactions*, a whole section in his description of Leeds, and a learned discussion with Gale over its ancient name. Thoresby's urns allowed him to discourse (as Woodward was just then doing) on their dates and on the ancient burial customs. The altars, molds, bricks, and urns were also described, together with tesselated pavements, many miscellaneous items, lamps, and especially a large Roman sepulchre unearthed at York, big enough for three or four whole bodies. There were spurs and swords and other ancient weapons but, most interesting of all, an ancient Roman shield that had been in his collection since 1697.

The shield was no rival to Dr. Woodward's; it was much plainer and more ordinary. But it anticipated the Doctor's by several years and was probably the first to be publicly described in England. Thoresby sent an account of it, together with an even earlier and somewhat differently shaped one in his collection, along with some general remarks to Dr. Lister, and they appeared in the *Transactions* in the summer of 1698.[56] In reading the ancient writers, Thoresby observed that the Romans had used three terms, "*Scutum, Parma* and *Clypeus*, for that defensive Weapon we generally English a Shield." But they had frequently used the three, especially under the Empire, much as the English did the Shield, Buckler, and Target. He described the form of the *scutum* and the *clypeus* as well as the votive shields that were usually affixed to palm trees and often illustrated on coins with inscriptions. Both the *scutum* and *parma* were "by very good Authors, as well more ancient as modern, positively said to be Wood covered with Leather" (here he referred his readers especially to Obadiah Walker's recent *History of Coins*). But neither of Thoresby's shields fit that description. The earlier was the *parma* kind. It was fifteen inches in diameter (only slightly larger than Woodward's), with a third of it taken up by the *umbo* "or protuberant Boss at the Naval." This was a convex iron plate, hollow on the inside, for the gladiator's hand. Within it was a lesser *umbo* where Thoresby thought there would have been fixed a *cuspis*, "or sharp offensive Weapon," for hand-to-hand fighting. Unfortunately he was unable to describe it, for both shields were here defective, "nor can I receive any Light from my Collection of Coins." The shield beyond the *umbo* was plain, with eleven circular equidistant rows of brass studs (two hundred and twenty-two in the outermost) set into circular plates of iron. "The inner Coat next to those Iron Plates (for I cut it a little open behind, that I might more distinctly discern the Work) is made of very thick hard strong Leather." There was then a second cover of the same, and plaited iron pins beneath to hold it together. The brass was purely ornamental, none of the iron showing. Finally there were two further coats inside, one of linen and a last of softer leather, all of which were bound together between two circular plates of iron. The other shield was a foot larger in circumference and differently shaped. Where the first was almost flat except for the *umbo*, the second was thoroughly concave, the studs and plates being also differently arranged. Typically, Thoresby concluded his full description with a conjecture: he thought these weapons had belonged to the *equites* rather than either the *velites* or *hastati*. Though he cited an anonymous *Roma illustrata* for a parallel description, he thought himself the first to describe the ornamental studs and ironwork which Camillus was said to have introduced against the Gauls. To confirm his description he supplied a drawing which was printed in the *Transactions*, and sometime later presented the flat shield to the Royal Society for its own collection of antiquities.

Thoresby's essay was a useful communication, exact in its description, plausible in its conjecture. Unfortunately it was quite wrong in its conclusions. When the picture was shown to the Royal Society, "a Scotch Gentleman present declared that it was Exactly of the Form of the Targetts used among the Highlanders with the same circle of Nails as this was described with."[57] Dr. Sloane purchased the draft anyway, and the essay was printed without qualification, so it is likely that most readers, including Dr. Woodward, continued to believe it to be Roman. It was natural enough then that Woodward should respond by sending Thoresby a copy of his own shield as soon as it was ready, together with a report on the urns discovered at Bishopsgate and a promise to "slip down some time or other, and surprise your collection." He also promised each of the two brothers Gale (sons of the famous Thomas, and antiquaries in their own right) copies of his icon. The replies are missing, but Thoresby's friend Richardson was able to report from personal observation that Dr. Woodward's shield was indeed "a very fine and valuable piece of antiquity." He could only compare it to the famous engraving in Spon. It was unimpeachable testimony indeed; only the day before, Richardson's friend Lhwyd had been elected to the Royal Society over the strenuous objections of Woodward.[58] Doubtless Thoresby was more than satisfied with the authority of this newest and most impressive of Roman monuments. In 1707 learned opinion seemed everywhere unanimous on the subject.

[6]

What Dr. Woodward wanted, of course, was to have his shield treated with the same attention as Thoresby's or Spon's. It seemed to him at least as great a prize and much more interesting historically. For a time he seems to have thought of attempting it himself. He filled the sheets of his notebook with annotations. He made long lists of classical citations and scribbled notes on points of detail and rebuttals to adverse comment. But he was too busy to put it all together. Even keeping up with contemporary scholarship was hard enough. In a letter to Dr. Hudson in 1705, he writes that he is reading "with abundant satisfaction" Montfaucon's *Diarium Italicum* and Fabricius's *Bibliotheca Graeca*, the latter alone a work of fourteen massive volumes.[59] When one Dr. Picenini appeared in London fresh from his travels in Asia Minor, Woodward immediately sought him out to obtain a transcript of the many inscriptions that he had heard about, "never yet published."[60] He purchased almost every book he could find about antiquities — Roman and otherwise — and took a special interest in the drawings and prints that might illustrate his collection, and his shield.

The library that Dr. Woodward built in these years was extraordinary. He saw that the objects which he prized could only be understood by setting them amongst others of their kind. No one man could hope to collect enough (any more than in natural history) to know them all. Fortunately, antiquaries like Spon had for a long time been publishing their discoveries, and Dr. Woodward set about assiduously purchasing and reading everything that he could obtain. He also bought pictures, drawings and engravings of every kind, but especially of classical antiquities. He left over four thousand volumes and hundreds of drawings at his death.[61] They dealt with many subjects, but a very large number concerned themselves with ancient Rome.

As Dr. Woodward considered his shield and sorted through his correspondence, it became increasingly clear that the single most illuminating classical object for his purposes was the Trajan column in Rome. Here was an ancient monument illustrating in great detail the warfare of the very period in which most antiquaries believed his shield had been made.[62] It was in any case the most authoritative set of illustrations in existence describing Roman weaponry and military costume. According to a recent authority, the reliefs are best compared with the accurate drawings made a century ago by the journalist-artists whom press photographers have now replaced. Thus when Robert Hooke wanted to confirm an opinion before the Royal Society about the galleys of the ancients, he turned inevitably to the famous column, and it was the natural recourse of anyone with an interest in ancient arms.[63]

Fortunately it could be studied in a number of published works, including a celebrated treatise by Raphael Fabretti.

The original monument itself, over a hundred feet high, was completed in 104 A.D. to commemorate the exploits of the Emperor, and it is surrounded by a bas-relief depicting the chief episodes of the Dacian Wars. Fabretti was a learned Italian antiquary who died in 1700 having won fame for his work on the column and for a remarkable horse, named Marco Polo, who had somehow acquired the unusual ability of standing still, and (like a hunting-dog) pointing, whenever he came near an antiquity. His master confessed that there was much that he would have overlooked were it not for his faithful companion.[64] The *Columna Traiani* was a magisterial volume that appeared first in 1683 and again in 1690 and won its author immediate fame and a quarrel with Gronovius. It brought together a wealth of information, especially from inscriptions, but from every other kind of antiquity also, to illustrate the Roman monument. Woodward owned a copy, as well as at least one other work on the column. Besides these, he owned five folders of drawings, with more than a hundred sketches of the monument. Moreover, he was interested in all the other Roman columns that survived. Thus he sought out the engravings of the *Columna Constantini* at the Earl of Pembroke's so that he could compare it with the Trajan and the Antonine monuments in his own collection. (The latter had just been discovered and published in 1705, raising in Woodward's words a "mighty noise" among the antiquaries.)[65] Set into this context, the shield, and even more the arms and setting depicted on the shield, could become really meaningful. Thus when someone complained about the balls atop the obelisks on Dr. Woodward's shield, the Doctor (after an uneasy moment) could show that that was how they were depicted on the Antonine column.[66] And it was plain from the Trajan relief that the soldiers at that time had never used a shield like Dr. Woodward's for combat, thus reinforcing his guess that it was a *clypeus votivus.*

One of Woodward's correspondents was the Kentish antiquary John Batteley. Batteley was the prototype of the leisurely scholar-clergyman of the day. He was Archdeacon of Canterbury from 1687 to his death in 1708 and, though busy collecting antiquities throughout his lifetime, published nothing beyond a single sermon. He had, however, assisted Bishop Fell by collating manuscripts for several editions of the classics, and he corresponded with a wide circle of antiquaries including John Strype, to whom he furnished much help. In 1707, with John Bagford and Humfrey Wanley, he met at the Bear Tavern in the Strand to begin anew the long-lapsed English Society of Antiquaries. After his death there was published his Latin tract, the *Antiquitates Rutupinae* (1711) – a pleasant description later translated, in an abbreviated version, as *The Antiquities of Richborough and Reculver.* These old

Roman sites in the south of England had long drawn the attention of antiquaries, and Reculver was said to have furnished more Roman coins than any other site in England. Batteley's advice to those whom he employed in searching for antiquities is interesting both for its enthusiasm about actually digging and the care with which Batteley tried to formulate some rules for archaeology. "I gave them," he writes of his assistants, "a strict caution not to clear such coins as were rusty, by rubbing them with sand or anything else; as I had seen many of the most valuable by that means quite obliterated; to sell no brass to the brasiers, but to reserve it for me . . . to break no urns or pots; if they found them broken and inscribed with any marks or letters to bring them to me." (This he notes was the origin of his own collection.) "In other places," he continues more generally, "let him who is curious in such researches, open barrows, let him explore encampments, trenches, and the places adjoining; let him examine the ancient public ways; let him without superstition or dread, open and ransack sepulchers . . . let him carefully explore the ruins of cliffs." When any traces of antiquities are discovered, "let him pursue them, and call in the assistance of connoisseurs; if he should discover any coins, either lying in a heap, or enclosed in an urn or pot, let him observe the latest, for they will nearly determine the time when they were buried." If there are inscriptions, they must be accurately copied. And all must be freely communicated to the world. "These" he concluded, "be the laws of Antiquaries."[67] No doubt the modern archaeologist would want to add much more, but the science had come a considerable way in the centuries since the classical revival began.

Batteley was very interested in the shield, which he thought undoubtedly genuine.[68] One problem particularly concerned him. Someone had suggested that the shield was really an *umbo* — that is to say, the center of a much larger Roman shield. This was made plausible by the fact that the edges of Dr. Woodward's buckler were rough and appeared to be cut down. There was also its small diameter. If so, he wrote to Woodward, it should have a little sharp cone in the center. Batteley had several among his collections and offered to send one. There was a problem, however, for if the shield were concave (alas, it was), he thought it could not be an *umbo* after all. Woodward was grateful for the information and for the "ancient spikes" that duly arrived. He was also grateful to Batteley for confirming that the colors and standards flown from the shield were the same as those on other authentic monuments. (The two men had been a little worried by the *vexilla* there.) Woodward found further confirmation for them in the engraving of a "Gem of M. Mason that exhibits the continence of Scipio." In another letter to Batteley, he explained why he had not placed the well-known passage of Silius Italicus at the bottom of his print of the shield. He had long known of the line there about one Crixus who claimed descent from Brennus and "displayed on his shield the Gauls weighing the gold on the sacred eminence of the Tarpeian hill." But he didn't include it there, "because it falls far short of a description of this, and besides that was a different shield." Naturally, it was important evidence in favor of Woodward's buckler, "indeed it had somewhat of the same design exhibited upon it," but it was only one among the many useful classical passages that Woodward was assembling. Meanwhile he was able to report that everyone agreed, both the continental antiquaries and the English, that the shield was "considerable for its antiquity, workmanship, and the story delineated upon it."

[7]

"I should be glad," Dr. Woodward wrote early in 1708, to see "the Letters with a further Dissertation upon the Shield . . . publish'd; 'tis what is called for on every Hand." The materials were ready, "but my Business will not allow me leisure for such a work." To Batteley he wrote that all these papers "must lie, amongst others of greater importance, for ever in the dark; unless some one who has more leisure than my business will allow me, will undertake to fit them for the view of the publick."[69] It was true; these were the years of geological controversy, the quarrel with Sloane, the urns at Bishopsgate — and always, the growing medical practice.

Two men at least had promised dissertations. One was the Englishman Abednego Seller, a non-juring parson who had accumulated a great library and a considerable reputation for classical learning, not perhaps entirely deserved.[70] The other was a much greater man, the Baron Ezechiel Spanheim, by all odds one of the greatest scholars in Europe. Somehow Spanheim had managed to combine an important career in politics with an abiding commitment to learning, and he was everywhere acknowledged to be one of the most formidable philologists and antiquaries alive. His great age and copious bibliography were meant to be climaxed by an enlarged version of a work that had early in life brought him fame, his *Dissertationes de praestantia et usu numismatum antiquorum* (1664). Here was a monumental treatise on the ancient coins, a synthesis of the numismatic learning that had absorbed the antiquaries since the Renaissance, a vast tome stuffed to the brim with every variety of classical erudition. Even so, Spanheim lived to see only the first volume republished in London in 1706, and died before he could attend either to the tract which he had promised Dr. Woodward or to a second volume of his *magnum opus* which had to be brought out posthumously.[71] It was too bad. Seller would not likely have added much to the discussion, but Spanheim might well have had the last word — if such were possible when scholars set no practical limits either to their own or to their readers' endurance.

Meanwhile there was yet another scholar who was busily writing about the shield and whose work seemed for a time to promise that fulfillment of which Dr. Woodward dreamed. Gisbertus Cuperus (or Cuypers, as he was known in his native Deventer) was second to none in Europe for either learning or fame. Indeed, he had some claim to being the most prolific of the learned correspondents of his day, and there was hardly a one in Europe who did not exchange information with him. Yet Cuper managed, like Spanheim, to combine a political career with his learning and a chair in history, first as burgomaster for Deventer, then as deputy to the Estates General of the United Provinces, and eventually as ambassador to the allied armies in Flanders.[72] His works bestrewed the field of classical scholarship — philology, epigraphy, numismatics, archaeology, and history — but he took an interest also in China and ancient Egypt, in theology, and even in geology. He was rewarded for his erudition by becoming one of the earliest foreign members of the French Academy of Inscriptions. When he first received a copy of Dr. Woodward's shield, he had already spent a lifetime in study and he was busily engaged in carrying out his ambassadorial functions in the midst of war — not once, however, overlooking an opportunity to examine a new cabinet or inspect a rare curiosity. Inevitably, all this activity was at the expense of that lucidity and simplicity which were the alternative virtues of the Classical Age. Like most of the *erudits,* his mind was cluttered with his learning and he found it a little difficult to steer a straight path to his destination. Yet while he often complained that his public duties interfered with his scholarship, he believed characteristically that his learning was an asset to his public service.[73]

That Cuper's opinion should be solicited was inevitable. He was already in touch with almost everyone who had anything to do with the shield, at least on the Continent; moreover, his skill in these matters was universally acknowledged. He had written already on many similar objects, identifying coins and statues and enlarging upon their iconography, conjecturing their dates and purposes, explaining their uses in understanding classical life and literature. As he wrote to Woodward later, his chief passions were sacred and pagan history and especially their conjunction, as illuminated by the coins, inscriptions, and other monuments of the ancient world.[74] His correspondence was like a great canvas spread across Europe to catch wind of the latest antiquarian finds, and he was ever ready with an opinion: to examine as a critic, to judge as a connoisseur.[75] By the time Valckenaer brought him word of Dr. Woodward's shield, he had already heard about it from Otto Sperling and in greater detail from his friend the Amsterdam burgomaster and virtuoso Nicolaas Witsen, and it was to Witsen that he sent his first letter on the subject.[76] Still he had more than enough to fill a second one and then a third, so that Valkenaer had no reason to be disappointed. In time all were sent on to Dr. Woodward, and this began a new correspondence which soon extended beyond the shield into other matters less amiable. Cuper thought that his letters on the shield would eventually be published. "I believe," he wrote confidently to the Abbé Bignon, "that I have sent to England all that one can say about the Buckler which exhibits the sack of Rome by the Gauls. I have tried to illustrate all the details. . . ."[77] Dr. Woodward was delighted, but somehow he found that even these large essays did not suffice for his grand intention.

Witsen had asked Cuper's opinion; not surprisingly, it differed little from the consensus of the virtuosi.[78] Like the rest, he did not believe it was contemporary with the events it described. The stone temples, the columns and obelisks, all put it unquestionably much later in Roman history. Cuper thought the workmanship outstanding, but further proof that it could not have been made during the Republic, when the arts were still primitive. He did not see how it could be dated more precisely except for the fact that it was iron, and iron was not very resistant. (Even ancient iron coins overlaid with bronze barely survived their rust and corrosion.) Since the shield seemed so perfect, he was inclined to place it as late in the Classical period as possible, perhaps the first or second century A.D. He wished he could see the object itself so that he could resolve some of the other problems, like how it had been worked into shape. "It is certain that hammering of iron was once practised by the Romans and it is clear from Julius Caesar that iron was much used by your Britons, as also that they offered iron in place of coins. If I knew where the shield was dug up and in what circumstances . . . I would be better able to satisfy your desire." Meanwhile, if he could obtain some more copies of the print, he would be happy to send them to his friends in France.

Valckenaer too had asked Cuper's opinion, and a month or so later he was sent the letter to Witsen along with a new and longer one in reply to his own questions.[79] He had, for example, asked whether Cuper thought that Dr. Woodward's shield was votive or combative. The great man was a little uncertain; the ferocious head at the center (probably the god Pan) made a military use possible. But he needed more information; like Batteley he supposed that if it was so, it must have had a hole in the *umbo* for a dagger point to be inserted. If Dr. Woodward would examine the reverse to see whether there was one, or whether there was any trace of the rings through which the soldier's left arm would normally be thrust, he would know for sure. If they were absent, then the shield must be votive, a reminder no doubt by some later Camillus of his ancestor's deeds. As to the figures, Cuper, like the others, was surprised to find the same costume for both Gauls and Romans, when they must at that period certainly have been different; here was another argument for the lateness of the shield. Like Gronovius, he thought that he spied the tribune Sulpitius, named by Plutarch, though he overlooked the cavalry commander L. Varenius.

Cuper's letter was characteristically discursive. The fact that the Celts and Romans appeared clean-shaven on the shield furnished an obvious occasion for an essay on the beard in ancient times. (Spon's remarks about Scipio no doubt suggested the theme.) The learned scholar combed the ancient writers to prove that beards were not customarily worn at the time of the sack of Rome. Among other things he showed that barbers had first come to Italy in AUC 454, a hundred years or so after the event. Clean-shaven soldiers were thus one more anachronism to show the late date and clumsy archaeology of the shield. The standards of the soldiers supplied still another opportunity for Cuper's learning. As he considered the *vexilla* on the shield, he was reminded of a correspondence he had once had with Otto Sperling in which they had argued about the form and material of the ancient banners.[80] It was an excellent example of that meticulous and inexhaustible attention to detail which looked to the rest of the world a little like pedantry, but which entirely absorbed the energies of the learned. No matter was too small to notice in a classical object. Sperling held that the Romans had borrowed their cloth standards from the barbarians late in the game, their own *vexilla* being spears decorated with silver or other metals. "I on the other hand argued that the Romans had early used flags and that the word *vexilla* always meant standards made from silk, wool, flax, or other cloth." Now, Cuper was satisfied that Dr. Woodward's shield confirmed his view. Just how this was possible when everything else upon it was anachronistic, he did not bother to explain.

But the learned man was by no means done. There was more to say about the handles of the ancient shields and about the other arms depicted on Dr. Woodward's monument. He was surprised that the geese had been left out. "The ancients indeed were very skillful, but far short in their organization of pictures and sculptures because of their ignorance of the art we now call perspective. Had they known this, they would certainly have represented this subject differently." Finally, Valckenaer had provoked him with another question, this time about the gold on the scales of the Gauls. Cuper saw at once that this must be still another anachronism, since gold had not yet been struck in Rome. This meant further explanation, but even Cuper at last began to falter. "No doubt you are tired of reading," he concluded to Valkenaer, "as I am of writing," and he was probably right on both counts.

Even so, he was still not finished. Valckenaer reported to him that Dr. Woodward had received both his letters gratefully, but with a number of questions. This required further reply, and so once again Cuper picked up his pen and set out what he had learned since his last.[81] Three appendices were

necessary now to deal with such matters as Roman helmets, swords, and
javelins, but the body of the letter was addressed to some new information
supplied by Dr. Woodward. Cuper was easily persuaded by the Doctor's
inspection that the shield must have been used for battle. (Traces of a handle
and of a place for a dagger point were both found upon it.) To clinch the
point, he saw that he must show that whole stories were often placed in relief
upon fighting shields, and he was able to quote here from the Roman
historian Ammianus Marcelinus, as well as from the famous passage in Silius
Italicus about the shield with Brennus upon it. ("Although the poet could
have invented this," he admitted, "it seems to me very probable that in fact
he saw this story set out on shields, either painted or in bas-relief.") A long
postscript, dated three days after the letter, developed the point still further
with a host of additional citations from Homer downward.

There was a last critical question that required attention. Dr. Woodward
wanted to know whether Cuper recalled any similar works in iron from the
time of Camillus, perhaps Greek or Etruscan. The Dutch scholar admitted
difficulty here; only a few scattered references, though he was certain, on
Pliny's authority, that the ancient smiths has been wonderfully skilled in
blending iron. His authority spoke, for example, of "goblets of iron dedicated
in the temple of Mars." Surely these might have been decorated in bas-relief,
since (in being offered to the gods) they must have had something remarkable
about them. If this was sensible, then Dr. Woodward's shield was just such
another. "I must admit that I can find no other example in Antiquity," Cuper
was forced to conclude, "yet later centuries were not entirely barren." Only
last spring, Cuper had seen in the cabinet of a friend a fine iron shield with a
relief upon it, and his second appendix consisted of an extract from a recent
letter to him describing it.[82] It was, he learned there, twenty-eight inches
long by eighteen inches broad, somewhat larger than Dr. Woodward's shield,
but like it with a hammered-out high relief. In the middle, was an oval, with a
battle depicted between mounted soldiers before the gates and walls of a city.
Four small side ovals surrounded the center, in which there were personifica-
tions of peace and war and between them figures of defeated soldiers and
commanders "in postures of wonderful art." The ornaments and arms were
all inlaid with fine gold and all the figures were antique. One of the small
panels, however, gave away its late date, for it depicted two cannons at the
feet of Mars, the god of war. With this hint and with the evidence of style,
Cuper's correspondent was ready to give his opinion. "It has been done
between the years 1500 and 1560, for the drawing of the ten big figures is
indisputably of M. Angelo, Raphael and Julio Romano. It seems to me
altogether likely that the same has been cast according to the drawing of the
best Masters and further wrought out by Aeneas Vico Parmensis because,

amongst my Papers, I find a drawing of Aeneas Vicus, afterwards engrav'd in Copper by himself of a piece like that of the middle great oval . . . which represents a Battel of the amazones." Since he owned a great number of Renaissance drawings, he thought himself an excellent judge of the different styles of art in different times.

Cuper had meant to reassure Dr. Woodward that iron could be employed at all times to produce a shield like the Doctor's. His example was produced as an argument in favor of the authenticity of the shield as an ancient iron object. But something had gone wrong; everything he had written, but most of all this alarming example from the cabinet of his friend, pointed the other way. It is true that there were no cannons in Dr. Woodward's shield and no drawings by Aeneas Vico that anyone knew resembling it. But surely some suspicion must have flickered through Cuper's mind as he penned these lines. As for the Doctor, it is hard to know just what he thought of them when at length he got the news, except that he did not print Cuper's letters and he never gave up his own conviction that his prized possession was a genuine Roman monument, obviously and conclusively authentic.

[8]

The war made traffic with France difficult. Dr. Woodward was frustrated to discover that the French did not know his work, and he set Scheuchzer and others to the task of establishing a Gallic connection. Cuper was a natural resource; for nothing, not even war, could inhibit his universal connections and vast correspondence. Already in 1708 he was describing Dr. Woodward's shield and his views on it to the great patron of French learning, the Abbé Bignon. Apparently the Abbé had some suspicions about it, since Cuper was not content to reject his argument that the shield was later than the time of Camillus; he now added categorically that it was not made "in the time of the tournaments" either. He thought that its iconography and the reference to Silius Italicus were enough to establish that. Cuper was able to inform the Abbé also about Dr. Woodward's views on the Deluge. "I have sent him some doubts about his System but I am still awaiting a response."[83] To his friend LaCroze he repeated his doubts, though he noted that there were many who approved.[84]

Needless to say, Dr. Woodward was not impressed by Cuper's geological skepticism. Of course on a matter like the Deluge, everyone must have an opinion, but the Doctor was convinced that only an experienced naturalist was entitled to evaluate his theory. Cuper knew the fossils in the collections of his friend Valckenaer, a cabinet he thought beyond comparison. But he had to rely on others, for the most part, for his knowledge of the subject. To

Dr. Woodward it all seemed very tedious, reviewing for the thousandth time
the arguments he had had to make against his critics, to explain the Deluge
and support his reading of Moses. "M. Cuperus gives me much trouble," he
complained wearily to Scheuchzer, "with long letters about my Nat. Hist. of
the Earth which is a subject he does not understand."[85] (As Dr. Woodward's
advocate, the Swiss naturalist had also to share some of the burden.)[86] For
Cuper, the Doctor's views were worth considering, but he did not find them
very persuasive; and for once the critic outlasted the naturalist, who lapsed
finally into disgusted silence.

It was not for some time that Dr. Woodward was put directly in touch
with the Abbé Bignon and the French men of learning. In 1710 he was still
employing Scheuchzer as his go-between and promising to send some fossils
and a copy of the shield. The Abbé was then probably the greatest patron of
erudition in Europe. He was the scion of a great house and rose to promi-
nence through his oratory and his family connections. When his uncle, the
Comte de Pontchartrain, was named controller general and secretary of state
with the department of literature and the various academies beneath him, the
young Bignon was brought into his service to help. In 1693, at the age of
twenty-nine, he became chaplain to the king, received the rich Abbey of
Saint-Aventin-en-l'Isle, and began to direct the activities of the three academies. He was largely responsible for the reorganization of the Academy of
Sciences and the Academy of Inscriptions, in both of which he was personally
active. When at length Pontchartrain became chancellor, Bignon became the
channel through which all the favors of the court were awarded to men of
learning and letters. He was nothing less than a kind of secretary of state for
science and learning: "Mecène de son siècle at l'ange tutelaire des sciences et
des savans."[87] There is a marvelous description of him in the splendor of his
island retreat (too long to quote here) which the Greek scholar, Ludolf
Kuster, sent to Dr. Bentley in 1714. "The friendship of such a man," he
properly concluded, "is not to be despised."[88]

Dr. Woodward understood that, but it was only in 1712 that he finally got
through to Bignon. The Abbé was delighted with his promise of fossils. When
they arrived, along with a copy of Dr. Woodward's book, he was even more
impressed. (He had not, he wrote to the author, had to read it in translation,
since he could manage the English language.) He then reported on the receipt
of his other gifts.[89]

I have communicated the copies of your shield to those among us who are
held not unskilled in such monuments. The majority of them believe that it is
not of great antiquity and that for two reasons. The force of the first is that
not one shield has survived among so many which must once have existed,
except for yours. None are to be found in Rome where an infinite number

ought to be seen nor in Italy or France. If one inquires zealously why this is so, the reason seems to lie in the nature of iron which cannot survive for many ages. If it had been less subject to injury of time many such monuments would have survived to our own day, particularly coins whose small bulk would give them a better chance of survival. Since none are now to be found in the cabinets of the learned, they infer therefore that your shield does not date back as far as you wish to believe. They further object, and this is the second argument, that the architecture and buildings which are shown in the shield [do not] belong in many respects to ancient Rome. No more can the arms and military standards cut on the shield be referred to that age. For these reasons many of our men of learning believe that it was made during the centuries when equestrian combats (*Tournaments* in the vulgar tongue) were held in high esteem. It is common knowledge that when such mock battles were the fashion, workshops produced many such, and those of wonderful workmanship, as may be seen from those that survive today.

Bignon was properly apologetic for an opinion which he feared would differ from the Doctor's. Yet it must have come as no surprise to Woodward, since Cuper had written to him already in 1709 about the objections of the French. When the Doctor first received the report, he "expostulated" after his usual fashion and Cuper had to write back hastily to assure him that the criticisms had not been his but only his French correspondent's. Woodward demanded to know exactly where anyone had seen a shield like his from the Middle Ages.[90] Cuper thought the whole subject needed further investigation, for while it was true that ancient histories were often displayed at tournaments, he was not at all certain that the art of sculpturing bas-reliefs in metal was then as exact and perfect as on Dr. Woodward's shield. (He thought it would be worth scouring the medieval romances to see.) On the whole, although with just a trace of hesitation, Cuper like Woodward preferred to believe it Roman.

But it wasn't only the French who were beginning to voice doubts about the shield. Unfortunately, however, the criticisms remained vague, either without author or without content, and suspicious because of personal animus. Thus Alexander Cunningham wrote to the Earl of Oxford in 1711 that he had heard that there had been a great "pother" about Dr. Woodward's shield and that both Hearne and Dodwell had written in its favor. "As I never was of Dodwell's mind in anything, so I suspend judgment in this, but in the questions sent from England I consulted the antiquaries at Rome; in short all of 'em say it is not ancient."[91] And even before that, Hearne had complained in his Diary that there were "some ill-natured men who run it down as Banter, particularly Dr. Gregory the Scotch man who understands just as much of Antiquity as he does of Greek." Indeed, it was a little hard to see just what the mathematician's credentials were for such a matter, except that he was an intimate of Woodward's old enemy Dr. Arbuthnot.[92] Perhaps there was more to these suspicions than appears, but it is unlikely that they were ever very fully developed, and they merely spurred Dr. Woodward more vigorously to the defense.

The Doctor did not miss his chance to reply to the Abbé and vindicate his weapon. Soon after he received his criticism, he sat down and drafted a lengthy reply in Latin.[93] His file of correspondence had grown heavy with authority, and he drew first of all upon its weight to impress Bignon. Everyone who had actually seen the shield, he assured the Abbé, agreed upon its authenticity. The greatest connoisseurs, the Earl of Pembroke and Sir Andrew Fountaine; the most learned scholars, Spanheim, Cuper, and Relandus; the different nations, Germans, Dutch, and Italians — all had joined to attest to it. As for the insinuations of the French that it might be from the time of the tournaments, Woodward offered three arguments in rebuttal. First, he insisted that the costume and the arms were all depicted on the shield according to ancient Roman usage. (They may not have been contemporary with Camillus, but neither were they medieval. The style of later times was, of course, very much ruder.) Secondly, while it was true that rust ruined iron and other metals, nevertheless Dr. Woodward himself possessed various antique items of iron, *stylis, claves et telorum cuspides,* and others still of brass, *urnis, pateris, simpulis, aliisque vasis.* Finally, one could see from some patching of the iron which was evidentally more recent than the shield itself that the weapon must be ancient. In a word, if any skeptical antiquary was to look at the object itself and compare it with something later, Woodward was confident he must change his opinion.

Indeed, it was hard to reply from the evidence of an engraving only, although clearly Bignon and his friends were not satisfied. Only one thing was left to Dr. Woodward. He had to find someone with the knowledge and authority — and access to the shield — for a proper rebuttal. Fortunately, the solution was close at hand. Dr. Woodward had found already in that most learned of Englishmen, Henry Dodwell, and his young friend Thomas Hearne, allies who were ready to join forces for that definitive treatise on the subject which alone might stifle all criticism. To this everyone agreed — Bignon who was doubtful, Otto Sperling who continued to think it Roman, and Cuper who wavered.[99] With the support of Dodwell and Hearne, Dr. Woodward might yet put his shield beyond all doubt.

Dr. Woodward. From a mezzotint by W. Humphrey (1774) after an original painting which was in the possession of Col. King. (Courtesy of the Trustees of the British Museum.)

Dr. Woodward. From an anonymous painting in the Sedgwick Museum. (Courtesy of the Sedgwick Museum, Cambridge; photograph by David Bursill.)

Gresham College. From a print by George Vertue (1739) in John Ward, *Lives of the Professors of Gresham College* (London, 1740). (Courtesy of the Society of Antiquaries.)

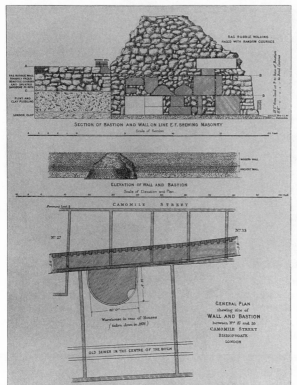

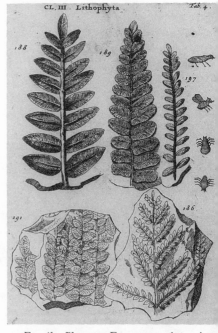

The Roman Wall at Bishopgate. Nineteenth Century excavations exposed the wall just where Dr. Woodward had studied it and confirmed his accuracy; from the frontispiece to John Edward Price, *On a Bastion of London Wall* (Westminster, 1880). (Courtesy of the Society of Antiquaries.)

Dr. Woodward's Cabinet of Fossils. The earliest of four cabinets built for Dr. Woodward and transferred at his death to the Woodward Museum Cambridge. (Courtesy of the Sedgwick Geological Museum.)

Fossil Plants. From a plate in Edward Lhwyd's *Lithophylacii Britannici Ichnographia* (London, 1699). (Courtesy of the Bodleian Library.)

SCIPIO RECEIVING THE KEYS OF CARTHAGE.

The Property of Gustavus Brander Esq.

Dr. Mead's Shield Depicting Scipio. From Frances Grose, *A Treatise on Ancient
rmour* (London, 1786). (Courtesy of the Society of Antiquaries.)

Dr. Woodward's Shield (front). (Courtesy of the Trustees of the British Museum.)

Dr. Woodward's Shield (reverse). (Courtesy of the Trustees of the British Museum.)

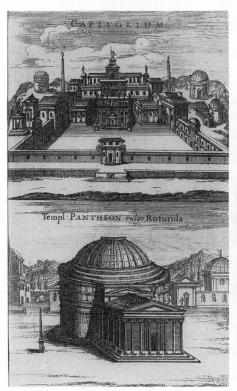

The Capitol and *The Pantheon.* From Basil Kennett, *Romae Antiquae or the Antiquities of Rome* (2nd ed., London, 1699). (Courtesy of the Bodleian Library.)

The Dress and Arms of Roman Soldiers. From Basil Kennett, *Romae Antiquae or the Antiquities of Rome* (2nd ed., London, 1699). (Courtesy of the Bodleian Library.)

Gem Depicting Scipio in Dr. Woodward's Collection. From Thomas Hearne's edition of Livy (Oxford, 1708), vol. VI.

The Shield of Dr. Spon. From Jacob Spon, *Recherches Curieuses* (Lyons, 1683). (Courtesy of the Society of Antiquaries.)

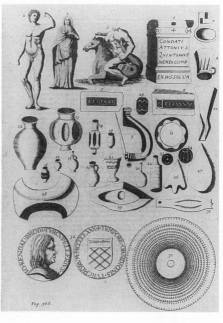

Roman Antiquities and a Shield [lower right] *in Ralph Thoresby's Museum.* From Thoresby's *Ducatus Leodiensis* (London, 1715). (Courtesy of the Bodleian Library.)

The Shield with the Life of Camillus in the Tower of London.
(British Crown copyright — reproduced with permission of the
Controller of Her Britannic Majesty's Stationery Office.)

The Death of Decebalus on the Trajan Column. From the atlas of plates accompanying Raffaello Fabretti, *Columna Traiani Syntagma* (Rome, 1683). (Courtesy of the Society of Antiquaries.)

The Helmet of John Kemp. (Courtesy of the Trustees of the British Museum.)

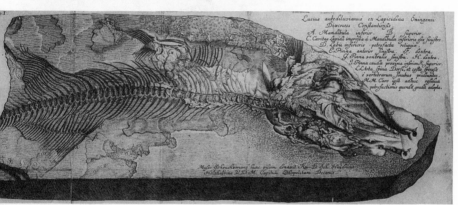

Antediluvian Fish in the Museum of J. J. Scheuchzer. From J. J. Scheuchzer, *Piscium uerelae* (Zurich, 1708). (Courtesy of the Bodleian Library.)

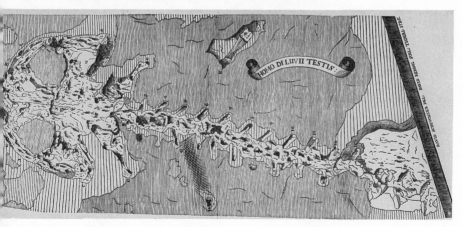

Antediluvian Man. From J. J. Scheuchzer, *Homo Diluvii Testis* (Zurich, 1726). Courtesy of the Trustees of the British Library.)

The Earl of Pembroke's Sculpture of Curtius Leaping into the Gulf. From Cary Creed, *The Marble Statues of the Right Hon. Earl of Pembroke at Wilton* 1730. (Courtesy of the Society of Antiquaries.)

X.

Enter Thomas Hearne

[1]

Nobody in Dr. Woodward's day is better known to us than Thomas Hearne. At the age of twenty-seven he resolved to keep a diary, and from 1705 until his death in 1735 he recorded in minutest detail all the events of his extraordinary mental life. Besides these daily entries, he kept almost every other scrap of paper that might serve his purpose, his reading notes and business records, the innumerable letters sent to him, and the drafts of many that he scattered over England. Once he wrote a brief autobiography, and always he carried with him a notebook in which he entered everything of interest. He was indefatigable, this poor son of a parish clerk, and he made up in sheer industry what he lacked in personal advantage. At twenty-seven he was already one of the most learned men in England; behind him lay an apprenticeship in historical scholarship that very few could match. And already half a dozen publications had flowed from his ready pen.[1]

Hearne was born in the village of White Waltham, not far from Windsor, on June 11, 1678. His father was an obscure clerk of the parish, but Hearne remembered him "in good Reputation for his Learning, more than ordinary considering his Education, and for his Skill in History and for his Love to Antiquities." He wrote a "firm legible hand" and gave to Thomas his first education. His uncle too, though only a tailor, was "a great Lover of Antiquities and . . . was never weary in talking about Subjects of that kind"; even as a boy he could be seen "pouring over the Old Tomb-Stones in his own Church-Yard, as soon almost as he was Master of the Alphabet."[2] No need to hunt further for the origins of that wonderful dedication that was to mark his life; Hearne's "natural propensity" and a devoted family seems to have launched him on his historian's career even before he had entered school.

The precocious boy was soon discovered by a neighboring Lord of the Manor, Mr. Francis Cherry of Shottesbrooke. Cherry was a splendid man, in Hearne's words "a man of great parts and one of the finest gentlemen in England."[3] A non-juror (that is to say one who refused to take the oaths and recognize the new monarchy of 1689), he was a country gentleman of great wealth, but also of extensive learning and genuine piety. He was a handsome man and a fine horseman, and it is said that when King William elected to follow him one day while stag-hunting, Cherry turned his horse at once to a dangerous precipice along the Thames, hoping that the "usurper" would follow him and break his neck. Later, when Queen Anne found out from Peachy her bottle-man that the rider in the distance was Cherry, she said, "Aye, he will not come to me now; I know the reason. But go and carry him a couple of bottles of red wine and white from me, and tell him that I esteem him one of the honestest gentlemen in my dominions."[4]

Cherry took Hearne under his wing about 1693 and sent him to school at Bray in Buckinghamshire, some three miles from Hearne's home; the six-mile daily walk soon fixed another of his lifelong habits. "I was the lowest boy in the Schoole, when I went first," Hearne recalled, "but in a little time (for I was not at the Grammar Scool above 3 or 4 Years in all) I got to be the Head boy of the School." Hearne's reminiscence was the result of the unexpected appearance of an old school friend in 1718. "Mr. Alexander, could not but observe Yesterday," he writes with obvious pride, "how I could very rarely be drawn to play, spending my time at my Book while other Boys were at play."[5] The curriculum was all Latin and Greek, but Hearne's interests already extended to English history also. Indeed, when an argument would arise among the boys in school, they turned to him as readily as the Master, "and they would often say they received more Satisfaction from him than they did from the Master himself, especially if a Point of some English History was mentioned, to the reading of which he was naturally addicted." Hearne kept an affectionate recollection for both of his masters there: Patrick Gordon, "a very learned man" who had been a professor at Aberdeen, and James Gibson. "I needed no spur," Hearne recalled, "and indeed (if I may be allowed to note this) I was never whipped by either of them, notwithstanding they were both severe enough to such as deserved correction."[6]

In 1695 Cherry took Hearne into his home at Shottesbrooke. The great manor house was alive with visitors, the discourse elevated and sober. "A great many Non-Jurors and other learned men used at those times to resort to Shottesbrooke, so that Mr. Cherry's house would be sometimes like a College."[7] There Hearne met Cherry's most intimate friend, the learned non-juror Henry Dodwell. Upon losing his Oxford chair as a result of declining the oath, Dodwell had been invited to the manor, where he lived out his remaining days under Cherry's patronage, writing massive volumes of historical erudition and walking once a day with the master (excepting only Sundays) in the garden, while they talked of learned and pious subjects. To Shottesbrooke frequently came one of the famous "seven bishops," since deprived, Thomas Ken, and in the summer of 1695 another deprived bishop and a greater scholar, the famous Dr. Hickes. The parish priest was White Kennett, beginning an important career in the service of the church and historical scholarship. Dr. Hickes had been forced into hiding at the time of the Revolution by his resistance to his deprivation, but had nonetheless embarked upon one of the great scholarly enterprises of the age, the restoration of the whole of Anglo-Saxon culture and history. For three months he lived at Shottesbrooke with Kennett, under the name of Dr. Smith and in ordinary clothes to avoid detection. There he could be seen walking upon Mr. Cherry's terrace, meditating. Unhappily, the news got abroad somehow and one night an attempt was made to apprehend him, the doctor just escaping by the back door. But Cherry's house remained a center of non-juring and of scholarly activity and left an indelible impression on the young Hearne.

While at Shottesbrooke, Hearne's education continued. Cherry himself and Dodwell instructed him. They read the classical authors together and the two scholars "explained the difficult places to him and always illustrated them with curious and useful Observations, such as have been wonderful Advantage to him since."[8] It was of course the usual training in "grammar," but through it Hearne could glimpse the larger and more appealing worlds of philology and history. Cherry gave him formal lessons in Latin and Greek and set him exercises, exactly as if he had been at school. It was preparation for an Oxford career, for Cherry was determined to send him (as he himself had gone before) to St. Edmund Hall. In the mornings, while Cherry perused the *Septuagint*, Hearne read him the Old Testament. And Hearne began to transcribe manuscripts for both his masters, the beginning of still another and even more central lifelong occupation.

In 1695 he was matriculated at St. Edmund Hall, though Cherry took him back to Shottesbrooke and sent him to Bray for another year. (It was a four-mile walk now each way!) But at last in 1696 he was lodged in Oxford to begin his university career. He could not know it then, but he was never again to leave his new lodgings in St. Edmund Hall, despite numerous temptations and even more frequent discouragements. As an undergraduate his most memorable experience seems to have been his discovery of the Bodleian Library, where he began to read voraciously and to copy furiously for Cherry and Dodwell. He still continued to spend time and vacations at Shottesbrooke (walking each way). For Cherry he transcribed Sir Henry Spelman's *History of Sacrilege*, some manuscript works by Sir John Fortescue, an account of the pensions given to former monks after the Reformation, and other pieces. (The first two eventually led to publication.) For Dodwell he copied Greek texts and twice prepared the learned man's *Paraenesis* for the press. The celebrated Dr. John Mill, principal of Edmund Hall soon picked him out to work for him as well; "finding the Young Man to be versed in MSS.," he got him to compare several for use in his magnificent new edition of the Greek Testament, then in preparation and one of the greatest works of scholarship of that learned age. (Hearne also remembered an expedition for Mill to Eton College, in 1699 where he spent three days collating a manuscript of Tatian and Athenagoras which was later employed in separate editions of the two Greek writers.) And finally, the erudite immigrant Dr. Grabe also found the young man useful in transcribing various pieces of patristic learning.[9] All in all it was perfect preparation under the most exacting masters for a lifetime of editing.

How Hearne loved the library! "He constantly went to the Bodleian every day, and studied there as long as the time allowed by the Statutes would admit."[10] He would not leave it or his beloved Oxford for anything. No wonder that the new keeper Dr. Hudson should choose Hearne as his assistant — especially since Hudson seems to have been lazy, and Hearne was as indefatigable as he was knowledgeable. He set about at once reorganizing the library and correcting and updating the catalog, which he found quite imperfect. In the course of this labor he examined every printed book in the collection. Then he turned to the coins, which were in an even more parlous state. Again he organized and cataloged them, becoming along the way adept in this new field of classical numismatics.[11] And all the time he read — not with the true variety of a virtuoso like Woodward, but with a breadth that is in its own right as astonishing. Hearne's interests were almost exclusively historical, but within that limitation there was no aspect of the entire past that was without interest. Even so, he had once had to catalog the collection of rarities in the Anatomy School and boasted his knowledge of nature over his rival Humfrey Wanley. ("H. Wanley is so little skill'd in Natural History that he knew not what a *Cornu Ammonis* was lately, when one was put into his hands and it lay with him a considerable while.")[12] His relish for manuscripts and books, coins and inscriptions, extended without limit to all times and places. And his passion was religious in its intensity. "O most gracious, and mercifull God, wonderful in thy providence," he once wrote in prayer, "I return all possible thanks to thee for the care thou hast always taken of me. I continually meet with most signal instances of this thy Providence, and one act yesterday, when I unexpectedly met with three old MSS, for which, in a particular manner, I return my thanks, beseeching thee to continue the same protection to me, a poor helpless sinner, and that for Jesus Christ his sake."[13] No wonder that in 1700 (after receiving his B.A.) he rejected a promising job in Maryland for an uncertain future at Oxford, against the urging of Cherry and Dodwell. The past was not very accessible in the forests of the New World, despite a fair allowance promised to him for buying books.[14]

[2]

In 1703 Hearne became Master of Arts and sometime soon afterward was offered a job as chaplain at Corpus Christi College. The principal, Thomas Turner, wanted to do Hearne a favor and allow him to continue his work at the Bodleian, but Hudson interfered and insisted that Hearne choose either one or the other. It was the beginning of a long and serious quarrel, but Hearne elected to stay at the Bodleian. In a similar fashion he had to decline another offer from All Souls College. For years he worked in the library on the merest pittance and without other support, until at length in 1712 he was appointed second keeper as well as janitor. By virtue of joining the two offices, he was able to keep the keys to the Bodleian "and opened the Door Morning and Evening and had the liberty of entering and going out as often as he pleased." It was enough to keep him from accepting (the following year) a post as librarian and keeper of the Museum of the Royal Society; "his circumstances," Hearne explained, "not permitting him to leave Oxford." [15]

No doubt the "circumstances" involved his work; Hearne knew the Bodleian as one can only know a great library after examining and cataloging its every book. In these years it was very much his own personal collection, and he displayed its treasures to visitors as great gentlemen revealed their cabinets. On the other hand, there is something a little pathological about Hearne's attachment to Oxford, for it continued even after the Bodleian was fully closed to him (the greatest disappointment of his life) and his friends offered all sorts of possibilities elsewhere. Hearne simply would not travel. Though the antiquities of distant parts of Britain interested him enormously, and the libraries and collections of friends and correspondents beckoned, he never even ventured as far as London. Once Sir Robert Throgmorton sent several invitations to him to come to visit him at his great house at Bucklands. [16] "The Person told him that I could not ride," Hearne reported. "I will send, says he, a Coach and six for him." But it would not avail. "He can ride no way, says the Person. He always walks." It was true; when Hearne did leave Oxford on rare occasions, it was to tramp the surrounding countryside, or to visit his father at White Waltham, always on foot. Throgmorton thought it a natural eccentricity. "Why the Duce is in it, says Sir Robert, so all Antiquaries use to do. I have known several, and they have all walk'd, Antony Wood not excepted. They are Men that love to make remarks, and they prefer walking to riding upon that account."

How Hearne loved to make "remarks"! His curiosity extended not simply to the past but to much in the present and particularly to the lives and thoughts of his scholarly contemporaries and to the little world of the university. His *Collections* are a mine, inexhaustible for the history of the Augustan spirit, at least on its weightier and more academic side. But what keeps them alive today in the eleven massive and closely printed volumes of the Oxford Historical Society (as much as their invaluable information) are the crotchets and prejudices of their author. Hearne was a touchy, humorless, and undoubtedly difficult man, with a pride very easily injured. He was also honest, conscientious, and loyal. He never married and he quarreled often with his friends as well as his innumerable enemies, but he was generous to the young (who often flocked to him) and kept the confidence of a growing number of patrons despite his uncompromising nature and dogmatic opinions. He felt passionately about religion and politics and he let it prejudice his every thought. The Whigs were contemptible; a scholar who was a Whig seemed to him a contradiction in terms. "Good Antiquaries have always been the best Friends to the Church", he wrote succinctly, "and have never proved Traytors to their Rightful Sovereigns."[17] His appetite for gossip was insatiable, particularly when it damned his opponents. The diaries furnished him the opportunity to revenge the wrongs, correct the errors, and condemn the wicked that was so often lacking in his real workaday life. He seems even to have suspected their eventual publication.[18] The one hundred and forty five notebooks of the original are thus an immense compound, like the man, of scholarship and exact information, rumor and prejudice, flung together just as they occurred to Hearne in the course of his daily rounds, in the library or his lodgings, the halls or taverns of Oxford, and the villages and fields nearby.

When Hearne had become famous, his enemies conspired against him (not for the first time) and contrived a way to embarrass him. They discovered in the Bodleian, as a result of a recent bequest by Francis Cherry's widow, a paper written by him many years before about the oaths, and they published it. In 1730 Hearne was a non-juror, and the Bodleian had been closed to him for fifteen years as a result of his refusal to subscribe to the oath of abjuration. But in 1700 it had been different. To take an Oxford degree meant to have to swear obedience. On June 11th of that year, Hearne remembered, "I put down in writing the Arguments that persuaded me to take the Oath, and this I did by way of Letter and afterwards lent it to Mr. Cherry, who wanted to be satisfyed about my reasons."[19] The arguments that moved him then, however, had been long abandoned. "I look upon them as weak and frivolous and I am so much dissatisfyed with this MS of mine that ... I revoke every Paragraph, line, word, letter and Title in it, and consign it over to the Fire." He was satisfied, however, that his enemies had been confounded in their attempt to embarrass him; all good men, he supposed, would say that they now had a better opinion of him than before, "declaring that 'tis an argument that I acted with deliberation and not rashly, when I formerly took the Oath of Allegeiance, and not as those do, who take Oaths without considering at all."[20] Consistency was after all not the only or the highest virtue.

This first work remains of interest (despite its repudiation) as the first piece of history, albeit polemical, from Hearne's hands. It is a pamphlet of some eighty pages. Apparently there had been a previous letter also, beginning the argument, with a defense of the notion that a King's breach of oath absolved the subject, and employing the histories of Henry VI and Edward IV as evidence. Now Hearne turns to the Empress Maud and Stephen, Earl of Blois, as additional confirmation. For twenty pages he tells their story from the sources, so that he can conclude his tract by showing that "those who took an Oath to Stephen as to their Sovereign Lord, notwithstanding they had before Sworn Allegiance to Her the Empresse, thought themselves, yea even were, loosed from their Obligation to her, because the Common Good of the Nation so required it."[21] To confirm his argument he quotes long passages from the manuscripts of Sir John Fortescue that he had been copying for Cherry.

The narrative was an unfortunate performance in more ways than one. Hearne's gifts did not lie that way. For one thing, it was halting and digressive, so that the story and the argument were equally obscured by irrelevant matter. And if that was a vice peculiar to the age (especially common to antiquaries and controversialists), Hearne was already lost to its temptations. Unhappily, a mind cluttered by learning was not likely to be improved by further erudition; neither Hearne's thought nor style became more lucid with time. Even more disturbing was his conflation of sources, which often seems haphazard and indiscriminate; he employed with equal authority contemporary writers and very much later and quite worthless chroniclers like John Harding. Hearne knew better; at least once he preferred William of Malmesbury to the alternatives because he lived at the time he was describing and was of unimpeachable character. But Hearne was not consistent, and he made his points with whatever was conveniently available. Of course there were different rules for historical polemic than for unvarnished history, when truth, as he writes here, "ought to Overbalance every thing else." But Hearne's powerful convictions and prejudices did not diminish with age, though they altered their direction. "Mr. Hearne," his editors wrote in 1731, "is a professed Non-Juror and a fiery Bigot to those of his own

Principles." [22] Alas, it was true (though as a bigot against differing principles Hearne was probably even more culpable), and the passion for non-juring was no more likely than his earlier convictions to improve his historical judgment.

[3]

To estimate Hearne's view of history in these early years, it would be fairer to turn to another early work. In 1702 he was approached by a London bookseller, Timothy Childe, to improve a popular introduction to history that had been published a few years before, entitled *Ductor Historicus*. It has usually been said that Hearne was the author of the original of 1698, but the W. J. who appears on the title page is not our man. In fact, the work was largely a translation from a French treatise by the Abbé de Vallemont, improved by two London writers, Abel Boyer and John Savage, and partly refashioned for an English audience. (Thus the fourteen epochs of the original no longer end with St. Louis and Louis XIV but with the downfall of Constantinople and "K. Charles II Restor'd.") [23] Its success led Childe to consider reissuing it enlarged, which he proposed to do by the addition of a new volume. Apparently Hearne had been considering something like this independently, and the two men quickly agreed to terms. Eventually the first volume was also revised by Hearne with additions, and the new work appeared in 1704-05. It was republished several times in later years but without Hearne's authorization, a tribute to its continuing usefulness.

The composite work belongs to a genre characteristic of the period. It was intended, Hearne wrote, as an introduction to history for students, "but also a compendious and (if I may be allowed to say it) a full Relation of all the most remarkable Events which have happen'd from the Beginning of the World to the Fatal Period of the Roman Empire." The anonymous preface set it among a host of similar tracts, largely continental, that had supplied the needs of students throughout the century. It combined the traditional *ars historia* with an epitome of universal history in a single convenient work.

The first volume was edited and published by Hearne in 1705 after he had completed the second. [24] In a "Praemonition" to the reader, Hearne claimed to have refashioned it so thoroughly as to have made it a new work. It comprised three books: a general chronology from Creation to the present; a consideration of the great historians, ancient and English; and "a Compendious History of all the Ancient Monarchies and States from the Creation to the Birth of Christ." "I have compared the whole with the Original Authors," Hearne claimed, "which I have all along quoted, Corrected divers Errors . . . made large Additions . . . and subjoyned an Account of the Foundation of Cities, etc. from the Beginning of the World to the Birth of Christ." The

Bodleian had indeed been "ransack'd for the Historical Matters here set down." The second volume, which was wholly his own, advanced the "compendious history" to the founding of the new Empire by Charlemagne. A third was promised and begun but abandoned, says Hearne, when the field was preempted by the new work of Samuel Pufendorf.[25] The whole retained the original framework of Vallemont and many of his sentiments. It was also necessarily arranged for a popular audience (Childe was very insistent here). Yet it may be fairly taken as Hearne's own. He certainly had *carte blanche* to improve it, and in 1711 he was still recommending it for the son of a friend.[26]

Apart from its information (the real justification of the volume), the work is replete with the commonplace observations of the day on history and chronology. Hearne was not very original here, or anywhere else, when he considered his discipline theoretically. Indeed, if ever a man lacked the gift of abstract thought it was he. But for all that, there were large areas of agreement about history in the early eighteenth century, and the *Ductor Historicus* is as good a place as any to seek them out. Thus the chapter "of the Benefits accruing by the Study of History"[27] is an epitome of Renaissance commonplaces about the value of the subject in teaching virtue by example, in furnishing diversion, in offering instruction in all sciences, and in defending religion. The chapter on writing history is frankly credited to the translator of Sallust (the Abbé Cassagnes of the French Academy, we learn from Vallemont) and recounts the traditional rhetorical advice of Antiquity on the composition of narrative. There is much on style and political reflection, nothing at all on research or criticism. But then, the *Ductor Historicus* (especially in its original conception) was not meant for the antiquary but for the young literate gentleman for whom history was narrative, an essential ingredient in the "polite" (i.e., classical or neoclassical) literature which was to furnish him his general culture.[28]

The chapter on the historians drew much more extensive revision. Even so, Hearne followed his original source in beginning with the Bible as the most authoritative of historical works. "The History of the Jews is contain'd in the Books of the Old Testament which is sufficient to convince any Christian that it is unquestionable, and will never admit of Doubt."[29] *Genesis* was the book that furnished the greatest difficulties and the fuel for deists and atheists. But Moses was an unimpeachable historian. It was true that he wrote long after the events. "But this ought not to gravel any Man of Sense, for when once he has conceived that Moses foresaw future things, through the Spirit of God which revealed them to him, it will not be hard to believe, that he was also Inspired with those that were past and before his Days." Even if the arguments of Revelation and the supernatural were put aside, Moses' authority could be demonstrated. Through a chain of testimony stretching from Creation itself and involving only a small number of men, it could be shown that Moses' account came directly from eyewitnesses. It was demonstrable that "Adam, Methusaleh, Sem, Isaac, Levi, and Amram, the Father of Moses, have successively convers'd with and instructed each other in the History of the World." (A table is annexed to make the relationships and chronology clear.) Doubtless Dr. Woodward would have been pleased with the argument. (I do not know that he read it.) He would have been even more interested in Hearne's proposed "Discourse concerning the Plantation of the World by the Sonnes of Noah," a subject dear to his own heart, or the essay in which Hearne planned to defend the Universal Deluge.[30] Unfortunately, the problems of the Pentateuch are dismissed in the *Ductor* as too complex and theological for an introductory work. But a final caveat is appended. There are rules by which to question ancient history, "But these are by no means to make use of in relation to the Canonical Writings; we must always acquiesce in them whatever difficulties we meet."

The rules for ancient history are given earlier, in the book on chronology, but they too are disappointing.[31] There are four altogether, and they caution the reader how far to accept the reliability of authors. Thus writers contemporary with the events, Hearne indicates, are to be credited, provided they are not contradicted by other contemporaries "of known integrity and ability." Next to contemporaries, the reader should prefer those nearer the age to those farther away. Anonymous or later works should be suspected, "especially if they clash with Reason against the constant Tradition of the ancients." Finally, we must distrust the moderns, especially when they disagree among themselves or with the ancients. In the age of Mabillon it all seems disconcertingly elementary. Fortunately, Hearne (who merely acquiesces here) was reading the great French critic directly, along with others like Montfaucon (or closer to home, Hickes and Wanley), all of whom were bringing vastly more sophistication to the evaluation of testimony. In any case, the rules enumerated here had been formulated expressly for the reading of classical histories where the rigor of the new criticism was still abated, and not for the scholar or antiquary intending a fresh investigation of the past.

This is borne out in the pages that follow, on the individual historians of the ancient world. Hearne obviously enjoyed this part of his task, rewriting more extensively than elsewhere. The bibliographer and editor was beginning to find his metier. Relying heavily on the work of Degory Wheare, who had furnished the standard English treatise on the subject, well-known now in Henry Dodwell's version, he reorganized the list so that it would form a single "thread of History from the beginning, digested in the Order they ought to be read." He then added his own judgment and remarks drawn from commen-tators like Lipsius and Casaubon.

One turns at once to Livy. In the original, the Roman historian was introduced with the assertion that "there are a world of Learned Men that cry up to the Skies, the Merit of Livy's History." Vossius and Alphonso of Castile are the two examples offered, the latter recovering from a very dangerous fit of sickness as a result of having read the historian. Hearne adds a long paragraph describing the author and the work. He follows it with some further testimonials drawn from Cremutius Cordus' *Tacitus*, from Quintilian, and from Casaubon. At this point the original author qualifies Livy's praise with a teasing paragraph about his reliability, and this is also repeated by Hearne. "Some think it strange that Livy, who was a Man of Wit, should relate so many popular Reports, which he did not believe at all himself, as he always seems to Insinuate." The history, after all, was filled with prodigious and incredible events: oxen speaking, chickens changing their sex, gods weeping or sweating blood. The French pyrrhonist La Mothe le Vayer is quoted to raise some further troubling questions. Unfortunately, both the original and Hearne pass on at once to a long quotation from the French critic Rapin which, instead of replying to this skepticism, extravagantly praises Livy for his style and expression. Both accounts conclude by acclaim-ing the Roman as "the great Master of this Art" and pass on to other authors.[32]

Hearne, at any rate, seems not to have been much taken by these questions. He begins his account of Roman history later in the volume with the problem, but offers nothing toward its resolution.[33] "It must be ac-knowledged," he admits, "that all the Account we have of Italy before Romulus is very Fabulous and Precarious and such as no Historian can rely upon." That does not stop him from making the attempt (with dates), beginning with Uranus. He shows some relief at arriving at Romulus, and scurries through the rest of the regal history without further cavil. But though Horatius appears heroically at the Bridge and Livy supplies much else, Camillus defeats the Gauls without any mention of Brennus, scales, or tribute, (though the geese are not overlooked). No doubt the omission was simply the result of the restrictions of a "compendious" history which would not allow embellishment, rather than any skepticism. Some years later, considering the similar problems raised by Geoffrey of Monmouth's suspi-cious history, Hearne took pains to defend the notion that the medieval chronicle, like the ancient histories, preserved a kernel of truth, despite some obvious difficulties. Hearne's argument in some ways recalled Perizonius, whose work he had read. "For my part tho' I believe that the Story in

Geffrey is corrupted with vast Variety of Fables, yet I am so far from thinking that 'tis all Romantick, that I believe tis well enough grounded." There was as much reason, he thought, to disbelieve "the early Part of the Roman History itself which we well know is attended with odd Relations, even in Livy and the most Judicious of their Writers."[34] The sensible student must try to discover what the reason was for this mixture of truth and legend and try to discriminate between the two. It was a useful idea, but Hearne was not very successful in applying it; he kept altogether too much of both the medieval chronicler and the ancient historian.

There is a great deal more in the *Ductor Historicus* to take the measure of the young Hearne. The final appended chapter on the foundation of cities was entirely his own, and shows his wide reading and interest in coins to some advantage. His talent for antiquarian digression is indicated by interpolated essays on the invention of chess (at anno 3340, with a passage from the Poet Lydgate), a history of silkworms and of the making of glass, and another on the founding of the University of Paris. Strangely, although Chichester appears preeminently, there is no mention at all of London.[35] The list of British historians (which succeeds the ancients) was, again, all his own compilation. It is with one exception limited to narrative histories and brief evaluations. The exception, which Hearne thinks a useful preliminary to any study of English history, is Camden's *Britannia*. The author was "the Prince of our English Antiquaries" and, with the earlier writer John Leland, one of Hearne's heroes. Most impressive is the very thorough bibliography of printed chronicles which is annexed. Hearne was already building the foundations for the special work that was to occupy him for most of his life.

Finally, the preface to the second volume offers the best view of Hearne's interests at the end of 1704.[36] Vallemont, he believed, had "abundantly proved" the usefulness of history, so he chose a subject somewhat more controversial: the use of coins and inscriptions for ancient history. And though his observations are still commonplace (they are drawn with credits from Spanheim, Obadiah Walker, and other writers on numismatics and epigraphy), they retain some interest. At least they show the new conviction in antiquarian method beginning to overcome the older allegiance to literary authority, though Hearne does not seem very clear about the consequences; inscriptions and coins, he argues, are really more reliable than ancient manuscripts. Their corruptions are less frequent, and the interpolations and alterations of ignorant scribes are quite lacking. For some historical problems like chronology they are unsurpassed. Moderns like Sigonius, Marsham, and Dodwell (to name only a few) "have made such considerable Discoveries in ancient History, as must be acknowledged to have far surpassed any thing of that Kind that had been attempted before." Moreover, inscriptions are essential to interpreting literary documents and to informing us about ancient scripts and alphabets. Coins add besides the names and likenesses of ancient men, with dates and places, as well as the "figures of garments," animals, etc. And their discovery furnishes clues to the whereabouts of ancient garrisons. They are therefore quite indispensable to the student of ancient history, and Hearne promises to illustrate their value in the work that follows.

[4]

If I have lingered unduly over Hearne's juvenilia, it is because they seem to me illuminating both for the man who took up the cause of Dr. Woodward's shield and for the age which set the framework for that discussion. Hearne grew beyond the *Ductor Historicus*, which was in any case a species of hack work, but he did not change many of the opinions he held there. He remained throughout his life a strange compound, rather more than his contemporaries a mixture of sophistication and credulity, at once old-fashioned in his tastes and convictions yet "modern" in his technical competence and enthusiasm for contemporary scholarship. Even as the young scholar was writing the *Ductor Historicus,* he was discovering his true vocation; he was becoming the learned editor, at once philologist and antiquary.

Hearne had learned from Cherry and Dodwell how to make appropriate observations upon a text. Now in the quiet of the Bodleian Library, as he poured over old manuscripts and learned editions, as he copied for great scholars like Mill and Grabe, he learned how to collate and amend texts, how to read obscure hands and decipher strange locutions. Among his first works were indexes, to Josephus, Cyril (1703), Dr. Edward's *Perservative against Socinians* (1704), and the Earl of Clarendon's monumental history of the civil wars (1704): hack work but good discipline. His first real publication was the *Reliquiae Bodleianae* (1703), an edition of some of Thomas Bodley's letters with a characteristic introduction on the history of libraries. The Bodleian, Hearne was confident, was superior even to the Vatican Library, especially since they employed there "Persons on purpose to forge old Hands, and so to transcribe the Fathers, and to put in and out, what they will make against their Religion." (His authority was the somewhat partisan Thomas James, *Corruption of Scriptures.*)[37] He also indicated here his convictions about the superiority of the ancients — the great number of books in Ptolemy's library proving "that the Ancients did excel the Moderns in Literature, notwithstanding what Mr. Wotton may say to the contrary." But it was his editions of various classical works that best displayed his growing technical competence and won the plaudits of contemporaries. In these same years they made their appearance, one by one, leading gradually up to his ambitious Livy: in 1703 the younger Pliny's *Epistles* and *Panegyric*, in 1704 Eutropius, in 1705 Justin. For each he collated several manuscripts, showed the variants, and contributed brief notes. Each edition won praise at home and abroad in the Leipzig *Acta* and in Le Clerc's *Bibliothèque Choisie*.[38] But his triumph was Livy, and it was through Livy that he drew the attention of Dr. Woodward and was compelled to take notice of the shield. In 1705, as he entered the first of his remarks in his notebooks, he had already embarked on that new venture. He had, in any case, come a long way from the early days at White Waltham.

[5]

Hearne intended his Livy to be his most ambitious work so far. In the summer of 1705 he was already reading assiduously for it. Cherry lent him his copy of Gronovius' edition with his notes on a New College MS. On July 7 he recorded in his notebook "Some Corrections of Livy in Pareus's *Lexicon Criticum.*" The next day he writes that he has heard well of Morhofius's Discourse, *De Livii Pativinitate*. Two days later he reminds himself to consult *Fabrettus contra Gronovium* and *Gronovius contra Fabretti*, as well as Scheffer's *de re vehiculari* ("where are Emendations"); also Boxhornius, and Freinshemius' edition of Lucius Florus. On July 15 he thanks Thomas Smith for help and notes some things in Tanaquil Faber's *Epistles* concerning Livy; and a little later he is annotating Vossius and Richter and remembering to look over Casaubon's *Epistolae* to see "whether there be anything in it concerning Livy." His reading was prodigious; he was determined to leave nothing unnoticed that might illustrate the text or the history of Livy's times. He read Perizonius, "concerning the Fidelity of Livy".[39] Among his many discoveries was, inevitably, Spon's Scipio shield.

The months passed and the notes accumulated. In March of the following year Hearne set down an account of his intended edition.[40] It was to be a handsome octavo on good paper and in good letter in four or five volumes. A chronology would be placed at the top of every page, and elsewhere as required. At the bottom of each page would go the various manuscript readings that had not been collated before, as well as the "Conjectures and Emendations of Learned Men (in a great Number) not yet taken Notice of in any edition whatsoever, together with select Readings, Corrections and Illustrations that are in other Editions." At the end of each volume, or perhaps at the end of the work, Hearne intended "such larger Annotations and Explications" as would not fit conveniently in the footnotes. Finally he expected to include such inscriptions and coins as would serve to illustrate the text, an accurate index and (perhaps) two or three maps. "The whole's intended to be with the utmost Exactness and Accuracy, as well as Beauty, in order to recommend it to the Youth of our Nation and others." In the end, to establish his text, he collated half a dozen manuscripts from English libraries, along with the printed version.[41]

It was some years before the labor was completed, though the first sheets were ready in August of 1706. Of course Hearne was busy with other things as well: a new edition of Spelman's *Life of Alfred*, for example. He divided his time equally between his two passions, classical antiquity and English history. Methodically, patiently, he pursued the sources of each, the manuscripts as well as the printed books, the documents, coins, and inscriptions as well as the literary works. Early in 1706 he drew up a list of writers to be consulted on classical antiquities (*de nummis, de epigrammatibus, de aedificiis, de statuis, de gemmis,* etc.), over sixty titles long. No detail was too small for this "exact and accurate" edition. When an English translation of Herodotus was announced, he wrote his regrets to Cherry. "I am sorry to see so many of our Classicks, both Greek and Latin . . . appear in English; which is certainly prejudicial to Learning. Young Gentlemen being by that means induc'd to neglect the Originals."[42] When he was still an undergraduate, he had read over Herodotus and Thucydides in Greek and found several mistakes and omissions in the great Dr. Gale's edition. "He was certainly a very Learned Man; but all his Editions of Books are full of Faults, he not taking due care (which is all the first and indeed the most considerable thing in an edition) to have them nicely corrected." He was determined at all costs to avoid such a charge himself. Perhaps it was at this time that Hearne composed his prayer to the Almighty, that He prosper his work, "and so to direct me in the printing of it, and in writing observations relating to it, that it may be a most accurate performance and be gratefull and useful to the learned world, and to all lovers of antiquity."[43]

It was Hudson who seems to have prompted Hearne to his classical editions. He was himself a considerable classical scholar and a collaborator with Dodwell on several imposing Greek editions. It was through Hudson that Hearne and Dr. Woodward first became acquainted and that Woodward began to take an interest in the new Livy. The two older men liked to share antiquarian news, and Hudson passed Woodward's letters along to the inquisitive Hearne. Thus early in 1706 the Doctor wrote to Hudson, promising to show him some interesting medals if he should visit Gresham College, and reporting on some new books and antiquities. There was a fine new print of the Theodosian column of Constantinople, engraved for the French king, like the Trajan and Antonine pillars, in the hands of the Earl of Pembroke. "Dr. Harwood of Doctors Commons," Woodward was also able to report, "shew'd me yesterday a Module [Model] of an ancient Roman Hypocaust lately discover'd at [W] Roxeter . . . very intire . . . and the Contrivance admirable."

Woodward, we have seen, persuaded Harwood to publish his account in the *Transactions*. Now Woodward did not forget to thank Hudson for some observations on the "sinking of a house" and any other information he might be able to furnish for "the History of Nature." Hearne was especially interested in the *Columna Constantini*. A few months later Woodward was excited to announce the appearance of the new edition of Spanheim's great work on coins. "As he is an excellent Scholar, so he has a mighty Mastery in that Particular science."[44] Woodward only hoped that they might have the second volume soon.

Sometime later that year (1706), Woodward promised help with the edition of Livy. He had two useful suggestions that might serve to illustrate the text. There was the shield which was just then being engraved, a print of which he promised to send as soon as it arrived from Holland. "I have also caused a Drawing to be made of the famd Action of Scipio, exhibited upon a large Antique Gemm which shall bear the former Company." The engraving was ready by the end of October; it only awaited Hudson's instructions. By the spring of 1707 it had arrived safely at Oxford. Hudson was very pleased; Woodward was anxious to have his thoughts on its iconography. The Doctor hoped that Hudson would pass a copy along to the Dean of Christ Church, and he offered to send further copies as needed.[45]

Hearne at once began to consider the engraving. He corresponded with Smith and Dodwell about it; he was concerned about some unaccounted details, particularly the horsemen without bridles. On August 26 he wrote to Woodward.[46] He had not yet had time to consider the shield fully, but even at first sight he thought it done long after the time of Camillus. Still he did not believe that it was as late as some were saying. "Without doubt 'twas done by one of the *gens Furia,* to revive the Memory of the Dictator's driving the Gauls from Rome; and none seems more likely to have been the Author of it than the Fabius Camillus who is mentioned by Suetonius in his Life of Claudius, who was descended from the Dictator and by his own Military Actions did add fresh Honour to the Family, as is observ'd by Tacitus in his Annals." It was exactly what the best foreign antiquaries were concluding abroad, and indeed Hearne had seen some of their opinions that very morning. Apparently Dr. Woodward had sent his file of foreign correspondence to his friend Dr. King at Merton College, where Hearne was able to read it and extract passages.[47] The absent geese, Hearne continued, one of the several details to bother him, could be accounted for by the late date of the shield. "And because the Numidians were famous for fighting on Horses without Bridles, the Horses on this Shield might be represented without Bridles on purpose to shew that the Romans were not at all inferior." There was another possibility. The Roman writers on "Strategycks" tell us that when charging, the ancients thought it expedient to use all the natural strength of their horses without the inhibition of bridles. "However," Hearne concluded, "if neither of these be the true reason why the Horses are thus represented, yet the thing ought not to appear more absurd than that the Romans . . . on Trajan's Pillar [are] fighting with the Dacians with their bare fists without any arms."

Hearne concluded his letter to Woodward with thanks for another favor. The Doctor had also sent him the drawing of Scipio. "I have by me the Draught of your Gem," Hearne replied, "which when the Text of Livy is off I will consider and compare with the votive Shield in Spon's *Miscellanea.*" He did more than that; a year later he printed it side by side with the illustration from Spon in the sixth volume of his edition of Livy, with appropriate thanks to the learned donor. Hearne saw the value of a contemporary picture.[48] It was not terribly useful perhaps, for it merely depicted Scipio on a pedestal surrounded by admirers, but at least it was original and had not been printed elsewhere. Thomas Smith, who seems to have read over Hearne's letter, urged him to translate his views on the shield into Latin, either for an appendix to Livy or as a separate little dissertation.[49]

Hearne was not yet satisfied with the shield, however, and the correspondence with Woodward continued. The Doctor, of course, was delighted to have the learned young man's interest. " 'Tis with great Satisfaction," he wrote on September 18, "I learne that the Iron of the Shield was so much to the Gust of a Gentleman of your Learning and good Sense."[50] If any of Hearne's friends wanted more copies, he need only ask. (Not long after, Hearne sent one to Smith.) As for the missing geese, it was not to be expected that they would be depicted on so small an object. Besides, their appearance would have been incongruous, since that episode was over before the appearance of Camillus. The shield, Woodward reminded Hearne, represents only the exact moment of the Roman's triumph. He agreed that it was done about the time that Hearne had suggested, and he concluded by promising to show him the Scipio gem itself.

A week or two later Hearne wrote again. He was still bothered by the lack of bridles. He had to dissent from Woodward; he had found the Romans using them as early as AUC 365. And the literary authorities all agreed that they were regularly employed. Even Livy, who tells of Numidians without bridles, remarks that the Romans thought it very strange and unusual. Indeed, Hearne was inclined to think that the ancestors of Romulus coming from Troy (he is of course thinking of the Aeneas story) brought them with them, since Homer tells of their use in the Trojan Wars. Both agreed that it was done in

Claudius' time but the designer could hardly be ignorant of bridles, "since he saw the contrary represented on Coyns (amongst which were those that exhibited the Triumph of Camillus in a Chariot of White Horses) and other Monuments of that kind." It was a puzzle. But whatever the solution, he concluded, " 'Tis certain the Learned World is highly indebted to you for the expence you have been at in the Engraving the Shield: and I am heartily glad that such a valuable Monument fell into so good Hands. Men of skill and Judgment will always set a Price upon such Curiosities, and will dispose the little, Trivial Objections ignorant Persons may make against their being genuine."[51] After all, the same nitpicking would work as well against Spon's Scipionic Shield or "any other Monument of antiquity."[52]

Woodward found time for only a hasty reply. He had meant only to say that it was some time after the Deluge before horses were tamed and trained, and later still when bridles were invented, and that they had been taken up at different times in different countries. Most ancient statues of horses showed no trace of them. The designer of the shield had probably had them in mind when he created the shield. Like Hearne, he believed that the work belonged to the early Empire, the workmen having "aimed at expressing a Manner more antique than that of his own Age." (They wouldn't have bothered then to research the matter.) He was, he added, most anxious to see Hearne's Livy.[53]

Hearne, however, was interested in more than the problem of bridles. He wondered just how and where the shield had been made. A long entry in the *Remarks* for November 1708 suggests his progress here.[54] He had found considerable information about the workshops of the Empire where armor was designed and made. There were, he wrote, "Divers colleges of Fabricenses in the Eastern and Western Parts of the Roman Empire and I believe there was one particularly assign'd to Britain." It is not clear why he thought so, especially since his source, the *Notitia Imperii*, mentioned eight *fabricae* in France but none in Britain. Their business was to make public arms, and they received a public salary for it. They had a governor of their own, the *Primicerius Fabricae* or *magister officiorum*, and were distinct from the *Barbaricarii*, whose special business was to adorn and beautify the instruments made by the *fabricenses*. "And 'twas by one of these *Barbaricarii*," Hearne was convinced, "that the Antient Shield of Dr. Woodward and others of that kind was beautified."

It was an interesting theory, but for the time being Hearne pursued it no further. His Livy had appeared, and he was ready for other enterprises. For the moment, anyway, he had done his service for Dr. Woodward and for classical antiquity; with the sixth volume of Livy, Dr. Woodward's shield had made a new public appearance. The fold-out engraving by Michael Burghers was less ambitious than Woodward's, but it was really quite handsome. Indeed, Hearne thought it superior to Woodward's own print. Now, annexed to the appropriate passage in Livy, it made its way throughout Europe into the libraries of gentlemen and scholars.[55]

XI.

Mr. Dodwell Discovers the Shield

Hearne never lost his attachment to Shottesbrooke, to his patron Cherry, and to his master Dodwell. In the summer of 1707 he reported his observations on the shield to the latter and received a characteristically learned reply. "I do not love venturing to give accounts of Antiquityes," Dodwell began, "till I be satisfied of their genuineness. If this you mention of Camillus be genuine, it should be further considered, whether it be of his own Age, or by some of the *geni furia* descended from him or some Prince who had an esteem above others for his memory who might represent the facts of his Age according to the fashions of their own."[1] Dodwell had not seen the shield and he was as yet unaware of the conjectures of other scholars. But he was willing to offer some suggestions of his own to Hearne, especially on the troublesome question of horses without bridles. He remembered reading, for example, that the ancients had thought it useful to eliminate bridles when charging, thus benefitting from the "full natural strength of their horses." Thus it had been at Marathon, long before the time of Camillus, and so it might well have been with the Greek colonies in Italy later. That was probably also how the Numidians took up the practice. "It is therefore strange if the Romans could be so long ignorant of it, till the second Punick War." (This, unfortunately, had been Hearne's notion.) However, there was another possibility. In Trajan's column, the Romans were shown fighting with bare fists, "perhaps not to offend the eyes of their fellow citizens with arms in the hands of citizens." If no arms, why must there be bridles? That the soldiers on Dr. Woodward's shield had, in fact, possessed arms (though no bridles) did not trouble Dodwell; learning, not logic, seems to have been his strong point. Anyway, he saw an advantage to doing without bridles if the horsemen had to employ both hands with shield and sword. Unfortunately, he concluded, he had not been able to see the coins of Camillus' triumph that Hearne had mentioned; nor was he entirely confident that Hearne's account of the shield was right.

200

It was the beginning of the great man's preoccupation with Dr. Woodward's shield. Perhaps it was inevitable that he should become concerned. Dodwell was probably the most learned man in England, at least in his combination of classical and Christian studies. Moreover, he had spent a lifetime specializing in scholarly puzzles, particularly problems of authenticity and dating. He was familiar as few in his time with the techniques of modern philological and antiquarian criticism. For years he had corresponded familiarly with the great continental scholars Graevius, Spanheim, Perizonius, *et al.* Hearne was not entirely wrong when he remarked reverentially that Dodwell was "of such an universal knowledge and profound Judgment in all Books and Sentences that his name is rendered famous all over Italie, France and Germanie."[2] His works were regularly reviewed and esteemed in the learned journals abroad.

Yet he was a complex man with an involuted mind and a liking for paradox. Even his friends sometimes wondered about him, while his many critics suspected the intelligence that lay concealed behind the obvious weight of his erudition. Thus Edward Calamy drew Dodwell's portrait in 1691.[3] He found Dodwell characteristically holding forth in a coffeehouse. "Nothing pleased him better," he wrote, "then to have a question proposed to him, upon a difficulty in chronology, a piece of History, either civil or ecclesiastical, or about ancient customs." Once launched, he would "pour out a flood of learning with great freedom." Calamy thought that Dodwell disliked controversy, though he noticed him chiefly in the midst of it. He remembered the older man seeking him out at a little table by himself, in preference to the noisy crowd of controversialists. "Then he would ask me if I had any questions to propose to him, with which I took care not to be unprovided. He would on a sudden, and off-hand, make returns that would sometimes be very surprising, though not always especially satisfactory. In order to the proof of a point that he had lain stress upon, he used to lay down a chain of principles, and if they were all granted him, his proof would be good, but if any one link in the chain failed, his whole scheme came to nothing. He was no great reasoner," Calamy concluded, "nor at all remarkable for his management of an argument."

Among his critics were those who had no patience for him whatsoever. Bishop Nicolson, for example, recognized Dodwell's great reputation as a scholar and was willing to admit that he might even deserve it. "Perhaps his Arguments may have their Depths too; for I could never see nor feel the Bottom of any of 'em. If Posterity understands either his Latin or English (which I much Question) he may possibly be as great a man in succeeding Ages as this."[4] Nicolson was prejudiced but he was not altogether unfair, though the contemporary verdict was sometimes more charitable. Nor was posterity more kind to Dodwell. Edward Gibbon allowed for his learning, but found his method and style "the one perplexed beyond imagination, the other negligent to a degree of barbarism." And Macaulay was completely baffled by him. "He had perused innumerable volumes in various languages and had indeed acquired more learning than his slender faculties were able to bear. The small intellectual spark which he possessed was put out by the fuel." Some of his works, Macaulay thought, seemed to have been composed in a madhouse.[5] The French biographer Nicéron was gentler but of pretty much the same mind.

On the other hand, he put many of his contemporaries in awe of his talents. "We have had the famous Mr. Dodwell," wrote one Cambridge man, "for about ten dayes." He was constantly at the coffeehouse, "and while he was there the coffee house never failed to be well entertained. He makes but a mean figure but I believe he is what I have heard said of him, a man of the most comprehensive knowledge in England. He talks very freely of any thing the company has a mind to put him upon and seems to do it with great exactness." The writer spent hours admiring Dodwell's performance and never left the coffeehouse without regret.[6]

Among the non-jurors and in the circle at Shottesbrooke, Dodwell was revered as a learned saint. One suspects that his controversial religious opinions prejudiced both his critics and his admirers. Cherry supported him and his growing family at his own expense for years. For his biographer, Francis Brokesby, he was "a great Pattern of indefatigable Industry" and an infinitely humble and pious man of principle.[7] But no one perhaps was more influenced by him, nor kept him in more continuous esteem, than his young scholar friend Thomas Hearne. "I never met with any who behaved himself more like a true Christian and sound Scholar. Such I find his Modesty, and his Humilitie, and his ready and quick Apprehension of Things, such his Reading and Skill and withall such his readiness of Communication and his desire of benefitting others, that I was transported." "I take him to be the greatest Scholar in Europe," he recalled after his death, "but what exceeds that, his Piety and Sanctity was beyond Compare." For Hearne, at any rate, Henry Dodwell was the one model he knew who was almost without imperfection.[8]

[2]

"Shottesbrook," Hearne recollected, was "a little village, pleasantly and delightfully seated, whether you respect its admirable conveniences for a recluse and retired life, or the great advantages here afforded for those who delighted in sports and pleasures, especially hunting." While Cherry followed the hounds, Dodwell took up the life of scholarly seclusion. In the end his example moved even his patron to emulation, and Cherry too took up scholarship, till he "studied almost day and night."[9] Dodwell's influence was profound on the circle that gathered about him; it was as much a result of his character as his learning. He was pious and ascetic; he "forebore the Use of strong Liquors and delicious fare," and for three days a week ate no meal till suppertime, "and then no flesh." Coffee was his only apparent vice. His life seems to have been comprised in equal parts of prayer and learning, and he lost no opportunity for either. Thus he, like Hearne, preferred to travel on foot, rather than by horse or coach, "that he might thereby be Master of his own Hours, often reading as he walked." For that purpose he carried with him little books to stuff in his pockets: the Hebrew Bible in four volumes, the Greek New Testament, the Book of Common Prayer, Augustine's *Confessions,* or the *Imitation of Christ.*[10] Hearne remembered that he liked to walk over to Henley to visit a friend at a local inn. "The first time he was there, being beforehand with his Friend, he ask'd if they had got a Bible. They brought him a great Bible and he read 'till his Friend came. Ever after this, as soon as they saw Mr. Dodwell at any time away, they carried in a great Bible." In the evenings, "besides what was employed in Religious Exercises" he would have his wife or daughters read some history aloud that they might all be benefitted as well as delighted. His piety and his learning gave him an unworldly air; even his biographer felt obliged to apologize for "his retired Life, and his Unacquaintedness with the World, his studying Books more than Men." At least it furnished a convenient explanation for why he expressed his thoughts "less clearly than he would have done, had he more frequently conversed with the Generality of Mankind."[11]

When Dodwell came to live with Cherry in 1691, he had already established himself as a formidable man of learning. No one (since his time) has ever taken the pains to reconstruct his career, and it remains an unlikely prospect. But something at least must be said here to describe the scholarship of this man who, late in life, took up the cause of Dr. Woodward's shield. Indeed it was Dodwell who came the closest of all of Woodward's many correspondents to writing the definitive treatise on the subject. Only death intervened and left Thomas Hearne to publish the incomplete manuscript.

Dodwell was about fifty when he came to Shottesbrook. It was the nadir of his career. Three years before he had been appointed Camden Professor at Oxford. This had been the first history chair to be endowed in England, and Dodwell was pressed to his appointment entirely on the merits of his great reputation for historical learning.[12] Like his predecessors he selected Roman history as his theme and plunged into his lectures with characteristic enthusiasm. They were a great success. Hearne's schoolteacher, James Gibson, remembered the occasion, "so thronged and crowded that he never either saw or heard the like, so that there was no standing with ease, which made him get out (otherwise much against his will) before the Lecture (an admirable one) was ended."[13]

Unfortunately the Glorious Revolution aborted his new career by introducing oaths — the same oaths that later troubled Thomas Hearne — and by leading to schism. Dodwell would not dissimulate. "If I may not keep my place without Sin or Scandal, I desire not to keep it at all."[14] He refused the oaths and was forced to retire, though hardship and penury were the immediate consequence. It was all the more painful to him because he had for years set himself stalwartly to defending the Church and had spoken unflinchingly against the Catholics, even under James II. His great fame abroad, his sanctity and severity of life, had gained him, wrote Anthony Wood, "a veneration very peculiar and distinguishing amongst all sorts of people." But it was to no avail. "He lives obscurely now (mostly in his cell in the north suburb of Oxford)"; it was left to him only to prepare his Camden lectures for the press.[15] It was about this time that Cherry found him, brought him to Shottesbrook, and built him a house where he could enjoy what Hearne called its "admirable conveniences for a recluse and retired life."

Yet Dodwell was anything but cloistered in this busy center of non-juring activity. In a way he simply resumed the life he had led before his appointment at Oxford, only more securely with the confident support of his friend Cherry. As before, he continued to divide his time between ancient and ecclesiastical history; but where before he had engaged in sporadic religious controversy, now he was leader of the non-juroring community and his pen was even busier in the defense of his religious convictions.

Nevertheless, the first thing he thought to do was to complete his Camden lectures for the press. Not all of them had been delivered; now they had to be edited, annotated, and furnished with tables. Eventually they were ready, eight hundred pages of Latin text.[16] Dodwell had chosen to illustrate that awkward period of Imperial Roman history for which the chief literary authority was the *Historia Augusta.* As we have noticed, ancient history was still being taught in his time largely by way of commentary on the classical narratives. Dodwell's lectures were an elaborate gloss on an obstinate and difficult text. In a way it was the perfect document for a display of critical· learning. Its date and authorship were dubious, its narrative suspect, and its chronology obscure. It remains so today. Dodwell lavished his scholarship upon the *History* to try to solve those difficulties and to illuminate the period. It was a characteristic humanist effort, wrong often in its conclusions, but learned and sophisticated in its technique. Across the Channel at Deventer, Gisbert Cuper read it and found it very erudite and comprehensive, but more than a little obscure and confusing.[17]

Perhaps Dodwell's inaugural lecture, delivered in May 1688, was the most interesting; it seems to have met with general applause.[18] It was devoted largely to a consideration of the use and difficulty of historical writing. Dodwell thought that chronology was the backbone of all historiographical activity. The chronology he admired was the new discipline launched by the humanism of the sixteenth century and associated with the great Joseph Scaliger. From the first publication of the *De Emendatione* in 1583, European scholars pored over the old texts with a new concern. "Hitherto," writes Mark Pattison, "the utmost extent of chronological skill which historians had possessed or dreamed of had been to arrange past facts in a tabular series as an aid to memory. Of the mathematical principles on which the calculation of periods rests, the philologians understood nothing."[19] Scaliger was the first to apply the new astronomy to the problems of historical chronology; he was "the first who undertook to give us a compleat Chronology, and solid Principles to bring History into an Exact Order founded upon Rules. Before him . . . no Body ever attempted the like. There was nothing but what was very confus'd and uncertain."[20] He tried to reconstruct the ancient epochs and to show the means by which they were constructed. Thereupon, the astronomers as well as the philologists took up the problem; among Dodwell's contemporaries, both Newton and Halley became passionately involved.

Essentially the task remained philological. Scaliger saw that ancient history could only be reconstructed as a whole, that classical and Biblical history had to be considered together, and that, particularly for the earliest period, the chronologists of the Empire had to be employed. Those ancient scholars, dry and difficult, frequently embedded in the works of others, had copied information from earlier sources, often misconstruing it but preserving facts and traditions that might prove useful. "He set himself, accordingly," Pattison writes, "to collect the distorted fragments of Berosus, Menander, Manetho, and Abydenis — names which he first dragged from the oblivion of more than 1,000 years, but which have ended by rivetting the attention of historical antiquaries." If they were too much beneath the dignity of the rhetoricians to serve the uses of imitation, their very obscurity awoke the critical possibilities of philological method and came to absorb the talents of seventeenth-century scholars. Furthermore, they seemed to furnish a new key to those earliest ages of history, to the primitive period before the times which classical narrative illuminates. But they had first to be rescued, understood, compared. In Scaliger's *Thesaurus Temporum* (1606), "every chronological relic extant in Greek or Latin was reproduced, placed in order, restored and made intelligible." It was indeed, a "prodigious Heap of Materials."[21] On this basis alone Scaliger is entitled to Niebuhr's praise as the greatest of philologists.

Now, for the benefit of his Oxford audience, Dodwell enlarged upon the difficulties of the new discipline. Choosing Greek history for his example, he showed that the early period was riddled with difficulties. There were no annals before Thucydides and little reliable history before the time of Darius. The Athenian *archontes* were a little earlier, but also lame and imperfect. Simply comparing the diverse accounts of Cyrus in Herodotus, Xenophon, and Ctesias was to cast suspicion upon them. No account of the events before the Olympiads was worth credence. Indeed, nothing could be known securely before the invention of letters and the keeping of archives in temples. It was the first task of the historian to ferret out from the later writers the earliest sources, to define their character and see their limitations. But here there was another difficulty. Each Greek city had kept its own reckoning of events and magistracies. The calendars did not coincide; the divergence of years was very great, and some kept to lunar while others employed solar calculations. Chronology was thus an indispensable discipline; through it alone could one furnish a reliable skeleton on which to flesh out a reconstruction of the past. Moreover, it was a unique critical instrument which, joined to the other philological disciplines, could eliminate much of the monstrous accretion of legend and later invention. No other technique was as likely to disclose anachronism and error.

In an appendix to his lectures Dodwell printed a fragment of the *Libri Lintei* (the ancient linen rolls mentioned by Livy), "intermix'd with many useful and learned Observations." This allowed him the possibility of expatiating on the earliest sources of Roman history and the problems of Roman chronology. Using later writers (including Livy), he described the earliest family records and pontifical annals and showed how they survived in the later works as the only ultimate sources for the history of early Rome. It was, however, but the beginning of Dodwell's interest in the problem; in the next years he set about his *magnum opus*, a study of all the chronological systems of antiquity, a work to be entitled *De Cyclis*.

[3]

Dodwell's scholarly interests dated from his university days. (He was Irish originally and attended Trinity College.) Even then he was concerned with those doubtful sources that pretended to fill up the gap in history before the Olympiads. Of these, perhaps the most important was the Phoenician writer Sanchoniathon. As Dodwell wrote to his new friend Thomas Smith (in 1667), the old Phoenician was "generally to be reputed for the faithfullest and ancientest and consequently the most usefull heathen author that was extant."[22] Scaliger himself and the other great men of the seventeenth century – Grotius, Bochart, John Selden, *et al.* – had all employed him in their chronological calculations. But Dodwell was doubtful; "the work itself is too full of the later vanity of those nations that pretended to antiquity, of arrogating to themselves the famous persons and instances of other nations." In a microscopic hand and at great length he set out his arguments and displayed his already amazing erudition in Latin, Greek, and Hebrew.

While in retirement in North Oxford, Dodwell began to arrange his papers and came upon the correspondence with Smith. As a result, he published that year (1691) *A Discourse concerning Sanchoniathon's Phoenician History.* He now resumed his arguments against the work. The original text had apparently been turned into Greek by one Philon Byblius in the first century A.D. and survived partially in the later works of Porphyry and Eusebius. Even as Dodwell was writing, a translation into English was being prepared by Richard Cumberland, the Bishop of Peterborough. Cumberland not only accepted the reliability of the document, he synchronized it more fully than before with the Biblical narratives and claimed to have restored the missing profane history from Adam to the first Olympiad. He supposed that Sanchoniathon "preserv'd the History of the idolatrous Line of Cain, as Moses did that of Seth." Since the Bible was the font of sacred history, and Sanchoniathon the font of pagan history, it remained only to combine the two to restore the complete early record of mankind. Cumberland's work was an improvement over any of his predecessors, "a Discovery that has hitherto escap'd the Inquisitiveness of all other Learned Men."[23] Fortunately it remained for a time unpublished, or it would certainly have occasioned controversy with the learned non-juror.

As it was, Dodwell was exercised enough by those who wished to use the Phoenician to "clear" the Old Testament of its obscure historical passages. He set about reviewing all the evidence for the authenticity of the text. Fortunately, there was a convenient model for this kind of investigation (familiar to Dodwell) in the remarkable work of Isaac Casaubon, who some years before had inquired into the fabulous Hermes Trismegistus, in the very same manner. After close critical examination, Casaubon had transformed Hermes from an ancient Egyptian into the forgery of a post-Christian writer. As Francis Yates has pointed out, "the weapons of textual criticism which Casaubon used for his dating – both from content and from style – had been developed long ago by the Latin humanists."[24] But it was long before they were applied to all the texts in all the languages that had come down from Antiquity, and here Dodwell and his contemporaries saw their opportunity.

Dodwell concluded, after a wide canvas of the sources, that all knowledge of Sanchoniathon depended entirely upon the testimony of Philon by way of Eusebius and the pagan philosopher Porphyry. There was simply no trace of the Phoenician beforehand, although he was supposed to have lived in very early times. Moreover, all the later writers who mentioned him were dependent upon Philon. As for Porphyry, he had employed the work, according to Eusebius, against the Christians, and Dodwell was at pains to show how "such disingenuities" were customarily devised by the pagan writers for their own purposes. Indeed, the forgery was transparent enough. After much further consideration he was ready to conclude: "I have shewn how little he favored of the Antient simplicity and how much of the modern Emulations . . . how little creditable he is in his pretended means of Information," and in general how little reliance could be placed on any similar heathen accounts. [25] Indeed, before the *Septuagint* had made the Bible known among the Greeks, there was nothing at all of the antediluvian world in the writings of the ancient historians. Where, for example, was the Flood in Herodotus, Xenophon, or Ctesius? It is only long afterwards that we hear of Berosus, Manetho, Menander, and the rest. As a friend wrote to Dodwell in 1680 after reading his arguments, it had been possible to suspect Sanchoniathon before, but now it was impossible to believe in him. [26] Dr. Woodward seems also to have been convinced by Dodwell's judgment. [27]

It was an impressive performance (though later discoveries must alter Dodwell's conclusions somewhat) and it showed Dodwell at his best. [28] In the years of his maturity he matched it often, though his conclusions were not always so persuasive. On one famous occasion, at least, he was quite wrong. It was the celebrated "Battle of the Books," the quarrel between Richard Bentley and his critics over the genuineness of the *Epistles* of Phalaris. Here, for once, Dodwell met his match with a man of deeper (if not broader) learning and a far more incisive intelligence. Here, for once, the skeptic was taken in by a forgery that had to be exposed by someone else. Richard Bentley belongs only with Scaliger, where Swift put them in his satire, as the greatest of philologists and critics, the two natural leaders of "modern learning."

There is no need to rehearse that famous episode here in any detail. [29] Suffice it to say that Dodwell was inveigled into a contest he would have preferred to ignore, pronounced on the wrong side of the argument in favor of the false *Epistles,* and then tried to retreat from the combat. He was able to hold his ground on some of the disputed chronological points but on the main issue he could only concede. In 1699 Bentley published his enlarged *Dissertations on the Epistles of Phalaris* and demolished the arguments of his opponents with a display both of learning and logic that showed once and for all that the letters were riddled with anachronisms and inconsistencies and that they must have been the invention of an obscure sophist writing hundreds of years after Phalaris had ruled in Sicily.[30] The Christ Church wits who combined forces to answer him chose to reply with satire rather than scholarship and their laughter proved an effective if temporary riposte. It was left to Dodwell to attempt a learned reply but all that he could do in the face of Bentley's arguments was to suggest that the letters were a translation of some lost original. That could account for why they were so long ignored in antiquity (a troublesome point), and it would also take care of most of the anachronisms and inconsistencies. Such an hypothesis, he wrote to Bentley, would "solve all your Objections as to words and some also as to things, if the Translator took the liberty of paraphrast, which you know was not infrequent."[31] It was an ingenious suggestion (Perizonius noticed that it was like his own view of the *Pentateuch*),[32] but in reality it conceded everything. How could one derive a date with any confidence from a late paraphrast? One might as well admit the forgery. But it is not easy for the scholar who has publicly proclaimed a position to retract. When Dodwell read Bentley's complete argument, he remained discreetly silent. He acknowledged receipt of the book, but it was several years before he returned once more to the subject.[33]

[4]

If I have lingered over Dodwell's criticism it is because it is so hard to arrive at a judicious estimate of it. Of his learning there was never a question; Bentley called him *doctissimus et integerrimus*; Nicéron compared him to Scaliger. [34] As Dodwell's treatment of Sanchoniathon showed, he could employ philological criticism with skill and originality. The great Dr. Mill told Hearne that he thought it better not to publish such works, "since 'twould do mischief to endeavor to overthrow an opinion which had been embrac'd for so many years."[35] But religious orthodoxy was not a bugaboo likely to frighten Dodwell from the truth. (In a few years he was to startle even his friends with his views on the immortality of the soul.) His general cast of mind was independent, even skeptical. Indeed, despite a genuinely pacific temperament he seems almost to have been embarrassed by too much agreement with others. He not only challenged most of the early Greek history in *De Cyclis*, but he also wrote some interesting paragraphs there against the early Roman history, radically criticizing the whole story of Romulus and the Alban kings.[36] He understood that through close examination of the literary sources it was possible to reconstruct the materials which Livy, Dionysius Halicarnassus, and the rest had employed for their narratives. But when this had been done, it was clear that the sources of the Roman historians were all too late to be reliable for the period they purported to describe. He was thus even more skeptical than his friend Perizonius, even more aware of the perils of early history.

Indeed, Dodwell's criticism, like that of the best of his contemporaries, had advanced beyond the possibilities of Renaissance philology. Scholars of 1700 were no longer content to discover and to expose legend and forgery; they had begun the task of reconstruction as well. Thus they understood that it was not enough simply to accept or to reject the testimony of the Latin historians, but that their authorities and the sources of their authorities would have first to be reconstituted and as far as possible criticized. Every scrap of information from all of classical literature that might help to describe those lost works would have to be employed. Behind Livy lay annals; behind the annals, the *libri lintei* and other possible authorities. Although almost all of this had disappeared, it might yet be painstakingly restored from the hints and fragments of later writers. The critical regression was infinite, but infinitely rewarding.

Needless to say, Dodwell's skills were neither perfect nor perfectly systematic. Philology, like any other science, develops its methods continuously and by slow degrees. Dodwell had grasped the latest philological accomplishment anyway, and he had attempted some of the most troublesome and elusive problems of ancient history – too many by far to admit of solution by any single scholar. But if his conclusions were often wrong, they were always sophisticated and interesting (though to be sure, it was hard at times to discover in their elaboration exactly what they were!). And with great patience he taught the circle at Shottesbrooke all that he knew about the useful, if somewhat arcane, disciplines of the new classical scholarship.[37]

Thus by 1700 Cherry was able to assist Dodwell in the writing of *De Cyclis* as well as to project some critical works of his own. And Thomas Hearne's apprenticeship under the learned scholar was long and rigorous. As we have seen, it began with the very rudiments of his Latin and Greek education. For some years afterwards he copied manuscripts for the Shottesbrook circle, and when he left for Oxford he remained in close correspondence with his two learned friends.[38] When *De Cyclis* appeared in 1701, Dodwell made a proposition to Hearne. He thought that he might take the tables from the end of the work and "collect a useful Manual for any who's conversant in reading the Historians themselves, but especially of Travellers as they may light on new Inscriptions that have new Archontes or Consuls or Victors."[39] Hearne was interested; he was already contemplating this kind of thing with the *Ductor Historicus*. He thought *De Cyclis* "one of the greatest and exactest Performances that ever was printed of its kind."[40] Although the scheme came to nothing – except as it was absorbed into the *Ductor* – Hearne was continuously learning from the older scholar. For years his notebooks were filled with the great man's suggestions, especially those dealing with criticism and chronology. "Mr. Dodwell has prov'd," Hearne wrote in 1706, "that Semiramis, that was five Generations before Nicotus according to Herodotus could not be the Wife of any Assyrian King that reigned at Ninieve, and Therefore not the wife of Ninus.... This he has done in a letter to Perizonius. He also some time agoe writ a Discourse about the Sicyonian chronology, which story he thinks to be fabulous. 'Tis writ in Latin." "Mr. Dodwell is of Opinion that Philostratus Life of Apollonius Tyanaeus is spurious; which ought to be consider'd by a Gentleman now engag'd in a new Edition of him." "Mr. Dodwell is of Opinion that Geoffrey of Monmouth's story of Britain is wholly to be rejected."[41] Hearne kept to his own opinions; he was temperamentally more conservative and credulous than the older man, but he was always impressed by Dodwell's judgments and he adopted many of the great man's habits. Many years later, after describing some ancient parchments, he noted in his diary that "Such Fragments are of good use in Critical Disputes. For such reasons I always take notice of them whenever I happen to light upon them. As I remember, the famous Mr. Dodwell used to do."[42]

When Hearne embarked upon his own editorial labors, he continued to receive advice from Dodwell. It was natural to consult the great scholar on all points of chronology and textual criticism. But sometimes Hearne was able to instruct the older man, especially where the collation of manuscripts (Hearne's own specialty) was involved.[43] Though Dodwell was primarily a literary scholar, he took a great interest in coins and inscriptions. Indeed, no chronologist could afford to overlook them. But he was as careful here to question their authority as any literary document. "Mr. Dodwell," Hearne recalled in his Diary in 1722, "did not care to write upon Coynes, or other Antiquities, unless he was pretty sure of their Genuineness." It was a useful notion, and Hearne kept it in mind.[44]

[5]

Hearne naturally hoped that he could get Dodwell to contribute the chronology for his Livy. In the end he was disappointed, Dodwell was content to advise him on how to do it for himself, using the tables he had lately set out for Dionysius Halicarnassus.[45] It was through the correspondence about Livy, however, that Dodwell learned first of Dr. Woodward's shield and was introduced to the last great chronological puzzle of his career. As we have seen, he was at once interested but characteristically cautious. He had to be sure of its authenticity before he could attempt to explain and illustrate it. He looked forward to seeing the shield for himself.

Perhaps he was still smarting from the recollection of Phalaris. In 1700 he wrote to Bishop Lloyd that he had "an unfinished Discourse concerning the time of Phalaris," in which he showed the inconsistency of some of Bentley's chronological arguments.[46] It remained unfinished and unpublished. At last in 1704 he returned to that awkward subject with two more dissertations, one on Phalaris and the other on Pythagoras, both dedicated to Baron Spanheim.[47] He was not anxious to reopen the controversy, he explained, but simply to enlarge on some useful points that had been raised in the quarrel. He still could not accept Bentley's chronology, and most of his work is a reaffirmation of the dates that he had earlier suggested, along with a general review of early Greek chronology. He was very tactful. On the main point (for the rest of the world, anyway), the authenticity of Phalaris, he sidestepped. He was willing to dispute one thing only, Bentley's last argument against the epistles: the fact that letters had been invented after the time of Phalaris by a Persian princess, Atossa. With characteristic ingenuity, Dodwell raised a host of objections to Atossa, to her sources, and to Bentley's notion of the invention of letters, while continuing to miss the main thrust of his rival's argument. When all was said and done, it remained that Phalaris' epistles could hardly have been written in the sixth century in the form in which they are received, regardless of when letters were first written. Nor were Dodwell's arguments for the existence of pre-Trojan and Homeric epistles very persuasive. What Dodwell lacked – but it was true of almost all the great scholars of his time – was an Occam's razor to apply to his learned proofs. In any case, having set out in the wrong direction on this one, it was very difficult for him to right himself.[48]

Meanwhile, in these years, Dodwell was busy at a host of other activities too numerous to recount. He was deeply engaged in religious controversy, defending the cause of the non-jurors and using his historical learning wherever possible. (His theological commitment was exclusively to the early church, which he was therefore obliged to examine.)[49] He continued to labor also in the service of classical literature. With the confident assurance of the Christian humanist, he believed that the classical historians were indispensable to the Christian gentleman and that the philosophy of Cicero was in nearly every way consistent with that of Christ; he even wrote a special discourse on each subject.[50] At the same time he held to his main scholarly task, illuminating the chronology of the ancient world, and he furnished a flood of treatises, several of them commissioned by Dr. Hudson, describing the dates of a host of ancient works and authors: Cyprian, Thucydides, Velleius

Patercullus, Quintilian, Caesar, Xenophon, Dionysius Hallicarnassus, the Greek geographers, and others still. Hearne knew of many other projects that remained in manuscript;[51] even failing eyesight did not slow Dodwell down. Through it all, he kept to his zealous quest for error and forgery. To some of his contemporaries he was the "last Appeal in matters of this Nature." And if he erred often, it was natural, as Hearne saw, that there must be some mistakes in so gigantic an undertaking.[52] But whatever else we may conclude about him, it is certain that Dodwell's errors were not the result of either ignorance or credulity. Indeed, he was as likely to eliminate the authentic as he was to accept unreliable testimony, out of an overabundance of suspicion and critical zeal.[53] Perhaps Nicéron's judgment was fairest: Dodwell was so exceedingly learned in ancient history that one could say of him, as of Scaliger, that he gave profit even when he was wrong, despite an obscure and digressive style. *Etiam cum errat, docet.*[54]

[6]

It was only in the winter of 1709-10 that Dodwell actually saw the shield. "I was with Dr. Woodward," he wrote to Hearne, "who was pleased among other privileges to show me the original of your shield."[55] He also showed Dodwell the learned correspondence that had accumulated, including Hearne's letter. "His design was, I perceived, to have my opinion concerning it." Indeed, Dodwell felt better qualified to offer one now that he had seen the original. He learned from Woodward that it could not be traced back very far; the ironmonger who had first noticed it was dead and could give no account of how he had got it. Characteristically, Dodwell proposed "whether or not it might not have been modern, upon the improvements or rather achievements of the manual Arts in Italy by the interest of some of the modern Camilli, a name frequently occurring in the late Histories of Italy." Woodward was annoyed; "The Dr. assured me that it had been twice patched with modern Iron, so that it could not be later than the time of Charles the Great." Dodwell hesitated; it was no use making a precipitous judgment when his eyes were so bad and the winter light so gloomy. "I reserve what I have to say for a second view."

A few days later Hearne replied. "Dr. Woodward," he assured Dodwell, "is a very ingenious, civil gentleman, one to whom I am highly obliged." He had never met him, though he had often been invited to London to view his curiosities. "I am glad you saw the Dr.'s shield," Hearne continued, "and hope I shall hereafter know your sentiments about it. I am still of the same mind that I was of when I writ my note upon it in the VIIIth volume of Livy, namely that it was made about the time of Claudius." Looking over his notes,

Hearne found that a friend, John Bagford, had told him about a year ago "that formerly there was a shield Gallery at Whitehall, in which was a great Collection of Shields, and other military Instruments as there is now at the Tower." His friend's opinion was that this had been one of them. Lately Hearne had received another letter from him telling of a new and parallel curiosity. It was an iron helmet with figures "chased and wrought rather better than those on the Dr.'s Shield."[56] It was in the famous collection of Mr. Kemp and seemed to be the very mate of Dr. Woodward's shield.

What was this helmet, and how had it gotten into the museum of John Kemp? Unfortunately, of all the important collectors of the period, Kemp is the most obscure. But the auction catalog at his death is eloquent enough about his efforts, and Bishop Nicolson has left us one very useful account of a visit in 1712 which (with some briefer notices) bears out the scope and magnificence of his holdings, "too valuable for a subject to keep in his own private possession."[57] Much of it had come into Kemp's hands from George, first Lord Cartaret, who had obtained them from his governor, John Gaillard, including apparently "the spoils of the late famous museum of Dr. Spon." Other objects — statues, altars, inscriptions, and medals — had been added from time to time. Nicolson saw two entire mummies "amongst his Innumerable Collection of Roman, Greek, Jewish, and Egyptian, Lamps, Vessels, Urns, Lares, etc." He saw besides many remarkable busts of Cicero, Aristotle, Harpocrates, Alexander the Great; seven hundred silver Roman medals alone, Greek inscriptions, "multitudes of antient Fibulae, Seals, and Rings," and coins and medallions of every kind and nationality. And finally he saw the iron helmet, "with fair figures of the very same kind with those on Dr. Woodward's shield." Kemp convinced him that it had been hammered out and not cast in a mold. But he learned that the owner was unwilling to make any extravagant claims for it. "Mr. Kemp is of Opinion with me that neither of those are Roman; it being inconceivable how anything of this metal, so slender, should have been preserv'd for so many Ages, either in the Earth or open Air." Unhappily, Nicolson would say no more, and Kemp seems never to have written down anything about his possessions. As we have seen, it was not the first time that someone was willing to express a doubt about the antiquity of Dr. Woodward's shield. Nevertheless, eventually, a subsequent owner united it with its mate, and believing them both to be Roman, bequeathed them fondly to the British Museum where they remain together still.[58]

Dodwell was delighted to hear about the helmet, though he doubted that its workmanship could exceed that of the shield.[59] As for Dr. Woodward's object, he was ready at last to formulate a judgment. The shield was not a *clypeus votivus* and it had not been done under the Emperor Claudius.

Characteristically, Dodwell had come to an independent judgment. The shield was a *parma equestris* and it was later. Hearne, however, was not ready to renew the question. For the time he let the matter rest (though unconvinced) and waited for the great man to set down his thoughts in a formal Latin treatise.

XII.

De Parma Equestri

[1]

While Dodwell was considering the shield, Hearne was faced with a difficult decision. Livy was finished; what was he to do next? He was torn in his ambition between his two lifelong interests, the classics and English history. Dr. Hudson and his superiors at Oxford urged him to the one, his natural instincts increasingly to the other. He had been trained by Cherry and Dodwell as a classical philologist, and his friend Thomas Smith thought he had a "natural genius" for Roman antiquities.[1] But the Shottesbrooke group had also a strong concern with English history, and Hearne's father and uncle had instilled in him a love of local antiquities almost from the cradle. " 'Twas always my opinion," he wrote afterward, "that the Greek and Latin and English Historians should be joyned together. For what reason I took this Method at my first coming to Oxford. I had the example of Mr. Dodwell and Mr. Cherry, besides my own interest."[2] He continued, therefore, to oscillate between the two, taking up as his next two projects a collection of all the writings of Cicero and an edition of John Spelman's *Life of King Alfred.*

The Cicero was a long-range project which became bogged down and was never published.[3] It continued Hearne's education in classical philology, and for a time he hoped that Dodwell would do the chronology. The Spelman, however, was quickly completed and determined Hearne on a new direction. Not that it was an easy decision; there was strong resistance in the Augustan Age to the idea of doing what was after all medieval, rather than ancient, history. Oxford, Hearne complained in 1709, had "little or no Relish for our English History and Antiquities so that at present, I cannot cultivate these Studies as much as my Inclination prompts me to, by reason of the Dependence, the Meanness of my Circumstances makes me have upon them." Still he determined to persevere, perhaps with a new edition of Camden's *Britannia.*[4] In the end, he turned to a more valuable enterprise, the manuscript notes of the great sixteenth-century antiquary John Leland. These occupied him for the next few years, after which he began that long series of editions of the medieval Latin chronicles for which he became famous. Yet even so, in the prefaces and appendices to all these works, Spelman, Leland, and the rest, he managed to stuff the latest discoveries of the antiquity of Roman England. In 1715 he could write honestly to Dr. Charlett that "I have joyned the Greek and Roman and our own Writers together and endeavored to spend an equal share of my time on each."[5]

Roman England was of course the nexus of his two interests, and it was no accident that he turned to Leland and Camden at this time. These were the originators of English antiquarian enterprise in the sixteenth century, and they too had each attempted both Roman and medieval English antiquities. It was the connection that Hearne insisted upon, the relevance of the one to the other. He saw what his predecessors had glimpsed and what others in his time were beginning to understand more fully, that English history had to borrow the techniques of classical philology if it was ever to have the authority and influence of antiquity. But the resistance was oppressive. "Young Noblemen and Gentlemen Commoners," Hearne wrote to Cherry in 1710, "think reading and perusing of middle-aged Antiquities improper for a genteel Education and the direct way to render their Manners unpolish'd." His new task was to persuade them of their error. "Mr. Hearne," wrote Canon Stratford from Oxford, "goes on without interruption in publishing those useful pieces of monkish antiquity by which he is to immortalize his own name and reform us all."[6] In the end he did win some encouragement, especially from the three great doctor-patrons, Sloane, Mead, and Woodward.

Hearne's edition of Alfred brought him once again into connection with Dodwell. He had transcribed Spelman's manuscript about 1705, but had been inhibited from publishing it because of the opposition of Dr. Charlett, the Master of University College. It was all due to Hearne's notion that King Alfred was the founder of the entire university and not simply the college. Hearne, of course, was wrong, as indeed was Charlett, but the episode started up two quarrels that continued for years: one with Charlett, the other with William Smith, who saw Hearne's error and exposed it.[7] The incident did little credit to Hearne's critical reputation (which was always weakest where his passion was greatest) or to Charlett's generosity. But Hearne's Alfred is more interesting to us — as it was to Dodwell — for the appendix, which contained a Roman inscription that had been recently discovered at Bath.[8]

Next to coins, inscriptions were the most plentiful and helpful of all the Roman antiquities that were available in the British Isles. They were like the coins, so a reviewer wrote of William Fleetwood's *Inscriptionum antiquarum sylloge* in 1691, in that "They have preserved [to] us very considerable events," many otherwise lost, "and serve very much to clear up or to prove the most important points of History or Chronology."[9] John Leland had set the fashion of transcribing them in England in the sixteenth century, and Camden recorded all those that had been discovered to his time in the *Britannia*. When the new edition of that work appeared in 1695, the number of inscriptions was vastly augmented, the most copious additions being those of Edward Lhwyd for Wales. Fleetwood's work (London, 1691) was a convenient compilation, drawn from all over the ancient world and arranged by subject to illustrate the religion and customs of both the pagans and early Christians. It was, however, only one in a series of similar treatises, many even more massive and ambitious, that had been drawn up abroad and were well known to antiquaries in England — not excepting Dr. Woodward, who once again seems to have owned them all. (The most impressive of these collections was undoubtedly the new edition in 1707 of Gruter, by Dodwell's friend J. G. Graevius, four great volumes of Roman inscriptions from throughout the empire.)[10] From time to time in England new finds were made, while travelers going abroad were expected to carry off or to copy out all that they could discover and bring them home. Their historical and literary value made them prizes for collectors like the Earl of Pembroke or Dr. Woodward.[11] Undoubtedly the most famous inscription in England was one that had turned up abroad and had been carried back to join the Earl of Arundel's large collection. It was the celebrated Parian chronicle, edited with great learning earlier in the century by John Selden. In Hearne's time, with the other inscriptions from Arundel's collection, it had been brought to Oxford, where the young scholar often contemplated it and considered re-publishing it more exactly.[12]

Hearne's new inscription was discovered just outside Bath and sent to him by Edward Halley. Although it had nothing to do with Alfred's Saxon history, "Yet because it has never yet been publish'd and because it illustrates the History of the Island during the time it was possess'd by the Romans, I think it may properly be allowed a Place here." It was a practice that was to become customary for Hearne.[13] Like most inscriptions, this one needed deciphering, and Hearne displayed his epigraphic skill in reading it and his historical learning in explaining it. It was not easy to make out, and Hearne soon had competition from other scholars who furnished different versions: Roger Gale in 1709, William Musgrave and Dodwell in 1711, Hearne again in 1712 (this time from copies by Gale and Oddy), and later still John Horsley (1732). Even Edward Lhwyd had found it interesting.[14]

Of course, the Bath inscription was only one among thousands that had been already deciphered, transcribed, and published. And it was by no means extraordinary. That it should excite so much curiosity and encourage so much "critical learning" was simply another sign of the immense enthusiasm for ancient Rome common to the Augustans. This one had come from a large stone slab and reads, in a modern version, "Julius Vitalis, armourer of the Twentieth Legion, Valeria Victrix, of 9 years service, aged 29, a Belgic Tribesman, with funeral at the cost of the Guild of Armourers, lies here." [15] Hearne understood it pretty well, though his reading of the abbreviations was not quite correct. (Using the inscriptions in Gruter, he read VV as Valens Victrix, for example.) He amended Halley's copy correctly, turning *Fabriciesis* into *Fabricensis*, and discoursed learnedly on their function of making arms. Dr. Woodward's shield had well prepared him for the task. Among other topics Hearne considered the problem of "pointing" in ancient inscriptions, citing (as always) a large number of recent epigraphic tracts, as well as the ancient sources directly. Thus Hearne displayed the full range of his antiquarian skills, though to be sure it seems not to have occurred to him to look at the inscription itself (Bath, like London, was apparently too far away). [16] He concluded his appendix by noticing that Dodwell also intended some remarks on the inscription. Apparently Hearne had sent him a copy, and the great man had determined also to set down his thoughts upon it.

[2]

The Bath inscription reached Dodwell amidst his new concern for Dr. Woodward's shield. It was not really a diversion — for as we have seen, both works raised some common questions, especially about Roman arms and armorers. Dodwell had necessarily become acquainted with inscriptions and the epigraphic science early in his career. He was not much of a collector, but a new Roman discovery always interested him, particularly if it could throw light upon his chronological or textual concerns. When Hearne turned to him for help with Livy in 1706, he replied with a description of his collection. [17] "My stock of Inscriptions is but small," he wrote. "Such as they are they are at your command.... I have a couple which I have explained in a letter to Mr. Goetz, who is now, I think, a Leipsick Professor." These he had sent to his friend Graevius. "I have also a Chester Inscription.... This I found I think since [16]90. I have another of Wales, perhaps from Mr. Lloyd of Jesus" (this of course was Dr. Woodward's old rival Edward Lhywd). Dodwell also had some inscriptions from the Roman wall in Scotland which he thought might have some value for the Roman history, if not directly for Livy. Cherry, he was able to report, had also a "Transcript of the Fasti Capitol immediately from the Stones." Although it had been printed, it was rare, and Dodwell thought it might be worth collating, "if there be any considerable difference from the Transcripts of Sigonius, Onuphrius, and Pighius."

It was not surprising, then, that Dodwell should find the new Bath discovery interesting. He seems to have been approached about it independently by an antiquary from Exeter named William Musgrave who had long been interested in local Roman antiquities. Musgrave (who was, incidentally, another medical doctor and Fellow of the Royal Society) had the advantage of being on the spot. When he saw the printed versions of the inscription by Hearne and Gale, he wrote to Dr. Sloane: "The Editors being not without errors, as being (I suppose) printed from Erroneous Copies, I have endeavor'd to procure one more exact and caused it to be engrav'd."[18] Musgrave was right; Dodwell, like Hearne, was dependent on others for his reading of the inscription. He was willing to set out his remarks, but he wrote anxiously to Hearne and to Musgrave to make certain of the details. The prospect anyway was inviting. "This being a great Rarity," Francis Brokesby reported, "has occasioned him [Mr. Dodwell] to consider the Offices of the Empire and other Diversities according to the several ages of the Empire and other Antiquities which will very much gratify the Learned Readers."[19] As early as January 1709 Dodwell seems to have finished his little work, which he thought to combine with his earlier remarks to Goetz. He assured Hearne a little later that the appendix to Alfred "has neither interfered with nor superceded mine on it."[20]

Dodwell's explication of the Bath inscription was complete and accurate. (VV is now correctly the legion called Valeria Victrix, identified by Dodwell from Dio.) Characteristically, he attempted to date both the epitaph and the career of the legion, to describe again the *fabricenses* and to say something about the *Belgae* in that part of England. Thus he meant both to explain the inscription and to illuminate the Romano-British history. It occurred to Musgrave, reading Hearne and knowing of Dodwell's work, that they might all three combine their efforts and publish together.[21] Dodwell was willing, and he urged the scheme upon Hearne. But it was to no avail; Hearne was envious and despised Musgrave both for his views on the inscription and (more importantly) his views on politics. In the end Dodwell's tract was published together with Musgrave's, and Hearne was left to complain to his diary.[22] Among other experts, Musgrave turned also to Gisbert Cuper for assistance, but Cuper was satisfied with Dodwell's work. On one point, however, a disagreement appeared between Musgrave and Cuper which led to a learned correspondence, some remarks in the *Philosophical Transactions*, and mutual respect.[23] Eventually Musgrave proposed Cuper for the Royal Society. He himself went on into old age publishing monographs on the Roman antiquities of the West Country and submitting occasional papers to the *Transactions*.

[3]

Meanwhile gloom was settling slowly upon Shottesbrooke. Dodwell's family had grown large but his circumstances remained narrow. Now his sight was seriously failing. Still he persisted in his studies, though his complaints grew. "I never loved writing on any Subject but on a prospect of respite for my own full satisfaction. Now it is a penance." He was exchanging letters with Hearne — about the inscription, about Dr. Woodward's shield, especially about the vexed question of the *cassides* (i.e., helmets) worn by the Roman soldiers. "Writing is so uneasy to me that I am loath to tell you what I have observed in the matter."[24] Hearne offered some advice. He was very pleased to learn that Dodwell was working on the shield, "But I hope you will take all due care and bring it into as little compass as possible." It was very hard to get long works printed, "especially on subjects that are curious." To his diary he was more frank. "Mr. Dodwell is now writing in Latin a Discourse upon Dr. Woodward's shield. But I am afraid he will be much too large, tho' tis certain what he shall do will be very curious and learned."[25] To confine himself was a hard prescription for a man whose pen, despite every complaint, was unceasing. But even Hearne, the most loyal of his admirers, could be wearied by the old man. When Dodwell visited him at Oxford in 1709 he showed him his latest two tracts, "one upon Theophilus Antiochenus and another upon Dionysius Periegesis." "There are in both (as there are in all his other works) abundance of curious, learned and usefull Remarks, such as will satisfy those that come with minds prepared for Instruction, not for Ridicule and Banter." Hearne was both interested and alarmed that Dodwell had (among other things) argued that the Acts of Ignatius were later than usually supposed. But he had to admit in the end that the dissertations were "too tedious" and that " 'tis the great unhappiness of this Excellently learned Person that he will neither be advised nor expunge anything he has once written."[26] Alas, it was true.

But now Dodwell needed assistance, and he received it from Hearne, from Brokesby who was now living at Shottesbrooke, and from Dr. Woodward. From Hearne he received help about the *cassides* in the form of collations of various classical tests at Oxford. Dodwell was anxious to discover just when they had gone out of use in favor of *galeae*, a point of possible value in dating the shield. "The Bucculae which are joyned with the Cassides," he reported to Hearne, "are Shields, as has been well observed by the excellent Du Cange, not visors which were unknown to the Antients. And the shield of the Doctor's I take for a *Parma Equestris* or such a *Bucula.*" He doubted Hearne's notion that it belonged to Camillus Scribonianus who had lived under

Claudian. He preferred the reign of Nero, for it was only then that the Romans were supposed to have equaled the Greeks in the art of molding iron. Hearne was not persuaded with Dodwell "that Cassides did not come into use amongst the Romans 'till the time of Tacitus." He sent references from Caesar and Plautus, among others. But he did not know how late to date the *cassides* or indeed how to distinguish them from the *galeae*, and was quite willing to leave that to the master. He was, however, pleased to inform Dodwell that the *Acta Eruditorum* had just given "an extraordinary Character" to his Livy and had reprinted Woodward's shield along with Hearne's opinion about its antiquity. He hoped it might be of some use.[27]

Woodward was more helpful. Not only could he furnish Dodwell with a view of the shield itself, he could give him the benefit of his vast correspondence and collections. Dodwell read the letters of the foreign savants as well as everything else that he could find relevant to the image displayed on the shield. He saw that it was essential for dating it to compare it with similar monuments. Thus he wrote to Woodward urgently in the winter of 1710-11, desiring to see for himself prints of the various columns of Titus, Trajan, and Antoninus, to see which offered the closest parallel to the shield.[28] He suspected that it would probably be Titus. "His Triumph was in the beginning of his father's reign which was the next settled one to Nero in whose time this Art of representing to the Life first attained to its perfection among the Romans." What he wanted to do was to see how close the shield was "to the Habits and Arms of the Horse when actually engaged as they are in yours." All the literary and iconological evidence was relevant here. Like Hearne (perhaps because of Hearne), he was disturbed about the absence of bridles. He agreed that the Roman generals forebore them when they thought it prudent. But the artist of the shield was not likely to have depicted a battle that way, since his usual custom seemed to be to represent the actions of Camillus according to the usages of his own time. Anyway, that was why he wanted to compare the shield with the columns, though renewed reference to his failing eyes and inability to get back to London made the prospect appear unlikely. Another problem was caused by the death of the illustrious Spanheim, who had promised Dodwell his support. Dodwell was pleased, however, to have read Woodward's discourse on the Egyptians, and he concluded his letter with praise for the Doctor's "great genius" in such matters.

Among Woodward's other helps, one was to put Dodwell in touch with a scholar named Theophilus Downes. "The said Mr. Downes," Hearne recalled, "was of Balliol College as a member of which he took the Degree of MA July 10, 1679. He was a very worthy learned Man, and wrote and published several things. After his Ejectment from the Fellowship of Balliol College (for he was a Fellow there) for refusing the Oaths to the Prince and Princess of

Orange, he travell'd beyond the Sea, and took care for some Years of several young Gentlemen, with great Success, he being a wise Man."[29] It was one of his pupils apparently who brought him a print of Dr. Woodward's shield and asked his opinion. Downes did what any other responsible antiquary would do. "I compared it then with books and prints of Antiquities I had by me."[30] But his conclusions were altogether unexpected. Unlike the majority of his contemporaries, he decided that the shield was not very ancient, and he set down his thoughts upon it in a brief Latin treatise. Somehow this got into Woodward's hands, and this started up a correspondence which became especially useful to Mr. Dodwell.

[4]

The *Stricturae breves*, as Downes' little Latin manuscript was known, was an antiquarian exercise meant solely to assist a young student. Downes had never intended it to be seen by anyone else, least of all Dr. Woodward, whom he esteemed, nor Dodwell, who meant to write a full-scale antiquarian treatise on the subject. He certainly did not mean to press his doubts against the learned man's opinions. "I have no prejudice nor inclination to bias," he protested. "I have intimated my opinion and my doubts as they are, and shall easily renounce them. I will not be a Heretic for a trifle." Still his opinion was worth considering, and Dodwell actively solicited it; he was impressed that Downes had consulted many collections of antiquities that were not easily come by at Shottesbrooke.[31]

Downes had summarized his thoughts in fourteen brief points to show that the shield was not a genuine antiquity.[32] His first arguments were arrayed against the scene depicted on the relief, which appeared to him manifestly not to be the Capitol in Rome. The hills portrayed there, for example, did not suggest either the Tarpeian rock or the general topography of the Capitol. There were two obelisks on the shield, but none that he could recall mentioned by any ancient writer. The temples depicted there did not appear like classical ones, and Downes was astonished at the omission of the most famous of them, the monumental *Jovis Capitolini*, one of the wonders of the Roman world. Only a very ignorant artist, he was convinced, would have overlooked it. Absent too were the fortifications around the Capitol described by Tacitus and mentioned by Plutarch, as also the gates and steps leading up to it. Even more inadequate, were the figures. Most of Downes' remaining points were intended to show that the arms and equipment of both the Gauls and the Romans were inappropriate. First of all, he thought that the enemy armies should have been differentiated and portrayed as Livy had described them. In detail, he raised objections to the shields and swords, boots and saddles, none of which resembled the literary descriptions of antiquity. Thus the swords on the monument were short, single-edged, and shaped like modern Turkish weapons, whereas the Romans had used double-edged weapons and the Gauls long and straight ones. Not even the sword cast dramatically onto the scales appeared authentic. Downes was vexed also by the horses without bridles and the swords without either sheaths or belts. Like Sperling, he was unhappy with the *vexilla*. And finally, he thought the golden coins lying in the scale and in heaps upon the ground were also anachronistic. Minted coins were impossible if Pliny was right that golden coins were first struck at Rome AUC 647. At the time of the Gallic invasion, gold was valued by weight, not by numbers — the ransom, according to Polybius, had been a thousand *weight*. Indeed, take all these arguments together, Downes was confident that the shield had to be made at a very much later time by a sculptor who was very ignorant of the ancient world.

In the spring of 1711, upon Dodwell's appeal for help, Downes renewed his objections to the shield. He took some further pains, apparently, looking over his extensive library and consulting with Dr. Woodward — even looking at the shield itself. He did not find the view very helpful, however. "To judg of it by the sculpture is beyond my Skill, Sculptors and painters and learned men of the greatest knowledg and experience are often mistaken in judging of Antiquityes and are often redressed."[33] While in this case he seemed to think that even "men of indifferent skill" might judge the shield by its style, he himself was unwilling. Instead, he offered a string of notes to confirm his original impression. Thus, to his objections to the Roman topography on the shield, he added now his doubt that there had ever been an amphitheater on the Capitol. (Among other things, he had read Hearne's note in Livy identifying one.) To his criticism of the armaments, he added some further remarks on the standards, relying upon Lipsius and the medals described by Augustinus (a sixteenth-century Spanish antiquary). "I have found square banners," he continued (those on the shield were forked), "from Augustus to Valentinus, many in coins, many in bas-reliefs, none like yours in the books I have, nor have I seen any that resemble them." In Vaillant's work on coins, he found Nero with a square standard; on the Antonine pillar, ten others; so also on the medals of Balla and Vespasian and on the Trajan column. It was true, however, that Woodward had shown him some prints engraved at Rome with standards like those on the shield. But were they accurate, and where were the originals? It was certain that prints of antiquities were often erroneous. (Here Downes summarized some of the mistakes in the published bas-reliefs of ancient Rome, including those even of Lipsius, who he thought had relied on bad drafts.) Once again he repeated his dissatisfaction with the Roman swords displayed on the shield, which were without parallel on any of the

"Arches, pillars, and bas reliefs." (Again, there were prints that showed them as on the shield, but they too were modern.) As for Dodwell's view that the shield belonged to the time of Nero, Downes was unconvinced. "I see nothing in the shield peculiar to that age." He concluded, however, by appealing to Dodwell not to attribute any of these opinions to him publicly, whatever use he might want to make of them.

Dodwell was impressed. Downes, after all, had the advantages of a splendid collection of books on the subject as well as a firsthand acquaintance with Rome. Clearly his objections had to be met. Earlier they had stirred Woodward to try to frame a reply — his notes are full of material rebutting Downes' objections. On May 12, 1711, Dodwell drafted a reply.[34] He began by regretting Downes' modesty. He was reluctant to conceal Downes' name, and even more reluctant to overlook the material so richly supplied by his letter. He was willing to agree upon one condition. Would Downes lend Dodwell his books? He promised to care for them faithfully. "A transient view at London," he wrote plaintively, "will hardly suffice me who have so weak eyes and cannot endure application of my mind for any long time at once." Indeed, he had begun to despair that he could make the journey at all. Especially did he want to see those places where Downes had mentioned the "Decursions" under Nero, as well as any other effigies of ancient horsemen that he might have. About the same time, he wrote urgently to his old friend Robert Nelson with a similar request.[35]

As far as Downes' arguments went, Dodwell did not agree. It was true that the scene on the shield was not the Capitol. But he did not despair of identifying the site. He thought that the theaters depicted on the shield would be helpful and would indeed give new ground for establishing the authenticity of the shield. That it was very old he had no doubt; it was "beyond the Skill of any late Author that lived within this thousand years and more." He was persuaded by a careful comparison of the particulars on the face of the shield with the ancient authors "and those few Copies of Monuments which I can meet with." He was very conscious of his limited resources and regretted that (unlike Downes) he had been unable to see the originals. But he was quite sure that the shield was "too ancient to be controlled by any effigial monument now extant." Nor was he going to be convinced by the "sayings or opinions of Learn'd men now living." Too many antiquaries gave their opinions too freely, relying on only a "transient View of Copyes of this Noble Originall." Dodwell at least had seen the shield. His proofs were too extensive to recapitulate in a letter but in the meantime he was grateful to Downes, and he closed his letter by appealing to him again for help. To Woodward he wrote at once reporting Downes' opinions but reaffirming his confidence in the shield.[36]

[5]

On June 7 Dodwell died. A week later Hearne wrote his obituary in his diary.[37] He thought Dodwell might have lived to a hundred had he taken care of his health. "He was of a small Stature of Body, but vigorous and Healthy, of a brisk, facetious Constitution, always chearfull even in the worst of times. He was humble and modest to a fault. His Learning was above the common Reach. . . . I take him to be the greatest scholar in Europe when he died, but what exceeds that his Piety and Sanctity was beyond compare. . . . Nothing could make him swerve from those good Principles of the Church of England . . . which occasion'd some People to call him an Obstinate Man; but Obstinacy in such Cases is always laudable. . . ." Dodwell apparently had not quite finished his dissertation on the shield; Brokesby thought that the manuscript might be published by Dr. Woodward. Though it was incomplete, it was vastly fuller than that of any other antiquary; no one, he was convinced, had taken a tenth of Dodwell's pains on it. "I doubt not, it will be welcome to Learned men, as discovering much of Roman Antiquities out of the Ordinary Road."[38]

It was natural for Brokesby to send the manuscript to Woodward, and for Woodward to offer to let Hearne see it. Both men were intensely curious. Unfortunately it was written out in that indecipherable miniscule that was Dodwell's peculiar script. Woodward could hardly make it out, and was typically too overwhelmed with other matters to do more than look it over.[39] Hearne on the other hand was only too eager to have a go at it; he even hinted to Woodward that he might be willing to edit it for publication. "Mr. Dodwell was one of the best friends I had and I was to have went but on Whitsuntide last on purpose to peruse the Discourse (for I never saw a word of it, tho he mentioned his Design to me several times) and to transact some other business with him. . . . I long to see what he has written upon this curious Subject and if you will be pleased to lend it me for a few days, I shall take it an extraordinary favour." If Woodward wanted to print the manuscript, Hearne hoped that he would use the University press and let him "give what assistance I can in the Publication by correcting the Press." Woodward was naturally delighted; the manuscript was incomplete and had not been corrected, "nor can any man do that better than yourself." Hearne could have it as long as he liked. By July Hearne had agreed to publish it.[40]

Indeed the two men were getting on splendidly. From the time that they had become acquainted over Hearne's Livy, their association had grown more cordial. In 1710 the Doctor could even offer a position to Hearne, who was still chafing under the restraints of Dr. Hudson. (Hearne declined.) Woodward became a subscriber for Hearne's Leland, got him the use of an important manuscript, and made his collections generally available.[41] Hearne responded with a flow of information and opinion and (eventually) an offer to publish Woodward's tract on the Roman walls.[42] The Dodwell project was thus sealed in mutual respect and common interest.

Only one incident temporarily ruffled feelings. In the midst of Hearne's other antiquarian activities, Ralph Thoresby sent him an account of some ancient instruments found near Bramham Moor in Yorkshire, and solicited Hearne's opinion. Hearne's reply was sent to the Royal Society and read by Hans Sloane.[43] The half-dozen brass instruments raised a familiar puzzle: What were they and who had used them? In some ways they were harder to identify than many of the other artifacts (or fossils) that were then coming to light. There was not much to compare them with, and without a conception of prehistory — without a Bronze Age — it was difficult to place them. Hearne and Thoresby surveyed the known English cultures, British, Roman, Saxon, and Danish, to find analogies. Dr. Woodward, who possessed quite a number of them in his museum, offered his help to both. Thoresby thought that they must be spearheads or walking staves of the ancient Britons. Hearne preferred to think them the chisels employed by Roman soldiers in cutting stones. He scoured the familiar collections of antiquities and coins: Augustinus and Gorlaeus, Gruter, Reinesius, Spon, Fabretti, and Graevius, ending inevitably with the Trajan Column, "one of the best Monuments we have by which to judge of the several Habits and Instruments made use of by them in their Military Exercises." The soldiers depicted there, he thought, were shown plainly using them.[44]

He was wrong, but that is not what disturbed Woodward. What bothered the Doctor was that Hearne had made passing reference to the shield in his discourse, insisting upon its authenticity but adding the phrase "notwithstanding the clamours without any proofs that have been made against it." Several of the Society, Woodward reported, were very "startled." Would Hearne, he urged, please direct Sloane "to blot the Passage out," since it was likely to be published? (He didn't mind if some youthful spirits wanted to quarrel with the shield, but surely it wasn't sensible to go ahead and publicize them.) Perhaps Sloane emphasized the "clamours"; what Hearne did not know was the bitterness between the Doctors that was then shaking the Royal Society. "I am sorry," he wrote at once to Woodward, "that my words in my Letter concerning the old Instruments found in Yorkshire should cause any of the Society *to startle.*" He wrote immediately (but apparently too late) asking Sloane "to expunge that Passage." He was shocked to learn of the "dissentions" in that august body.[45] Woodward's temper cooled and their friendship was resumed.

[6]

Hearne was not surprised to read Dodwell's tract. "Tis full of admirable Learning," he wrote to Cherry, "but too tedious."[45] There were several mistakes in it which he thought that Dodwell might have corrected had he lived. On the main point, he remained unpersuaded. Dodwell thought the shield a *parma equestris*; Hearne was convinced it was a *clypeus votivus*. "But Conjectures may be allow'd in things of this Nature where there is no Inscription to direct, and a greater Liberty of Fancy is allowable in such Cases than where we have plain History to guide us." He never doubted that the treatise should be printed as it was, if only out of reverence to the great man, although he did not think that any ordinary bookseller would be interested.

In short, the work required a subscription. Hearne was willing to edit it, he wrote to Woodward, provided he had assurance that he could sell enough to justify an impression. The trouble was Dodwell's prolixity. "I wish he had been less tedious in some Things. 'Tis what I took occasion to caution him about when I was inform'd of his undertaking. But his Learning was so diffusive that he could not confine himself."[47] No matter; Woodward furnished the guarantees and Hearne began to edit the text.[48] Woodward was willing to employ every help: the letters of his correspondents, the treatise of Abednego Seller (if it could be found), and his own corrections and remarks. "I have myself, a great Collection of Reflexions and Notes, founded partly on the antient Greek and Roman Writers, and partly on Monuments of Antiquity . . . but I have no prospect or Leisure ever to draw them into Form."[49] Even so, he sent to Hearne a long draft entitled "An Account and Description of an Antique Clypeus Votivus."

It is a good summary of Woodward's view of the shield in 1712, very like the similar Latin description he sent to the Abbé Bignon about this time. [50] In most respects it was simply a recapitulation of his original notions as outlined nine years before in his letter to Otto Sperling. The latter, he thought still, "one of the best Antiquaryes now living," and he was glad to report to Hearne that Sperling believed that it was a *clypeus votivus* and not a *parma*. (This was Woodward's opinion also.)[51] In this essay Woodward once again identified the details of the picture and glowingly described the shield. "It has all the most unquestionable Marks of Antiquity upon it; and yet 'tis very little defaced or broken, which makes it the more valuable and extraordinary." That it was of iron he thought probable due its "Strength and Durableness." It had once been gilded all over "when first finish'd or afterward, is uncertain," but most of it was worn off. "It appears to have had an Ansa, wherewith to wield it, on the Inside." He still thought the design beautiful and exact, the "Manner very bold and masterly and with Abundance of Life and Spirit. The Work is truely Good, a great deal of Art and

Skill appears in each Part of it; everything is free, easy, and natural, and the whole put together finely and with great Judgment."

How helpful all this was to Hearne does not appear. From time to time Woodward sent other suggestions. One oversight he attributed to Dodwell "towards the latter End, about the Capitol which you will see is plainly exhibited in the shield towards the top on the left Hand."[52] Downes' objection had apparently not registered. Dr. Woodward had a new suggestion also for its origin. He thought the shield might well be a British work, "made in Reproach to the Romans." It could even have been by a British artist; after a while they had developed a considerable art, though a Roman employed by the British would do as well. Hearne was not persuaded, and Woodward did not press the idea. The main thing was to get Dodwell into print, even if the great man had nodded once or twice. Woodward had no doubt that the treatise would be admired by the learned world, "and if it bear not its own charge I will take care you shall be no loser."[53]

The first sheets of Dodwell's treatise were ready by December 1712.[54] Woodward read them over and noticed some errors. Unfortunately, Hearne could not give the task his full attention. His great labor remained his edition of Leland, now coming forth volume by volume. It was not easy to put the rough notes of this sixteenth-century antiquary in order, especially when Hearne insisted on printing a perfectly faithful transcript. His standard of accuracy was far beyond anything that had ever been employed for the work of an English writer. "I am so religious in that Affair," Hearne claimed, "that I transcribe the very Faults." When Thoresby saw the edition, he compared it with a transcript and was pleased to discover that Hearne had "by your accuracy and diligence retrieved some words that could not be made out . . . from the autograph above fifty years agoe."[55] But Hearne also thought to add some comment of his own and a number of short pieces by other antiquaries: Gale (on the Roman roads), Woodward (on the urns), Bagford (on Roman London), etc. One piece of his own particularly preoccupied him; someone had discovered recently near Oxford still another Roman monument, a mosaic pavement with some figures on it. What with Hearne's official activities at the Bodleian, his edition of Leland with all its appendices, his correspondence and diary, and now this latest antiquarian puzzle, it is no wonder that work on Dodwell's manuscript lagged.

[7]

The pavement was discovered by a plow near Stonesfield, north of Oxford, on the 25th of January 1712. In the next six months Hearne walked over to visit it nine times.[56] At first the antiquary was uncertain, but he was swiftly persuaded that it was genuine Roman work. Indeed, "tesselated" pavements (so-called from the *tessurae* or colored tiles with which they were made) were not uncommon in England. They were naturally of exceptional interest to antiquaries. Whether plain or elaborately decorated, they furnished graphic evidence of the Roman habitation in Britain. The Stonesfield pavement was typically ambitious, about thirty feet long and twenty feet wide. In many ways it resembled Dr. Woodward's shield. It was immediate and vivid testimony about the ancient Romans, but it had first to be examined and classified and then subjected to criticism. What was its purpose and who had constructed it? When had it been made? What did the figures illustrated on it represent?

Hearne went immediately about the task, even to having a careful engraving made of the new discovery. He inspected it closely, talked to everyone who might be helpful, read voraciously, and formulated his opinion. His examination of the pavement was almost exactly analogous to his renewed consideration of Dr. Woodward's shield — and coincidental. There is not space to treat it fully here. Suffice it to say that Hearne arrived at an independent judgment; he declared the pavement to be an authentic Roman monument, the floor of a Roman general's tent, made late in the Empire and with a representation on it of the god Apollo Sagittarius. Most of Hearne's contemporaries, including Dr. Woodward, preferred to think that it portrayed Bacchus, and they were right; moreover, it was not a military monument. By March 1712 Hearne had finished his discourse, which was published soon afterward, both separately and in the *Itinerary*, volume 8. It provoked a bitter controversy in which I think we may say fairly that Hearne was worsted by a lesser man, though characteristically he never conceded.[57] Dr. Woodward, at any rate, was persuaded, and Ralph Thoresby was among those who were impressed.[58]

To be right or wrong is not perhaps as important as to have made the proper effort. "I desire that what I have observ'd in the Discourse should be looked upon as nothing more than uncertain Conjectures."[59] Hearne brought interest, even fame, to the new antiquity, however misguided his conclusions. "The Learned World," wrote Samuel Gale, upon receiving his copy, "is indebted to you for your Sedulous Preservation of so many antient Monuments which otherwise in a little Time must have utterly perished."[60] (The pavement itself was in fact eventually destroyed.) Hearne understood how to go about his job, and his essay was a praiseworthy effort to understand the meaning of an important Roman ruin and through it to gain a glimpse of Roman Britain. Though there were many, in Hearne's words, "ready to run down Antiquities," it was his task, with his friends, to show how pavements, coins, inscriptions, and shields could be made to illuminate the past, to flesh out the literary accounts, to restore the *lacunae* in the narratives, above all to build piece by piece a picture of the whole of ancient life and thought. Antiquities, as Hearne wrote elsewhere, were "the most uncorrupted Monuments of History." Where written documents were usually altered through the ages by generations of scribes and readers, the nonliterary remains were only changed "from the Moistness of the Soyl, the Badness of the Air . . . and from other accidents of that Kind, and they are therefore the best Authorities for correcting such Writings as have been corrupted." [61] That anyone might decry an interest in antiquities seemed to Hearne ludicrous. Yet after all, there had been some ready even to run down "the famous Shield printed in the Oxford Livy in which there are evident Tokens (in the opinion of Baron Spanheim, Dr. Thomas Smith, Mr. Dodwell, Cuperus, and several other very learned Men) of the Roman Art." One might as well deride the famous shield of Dr. Spon, or indeed "any other confessedly authentick Piece of Antiquity." [62] Hearne did not fail to include the Bath inscription in his review of recent and useful antiquarian discoveries.

[8]

Thus one way or another, from Livy to Dodwell's *De Parma*, Hearne kept alive his concern for Dr. Woodward's shield. Although he was busy with many other things, he never lost interest in the Doctor's prize. Nor did he swerve from his original opinion. It was, he repeated in his Leland, a *clypeus votivus*, one of the Roman *Bucculae*, "lodg'd in Temples, especially such as were consecrated to the honour of Juno Lacinia." Despite the "clamours" that had been raised against it, he was more than ever certain of its antiquity. [63] Relying upon Bagford, he was able to trace it from the Royal Shield Gallery, which "I am informed was inclosed in the days of King Henry the VIIIth who being a Prince ambitious of Glory . . . spared no costs in furnishing the Gallery with Shields, Helmets, etc." But Dr. Woodward's shield did not date from Henry's time, as some thought, "for tho' divers of them were made at that time, and there is one now being printed by Hans Holbein which has the Story of the taking of Bulloigne, yet others were more antient and 'tis probable this was brought to sight then." [64] Nor was he impressed with Downes' arguments. Though he was a "very ingenious man and a good divine," he was "not a very good Antiquary." The *Stricturae* were very weak. "The age of the Shield hath been fixed, both by Mr. Dodwell and myself. Mr. Dodwell (after I had given my opinion, little different from his) assigned it to the time of Nero." It was not enough to point to its anachronisms. "No judicious man ever believed it to be of the age of the Action represented." On the other hand, everything pointed to the age they had assigned.

All the *Stricturae* demonstrated was that the shield had not been made in the time of Camillus, "what no skilful man ever contended it was." Hearne thought that even Downes himself had been satisfied with this.[65]

Now, as Hearne transcribed Dodwell's work for the press, he resumed his researches. In the Bodleian he discovered a silver coin of Augustus with a *clypeus votivus* on the reverse. The ancient medals were invariably helpful to the student of antiquity. The Roman coins, Hearne noticed, "were always adapted to the particular Action. That makes Coyns of such excellent use in explaining and settling old Histories."[66] Dr. Woodward had promised some medals for use in explicating the Stonesfield pavement.[67] Meanwhile Hearne pored over the ancient histories, accumulating references and allusions to *clypei* in order to collate them with the archaeological evidence. Throughout, he confided to his diary his doubts about Dodwell's notion that it was a *parma*.[68]

Of course Dr. Woodward offered his advice, some of it useful, some otherwise. There were moments of suspicion, not unexpected with such irritable, indeed irascible, personalities. Apparently Hearne feared that Woodward might reprint the Dodwell work after Hearne's edition appeared. It was a natural concern, since Hearne insisted on printing only two hundred and forty copies, a restriction that he had already decided upon for his Leland and that was to become habitual in all his later publications.[69] Woodward was annoyed. The whole idea behind the work was to advertise the shield. Hearne threatened to withdraw; Woodward asked to have the manuscript back. The moment passed; trust was restored, although uneasily. Woodward began to correct the sheets as they appeared. He thought Hearne should put some compliment to Cuper in the preface, especially since Dodwell had taken issue with him, and since he was Woodward's friend and correspondent. He sent Hearne the exact weight and dimensions of the shield as well as a new print of it that had recently appeared in an edition of Baldwin de Calceo. Hearne thought it "miserably done" and Woodward agreed.[70] At least it was another indication of the popularity of the doctor's monument.

In March of 1713 Hearne sent a copy of the book to Woodward; he promised publication in a fortnight.[71] Unfortunately, Woodward's worst fears were realized. Hearne proved indiscreet, against the advice of the Doctor. The book was officially suppressed after only forty-three copies had been distributed to the subscribers. The trouble was with Hearne's introduction; in praising Dodwell he had introduced the ever-sensitive religious question. He had praised his hero as a conscientious non-juror and he had slighted one of Dodwell's opponents (*vir quidam mediocriter doctus*), insinuating that he had changed his politics to procure preferment. Woodward was bitterly disappointed. He had written long ago advising him "not to mingle any thing of Mr. Dodwell's Theological Studies with a Work of that Nature. I did it to prevent . . . a Censure and I wish that Caution had been observed." [72] Now, alas, it was too late.

Hearne had his own view of the matter. Indeed, it obsessed him, and besides writing explanations to many subscribers and friends he recorded his version at length in his diary, possibly intending to publish it. According to Hearne, the troubles began with a visit to the Bodleian by one Mr. Mollineux, "an Irish Gentleman" who had been recommended by the great coin collector Sir Andrew Fountaine. [73] As Dr. Hudson was occupied, it was Hearne's task to show the visitor around. It was a job he thoroughly enjoyed. Unfortunately no one told Hearne that his visitor "was a Person of but ill Principles" — nor, indeed, even his name.

For a time they looked over the "curiosities." The visitor "talk'd much both about MSS and Coyns, and, by his Discourse, one would have thought that he had spent much of his time in these Studies." They argued about one point which led Hearne to see that Mr. Mollineux was "a Man of some Confidence," though he was willing to allow his visitor the benefit of the doubt. At last they came to the Cabinet of Coynes in the Gallery. Mr. Mollineux asked to see the scarce coins of the ancients. Hearne was agreeable and produced "two Drawers in which are several very extraordinary ones, and which an Antiquary might immediately judge to be good and of very great use in explaining and illustrating History." These, of course, were the coins which Hearne himself had cataloged and arranged some years before. Mr. Mollineux immediately gave himself away, unable to say which were really scarce or to see "what considerable Part of History might be explained from any one of them." Instead, he asked if they had any spurious coins among them, a brass Otho, for example, or a Pertinax. Hearne told him that they had some, but had removed the counterfeits and kept them separate from the genuine ones. Hearne was offended; "me thinks Mr. Mollineux should not have inquired so strictly after those, and neglected to look upon other coynes I shew'd him, from which so much Light may be drawn for ancient History." Obviously Mollineux was a person "tho of good ready Discourse, yet of small Judgment in Matters of Antiquity."

If that had been all, it would have been no great matter. The interview, though uneasy, was undoubtedly typical enough and nicely illustrates the divergent interests of the dilettante collector and the earnest antiquary. And though Hearne was disturbed by Mollineux's "confidence" and his ignorance of antiquities, no harm had been done. Hearne had not yet discovered his guest to be "a Man of Republican, ill Principles, and of a Malignant Temper." It was only when they reached the Anatomy School that that truth became gradually apparent. There they spied a picture on the wall which some said was of Benjamin Hoadly. In his usual way Hearne spoke out forthrightly, mentioning no name but describing it as the picture of "one of the greatest Presbyterian, Republican, Antimonarchical, Whiggish, Fanatical Preachers living in England." Mollineux said nothing; but another mild encounter ensued over a tobacco stopper tipped with silver that belonged to Hearne, who believed it had come from an acorn planted by King Charles II. It was the Irishman's turn now to revenge himself by calling it "a Bawble." With both men inwardly seething, they passed on to a picture of a beautiful young man over whose head were the words *Eikon Basilike*, and beneath *Quid quaeritis ultra.* Hearne did not identify it but merely praised its beauty, and even Mollineux admitted that it was an exquisite performance. Hearne continued to avoid identifying it, and one of the party laughed nervously, telling the visitor "they are all Rebells, Mr. Mollineux, they are all Rebells in this Place." At last the group broke up and the visitors departed.

The picture, of course, was of the Pretender, and when Mollineux learned that, he was incensed. At once he spread the story all over town. Dr. Charlett was delighted to gain new evidence against his young adversary, and soon there was a growing conviction that Hearne had disgraced the university by showing off such a picture. The Vice-Chancellor himself came to visit Hearne and look at the gallery. Hearne defended himself. He had not identified the pictures; anyway there were lots of other "Pretenders and usurpers" on the walls, and no offense usually given. The Vice-Chancellor was not satisfied; a meeting of the library curators was called.

It was in this context that Dodwell's *de Parma* was printed at the university press. According to Hearne, the Vice-Chancellor saw it as the means to prosecute him, especially since Mollineux' case was not very strong. Hearne was never able to concede that the conscience or opinions of his foes might be as honest or at least as sensitive as his own. On the very day of the Vice-Chancellor's visit, he had sent him a copy of the new work "bound in Turkey leather." Though others found it perfectly acceptable — including, as Hearne noted, some reputable jurors — the Vice-Chancellor and some others "could not brooke what I had said in Commendation of Mr. Dodwell and the other Non-Swearers. They thought I had reflected by that means upon them and that I had published the Faults of the Persons that deprived Mr. Dodwell of his History Professorship to the World, for which I ought to be severely taken to task." [74]

Hearne's relish for this kind of contest was immense. He saw himself as the victim of a conspiracy hatched by his enemies out of personal and political malignancy. Perhaps he was right. But there is no mistaking his enjoyment over the role of aggrieved and innocent victim. It was for him and for the world a measure of the integrity of his conscience that he suffer for his opinions. Hearne was a Christian martyr though the arena on which he played out his role was a little less dramatic than his models. He survived the Dodwell incident, but almost at once provoked another. In the end he paid the price for his virtue (which was certainly genuine) by being forced to retire from his beloved library and to live in cloistered seclusion in St. Edmund Hall for the rest of his days.[75]

The Dodwell incident resulted in a hearing before the heads of all the Oxford Colleges, at which the charges were repeated. The President of Corpus Christi put the issue succinctly. "We shall be looked upon all as Jacobites said he, and very severe Censures will fall upon us, if we permitt Books of this kind to pass from the Theater Press unregarded."[76] Hearne defended himself; he had said nothing but what was demonstrably true. The heads of the colleges agreed that Hearne must retract his opinions or have the book suppressed. When Hearne categorically refused, an order was issued declaring that the book had been printed without license with "severall Offensive Expressions" and that Hearne was forbidden to sell or give away any further copies. When the order was read to the irascible antiquary, Hearne was nothing daunted. He immediately spotted an error "*operam* being written for *operum*" and was thus able to discredit the Registrar "as a very great Blockhead," as also indirectly the heads of the colleges and the Vice-Chancellor. Even so, he was able to induce the Vice-Chancellor to give him permission to dispose of the book without the offensive preface. In his own heart he was certain that he had been the victim of general injustice and that, among other things, the heads of the colleges did not have the legal right to suppress the book. And so he simply disregarded the order (though he was careful not to advertise the fact.)[77] For some time he sold copies without the preface; "then at length, upon consulting with some Friends whom I could trust, I got the Preface transmitted to those that had receiv'd Copies without it, and by the help of three or four faithfull Persons I dispersed other Copies with the intire Preface."

For Hearne it was a victory, though his enemies, Charlett, Hudson, and the rest, must have seen it differently. For Woodward it was another frustration. How hard he had labored through the years to advertise his beloved shield! How many had been the attempts to publicize it, thwarted by death or accident or by the press of other business! Now to have seen the immense labor and prestige of Mr. Dodwell greeted by controversy and derision, sidetracked by politics and religion, and finally suppressed! If there was any consolation, it was in those few copies that survived and circulated. There, at least, one could see the shield finally treated with appropriate seriousness and with all the learned resources of the age. If it had only not been accompanied by the mocking of the ignorant, stimulated by the imprudent conscience of Thomas Hearne.

[9]

Apart from the noxious matter, Hearne managed to introduce Dodwell's text with a minimum of fuss. There was only a brief preface (with a nod to Cuper), Hearne's earlier note to Livy on the shield (with the argument about the bridles), and Burghers' fine print. Hearne did add one further note of his own, on the *vexilla*. For those who were worried about the misshapen banners on the shield, he now offered the testimony of Vegetius and other ancients – as well as a coin of Antoninus Pius in the Bodleian – to show that streamers had been employed by the Romans from the earliest times. [78] Everything else was Dodwell's.

The argument consisted of forty-two substantial sections, one hundred and twelve pages, the last interrupted in mid-course by Dodwell's death. Much of it was recapitulation; Dodwell built cheerfully on the labors of his predecessors. Thus he began by crediting Hearne for the promising suggestion that the shield had once belonged to Henry VIII's gallery. Much ingenuity was naturally expended on establishing the chronology of the shield. Like the other antiquaries, Dodwell found the habit and costume of the soldiers on the shield most helpful here. As everyone had seen, they were not contemporary with the event but rather from a later time, when the arts and the Empire were in their most flourishing state. The shield could not therefore well antedate Augustus; on the other hand, the absence of either crosses or the ensigns of the pagan princes who persecuted the Christians showed that it could not be much later, either.

For more precision, Dodwell turned to the workmanship on the shield. Here Pliny was especially helpful. That useful writer had made it clear that the art of working metal had not been perfected until the reign of Nero. Nero's reign began about 54 BC; Pliny wrote about 77 AD. During that brief time Pliny remarked that the art had already declined. What more likely than that the artist who had made the shield lived in those few years? Indeed, Silius Italicus had described a shield very like this one showing Camillus, and he had been consul in the last year of Nero's reign.

Dodwell next considered the iron composition of the shield, quoting freely from Aristotle, Pausanias, and Pliny on its durability. He did not find it unlikely that the shield could have come down through the centuries in such excellent condition. At length, Dodwell was ready to consider the figures and to establish the iconography of the shield. Again, most of the work had been done. But there were many details and a few puzzles left to consider. Thus Dodwell enlarged at great length on the figure of Camillus, whom he found appropriately aged, bearded, and properly accoutred. The fact that he carried a single spear seemed perfectly consistent to Dodwell. In such a way had the other Roman dictators ridden, on like occasions. (Camillus' business was after all giving orders, not fighting.) Everything he found agreeable to Livy's account, but startling in its lifelike character.

Among the puzzles, Dodwell admitted to some uneasiness about the missing scabbards and belts for the soldiers. Elsewhere he identified the two figures riding behind Camillus as heralds. He showed how the bare forearms of the soldiers was customary; how the Gauls wore the same mail as the Romans (never below elbows or knees); that the boots were just as Tacitus had described of Caligula; and in general, how the costume of the soldiers accorded with the customs of imperial Rome. Then he turned to the difficult question of the scene. Though some had argued for the Capitol, Dodwell dissented. The theaters and circus suggested another setting, and the flames still flickering also argued against it. (The Capitol had been spared.) Unfortunately, as Dodwell was warming to his task, he was interrupted by death. How much more he would have had to say is unclear. As he tried to measure the distance between the Capitol and the Tiber, the pen fell from his hands and his lifework was done. Dr. Woodward's shield had received its final (if still incomplete) apotheosis.

XIII.

The Shield of Martinus Scriblerus

[1]

D r. Woodward's trials were not over. By an odd chance, at about the very time that Dodwell's *De Parma* was belatedly appearing, the leading wits of Augustan England were combining briefly into a new club. Whether it was the inspiration of Pope or Swift or Dr. Arbuthnot is a little obscure but by the spring of 1714 the three men had joined together with their friends John Gay and Thomas Parnell and their patron Robert Harley, Lord Oxford, to become the Scriblerians.[1] The association was brief, for the members had scattered long before the end of the year, but it was promising. In the end, some of the most ambitious achievements of Augustan satire — *Gulliver's Travels* and the *Dunciad*, for example — seem to have been born out of that association; while the one collective work of that spring, the *Memoirs of Martinus Scriblerus*, set down the satirical principles which continued to inspire them. Not least among their first targets was Dr. Woodward and his now notorious shield.

It could hardly be helped. The company was formed hard upon the heels of a project announced in the *Spectator*, probably by Pope, suggesting a new burlesque journal to be entitled *A History of the Works of the Unlearned*. It was to furnish satire each month on the follies of erudition and criticism by means of a series of ironic reviews like those supplied in more authentic journals such as the *Bibliothèque Choisie* or the *Acta Eruditorum*, or indeed the *bona fide*, *Works of the Learned*. Pope turned naturally to his friend Swift, who had already become famous for his satirical shafts at the new learning, and it is probably out of this suggestion and their new association that the Scriblerians were organized. Inevitably, Swift brought along his friend Dr. Arbuthnot.[2]

By all accounts, the Scotsman was a very attractive man. Dr. Arbuthnot's career had prospered handsomely since his first encounter with Dr. Woodward and his geological opinions. It must have been very exasperating to our

238

hero to find his rival grown famous as a physician, ingratiating himself with the court, and entering as a colleague first the Royal Society (1704) and then the College of Physicians (1710). In the next years, until the death of Queen Anne (1714), Arbuthnot was an influential man, prominent in the Tory interest, and on intimate terms with the great. Was this why Woodward refused to challenge him publicly? Or was it the disarming amiability and insouciance of the man? Arbuthnot was, wrote a contemporary, "a man of honour, whose mind seemed to be always pregnant with comic ideas and turned chiefly, if not only, to that which is ridiculous." To Arbuthnot every character or event had its comical aspect; he had a natural gift for satire, the rival at least of any of his contemporaries.[3]

It was Dr. Johnson who placed him first among the wits, "the most universal genius and excellent physician, a man of deep learning and a man of much humour."[4] Yet it was then and remains still impossible to estimate his wit exactly, for on the whole Dr. Arbuthnot chose to write anonymously and he took little proprietary interest in the comical ideas that he seems to have lavished upon his friends, and which they cheerfully appropriated. We shall probably never know just how much he contributed to Gay's *Three Hours after Marriage*, or whether he wrote the *Life and Adventures of Don Bilioso de l'Estomac* or Dr. Technicum's *Account of the Sickness and Death of Dr. Woodward.*[5] His carelessness is recorded in a well-substantiated anecdote. "No adventure," we are told, "of any Consequence ever occurred in which the Doctor did not write a pleasant Essay in a great folio Paper-Book which used to lie in his Parlour. Of these, however, he was so negligent, that while he was writing them at one End he suffered his Children to tear them out at the other for their Paper Kites."[6] In the same way, at the coffeehouse or in his rooms, his cheerful fantasies were scattered about to be employed by any who chose to use them.

Whatever the origins of the new club, Dr. Arbuthnot was probably the seminal figure among the Scriblerians. They chose most often to meet in his rooms at St. James' Palace. Their purpose, as it developed that spring, was "to write a satire in conjunction, on the abuses of learning; and to make it better received, they proposed to do it in the manner of Cervantes . . . under the history of some feigned adventures." They had observed that "those abuses still kept their ground against all the ablest and gravest Authors could say to discredit them."[7] (Since *A Tale of a Tub* had been insufficient to stifle Dr. Bentley and his friends, perhaps the *Memoirs of Scriblerus* could do better.) The design of the *Memoirs*, remembered Pope, "was to have ridiculed all the false taste in learning, under the character of a man of capacity enough [i.e., Martinus Scriblerus] that had dipped in every art and science, but injudiciously in each." Of course there were candidates enough for that role, but Pope recalled that "the adventure of the shield," anyway, had been specifically designed against Dr. Woodward and the antiquaries.[8]

The Scriblerian theme was commonplace enough. We have seen the virtuoso in either of his capacities, as natural philosopher or antiquary, become the object of fun. The quarrel between the Ancients and the Moderns had laid the issue bare: What was the use of learning to a gentleman? For the man of the world, the object was a smattering of polite learning, a classical literary education that would furnish the necessary polish to allow him to shine in the fashionable circles of the great world. It was all a matter of *style.*[9] To the virtuoso, to the modern savant, learning was much more an end in itself, a philosophical undertaking that might unlock the mysteries of the natural and human universe. Moreover, it was a passion and − like all such, to an age of common sense − more than a trifle ridiculous. Yet the Ancients and the Moderns, the wits and the virtuosi, were all locked into the same small snug world of Augustan London, too closely to be entirely distinguished, too cramped together completely to ignore each other. However they might decry it, each of the wits had his own flirtation with learning − as each of the learned continued to advocate polite letters. However they might try to laugh only at its excesses, the wits found the allure of the new learning too hard completely to resist.

[2]

Each of the Scriblerians could make a special contribution to the satire. According to Warburton, "Dr. Arbuthnot was skilled in every thing which related to *Science*; Mr. Pope was a master of the *fine arts*; and Dr. Swift excelled in the *knowledge of the world.* Wit they all had in equal measure."[10] His division of labor was, however, a little too pat. Swift, for example was hardly alone in his worldly pretensions among the wits − and if he detested natural science and philology more resolutely than either of his friends, even so, he kept beside him until his death the two dozen folio volumes of Graevius and Gronovius.[11] As we shall see, Pope was even more obviously enamored of the new classical learning, whatever suspicion he may have had of its excesses. Among other things, he once told Spence that he had written a Latin treatise on the buildings of ancient Rome.[12] And Dr. Arbuthnot, much more than his friends, could pretend to the whole range of virtuoso learning: natural philosophy *and* the new antiquarian science. The Scriblerians were thus not ignorant of the learning they attacked; each of them dabbled in it more or less, and at times they even pretended to contribute to it. But they found ridiculous what they took to be its excesses; it was the abuse of learning more than learning itself that they meant to satirize.

In this, perhaps, Dr. Arbuthnot's career is most instructive. He first established credentials as a natural philosopher with his criticism of Dr. Woodward. A few years later he published a tract on *The Usefulness of Mathematical Learning*. After he had become a celebrated physician, he launched his career as satirist with a series of political pamphlets attributed to John Bull (*Law is a Bottomless Pit*, etc.), each of them enormously popular. By 1713 the philosopher George Berkeley could describe him as "the Queen's physician, in great esteem with the whole Court. Nor is he less valuable for his learning, being a great philosopher, and reckoned among the first mathematicians of the age." His reputation, he noticed, included a "character of very uncommon virtue and probity." There seem to have been few reservations about either his learning or his character, either in his own time or ours.[13]

Yet his reputation for philosophy and learning rested on rather flimsy foundations. He was in no sense a contributor to the new scientific knowledge; indeed, he did not pretend to be. His criticism of Dr. Woodward was not based on any acquaintance with the materials of the Doctor's speculations, the observations and experiments which Woodward had boasted were the foundation of his theories. He had simply, though ably, dissected Woodward's argument to discover its flaws on the mathematical side. But his contributions to his own discipline were not more profound. He followed the mathematical exercises of his contemporaries with ease and sympathy, but he seems to have made no new discoveries himself. A very early essay on games was borrowed largely from Huyghens (Dr. Arbuthnot was an inveterate card-player).[14] A later contribution to the Royal Society *Transactions*, entitled "An Argument for Divine Providence, taken from the constant regularity of both sexes," was, for all its ingenuity in "political arithmetic," similarly derivative. When at last he elected to write directly about his subject, he was content to limit himself to extolling its usefulness, in a pleasant essay published in 1701.[15] It contains hardly any mathematics at all. He tried to show the great variety of purposes for which mathematics was valuable, not least as a tool for natural philosophy. In a passage that could well have alluded to Dr. Woodward, he drew the conclusion "that a natural Philosopher without Mathematics is a very odd sort of a Person." Undoubtedly Dr. Woodward was vulnerable here, especially where he had tried to explain the dissolution of the earth by means of "specific gravity." "I must needs say," Dr. Arbuthnot continued, "I have the last Contempt for those Gentlemen, that pretend to explain how the Earth was framed and yet can hardly measure an Acre of Ground upon the Surface of it."[16]

More startling perhaps was his allusion to mathematics and the new historical learning. Arbuthnot took for granted the value of history to the gentleman; he argued too that chronology and geography were indispensable preparation, "a relation of matter of fact being a very lifeless insipid thing without the circumstances of time and place." Chronology, particularly, was necessary "to unravel the confusion of Historians." As Arbuthnot saw, each of these disciplines had its technical side and required some mathematics. "I remember Mr. Halley has determin'd the day and hour of Julius Caesar's Landing in Britain from the circumstances of his relation."[17] But there was an even more startling example. "Everybody knows how great use our incomparable Historian, Mr. Dodwell, has made of the Calculated times of Eclipses, for settling the times of great Events, which before were as to this essential circumstance almost fabulous." It was odd that a satirist of learning should be impressed by this, the most learned and not the least ridiculous of the virtuosi.

It was odder still, perhaps, that Dr. Arbuthnot should elect to contribute to antiquarian lore directly himself. Nevertheless, in 1705 there appeared a brief work under his name entitled *Tables of Ancient Coins Weights and Measures Reduc'd to the English Standard.* For two decades he continued to develop his interest in the subject, until in 1727 he was able to republish it, greatly enlarged by "several Dissertations." The *Tables* were now no longer merely enumerated; they were framed by an elaborate survey of the prices and customs of the classical world. Here was a wit who took the antiquarian sciences with obvious seriousness. "I have always been of opinion," he wrote, "that young Gentlemen of an Age to consider more than the mere Words of an ancient Author, ought not only to take along with them the Chronology, Geography, and a clear Idea of the Antiquities form'd by ocular Inspection on Models and Figures; but likewise to exercise their Arithmetick in reducing the sums of Money, Weights and Measures mention'd in the Author, to those of their own Country."[18] As we have seen, the translation of the ancient measures to modern equivalents was *par excellence* a "modern" scholarly exercise, the nexus of several philological and antiquarian concerns. "I believe," wrote Dr. Arbuthnot about the classical authors, "it will be readily own'd that the knowledge of the Value of Money, Weights and Measures of the Ancients, is necessary to an understanding of their Writings." The poets, orators, and historians of the ancient world were otherwise quite unintelligible.

There was surely a paradox here. The Ancients, like Swift and Pope, insisted upon the imitation of classical authors. Such imitation depended, however, as even the most reluctant of them saw, on a knowledge of the meaning of the ancient writers. *Some* philology therefore was requisite, however mean the discipline. The unresolved problem of the Ancients was how to achieve this knowledge without losing the polish of the gentleman, how to employ philology without becoming a philologist. Presumably this was meant to be Dr. Arbuthnot's objective. "I believe I need not advertise the Reader," he apologized, "that in a Work of this Nature it is impossible to avoid Puerilities, Trifles, and joyning things naturally incoherent, it having that in common with Dictionaries and Books of Antiquities." In any case, Dr. Arbuthnot would not admit to being an expert; *that* would be to forsake the ideal of the gentleman. "I do not value my self on any Skill either in Languages History or Antiquity."[19] It was an honest, if somewhat disconcerting, statement.

If the antiquaries like Dr. Woodward had been a little more disposed to the wit of their critics, they might thus easily have turned the tables on them. Once admit the necessity for some learning, could there really be such a thing as too much? Unfortunately, Woodward was dead when Dr. Arbuthnot's enlarged work appeared. It is doubtful, however, whether he would have spotted the anomaly in his rival's position. It was the case of the *Transactioneer* again. However much Dr. Woodward might despise Dr. Sloane, or in this case Dr. Arbuthnot, he could hardly endorse a criticism of his enemy which impugned the very kind of thing to which he himself was committed. What is more curious is that Dr. Arbuthnot failed to see the same comic aspect to his own work that the Scriblerians had discovered in Woodward or Bentley. The wits, for example, had been able simply to quote from the *Philosophical Transactions* for satiric effect. (Just enumerating the titles of some of the papers could be hilarious.) Yet surely Dr. Arbuthnot's *Weights and Measures* afforded the same prospect. "They had a great variety of Cakes," he writes characteristically of the Romans, and in apparent seriousness, "Each of which may make a good Subject of a Dissertation for an Antiquary; as also whether they had Pyes."[20]

The paradox, of course, lay in the fact that it was only through a concern for such *minutiae* that a true understanding of the ancients could be achieved. Neither scholarship nor natural philosophy could afford to overlook precision and detail. Yet the development of learning and knowledge had extended too far for this to be lightly won. It was scholarly learning and exactitude that had, for example, undermined the very existence of the letters of Phalaris, but it had needed a Bentley to prove it. Dr. Arbuthnot and his friends never quite understood that, as indeed they never gave up on the letters.[21] But at times they came close. "The Curious have thought the most minute affairs of Rome worth their notice," remarked Arbuthnot, "and surely consideration of their wealth and experiences is at least of as great importance as Grammatical Criticisms, Rites, Ceremonies, Figures of Vases, Instruments, various Shapes of habits, etc., upon which the Learned have perhaps taken too much pains. . . ."[22]

"Too much pains" — surely no one could accuse Dr. Arbuthnot of that!

There is learning in his treatise, but it is not his; he takes what he needs from others, haphazardly and casually. Moreover, he shares with his fellow Scriblerians the conviction that the practical life is superior to the scholarly, even in scholarly questions. Thus when he comes to the price of Roman clothes, he complains that "the Antiquaries being but indifferent Taylors, they wrangle prodigiously about the cutting out the Toga. I am of opinion a Mantua-maker could decide these disputes better than the most learned of them."[23] Just so had Temple argued that a statesman like himself could better determine the authenticity of a prince's letter than a mere scholar like Bentley.

Thus Dr. Arbuthnot's treatise is disappointing as a work of scholarship. (Needless to say, that did not affect its popularity.)[24] It is disappointing in everything, except perhaps its mathematics. But to determine the value of the ancient coins, much more was needed than arithmetic. It was necessary to interpret many passages in the classical authors; to examine closely the classical objects themselves; to weigh and measure the coins; even to analyze their composition. And it was desirable to review the elaborate scholarly literature that had developed over two centuries. In short, it meant employing the whole apparatus of philological and antiquarian learning. But in none of these respects was Dr. Arbuthnot very ambitious, and it was alleged by a later writer that one could hardly tell from his treatise whether he had ever even looked at a Roman coin.[25] If the work was useful, it was as a casual epitome of some earlier efforts. It was, as a critic saw, "the first Collection of Foreign Prices, and of the Roman Wealth and Luxury that has appeared so largely in the English Dialect to this Hour."[26] If its reliability remained uncertain, that was a matter for the antiquaries to dispute;[27] it was not likely to bother anyone else. So while Dr. Arbuthnot was read and taught in the schools for over a century, Graevius and Gronovius tended to lie undisturbed on the shelves, except perhaps for an occasional scholar with the time to scan their tedious pages.

[3]

On July 3, 1714, Swift wrote to Dr. Arbuthnot. "To talk of Martin in any hands but yours, is a folly. You every day give better hints than all of us together could do in a twelvemonth." Pope, he thought, though the first begetter, had no genius for the task; Gay was too young and Parnell too idle. "I could put together, and lard, and strike out well enough, but all that relates to the sciences must be from you."[28] Take the "sciences" broadly enough, there was little else to do. Perhaps it was Swift who had first thought of the narrative form; it seems to have been Dr. Arbuthnot who stuffed it full of ideas, as it was Pope, undoubtedly, who polished it in the end. Parnell contributed certain episodes, and Gay, who acted as secretary, may have added something also. (*Three Hours after Marriage* was one of the first public fruits of the enterprise.) But there can be little doubt — although the evidence is entirely circumstantial — that it was Arbuthnot who called attention to Dr. Woodward and his shield.

The narrative scheme was simple enough. The preface introduces the reader to a decrepit Spanish grandee, the author of a Latin treatise *Codicillus, seu Liber Memorialis, Martini Scribleri.* With a mock oration he urges on the narrator the printing of his commentaries, and immediately the story of Martin begins. It is the biography of a pedant from birth to maturity, and on this flexible peg the Scriblerians thought that they could hang a limitless fund of satirical anecdote.

The first chapter sketches in the ancestry of Martin, "how he was begot, what Care was taken of him before he was born and what Prodigies attended his Birth." We meet his father Cornelius and his mother and are treated to a pedigree which invokes the names of Paracelsus and the Scaligers on the one side, Cardan and Aldovrandus on the other — natural philosophy joined to antiquarian science. Cornelius is compelled to journey to England in order to secure an inheritance, and determines to set aside a considerable sum "for the recovery of Manuscripts, the effossion of Coins, the procuring of Mummies, and for all those curious discoveries by which he hoped to become . . . a second Peireskius."[29] At last he is blessed with a son, Martin, whose birth is attested by prodigies — among them Mrs. Scriblerus "dream'd she was brought to bed of a huge Inkhorn, out of which issued several large streams of Ink, as it had been a Fountain." The child was destined obviously to write. Another appeared in the form of a monstrous fossil with black feathers spelling out the several branches of science. In the next chapter, Cornelius expounds on his son's future education into the "works of Nature under ground and . . . in the nature of Volcanoes, Earthquakes, Thunders, Tempests, and Hurricanes. . . ."

If Cornelius had begun to sound rather like Dr. Woodward, by the third chapter he was unmistakable. Here we learn "what befel the Doctor's Son, and his Shield, on the Day of the Christ'ning." (Presumably Dodwell's *De Parma* had made its appearance not long before these words were written.) Dr. Woodward's enthusiasm and Mr. Dodwell's erudition epitomized just that "false taste in learning" that the Scriblerians so despised. In any case, the first major incident of the satire was devoted to the shield of Cornelius Scriblerus. The latter, it seems, had read in Theocritus that the cradle of Hercules had been a shield, and "being possess'd of an antique Buckler, which he held as a most inestimable Relick, he determined to have the infant laid therein, and in that manner brought into the Study, to be shown to certain learned men of his acquaintance." Cornelius was so devoted to his shield that he had even composed a dissertation on it, "proving from the several properties, and particularly the colour of the Rust, the exact Chronology thereof." He thought he would entertain his guests with the Treatise and a modest supper. He therefore took the weapon from its case (where it had been secreted to keep it from any modern contamination) and gave it to his housemaid. She was to place the baby in it at the appropriate moment and cover it with a satin cloth.

The stage was set for the great moment. When at last the learned guests were seated, they entered naturally upon an antiquarian debate "about the Triclinium and the manner of Decubitus of the Ancients." (Arbuthnot, incidentally, had discussed both matters in his *Tables* of ancient coins.) [30] They were, however, soon interrupted by Cornelius. Today, he told them, I present my son to you, not in the ordinary vulgar manner but "cradled in my Ancient Shield, so famous through the Universities of Europe." He reminded them how he had purchased it at great cost and had it transported thither. But the incident is worth repeating verbatim.

Here he stopp'd his Speech, upon sight of the Maid, who enter'd the room with the Child; He took it in his arms and proceeded: "Behold then my Child, but first behold the Shield: Behold this rust — or rather let me call it this precious AErugo — behold this beautiful Varnish of Time, — this venerable Verdure of so many Ages —"

In speaking these words, he slowly lifted up the Mantle, which cover'd it, inch by inch; but at every inch he uncovered, his cheeks grew paler, his hand trembled, his nerves failed, till on sight of the whole the Tremor became universal: The Shield and the Infant both dropt to the ground, and he had only strength enough to cry out, "O God! my Shield, my Shield!"

The Truth was, the Maid (extremely concern'd for the reputation of her own cleanliness, and her young master's honour) had scoured it as clean as her Andirons.

Cornelius sunk back on a chair, the Guests stood astonished, the infant squawl'd, the maid ran in, snatch'd it up again in her arms, flew into her mistress's room, and told what had happen'd. Down stairs in an instant hurried all the Gossips, where they found the Doctor in a Trance: Hungary water, Hartshorn, and the confus'd noise of shrill voices, at length awaken'd him: when opening his eyes, he saw the Shield in the hands of the Housemaid. "O Woman! Woman! (he cry'd, and snatch'd it violently from her) was it to thy ignorance that this Relick owes its ruin? Where, where is the beautiful Crust that cover'd thee so long? where those Traces of Time, and *Fingers* as it were of Antiquity? where all those beautiful obscurities, the cause of much delightful disputation, where doubt and curiosity went hand in hand, and eternally exercised the speculations of the learned? All this the rude Touch of an ignorant woman hath done away! . . . Behold she hath cleaned it in like shameful sort, and shown it to be the head of a Nail. O my Shield! my Shield! well may I say with Horace, *non bene relicta Parmula*."

The Gossips, not at all inquiring into the cause of his sorrow, only asked, if the Child had no hurt? and cried, "Come, come, all is well, what has the Woman done but her duty? a tight cleanly wench I warrant her; what a stir a man makes about a *Bason*, that an hour ago, before this labour was bestowed upon it, a Country Barber would not have hung at his shop door." "A *Bason*! (cry'd another) no such matter, 'tis nothing but a paultry old *Sconce*, with the nozzle broke off." The learned Gentlemen, who till now had stood speechless, hereupon looking narrowly on the Shield, declar'd their Assent to this latter opinion; and desir'd Cornelius to be comforted, assuring him it was a *Sconce* and no other. But this, instead of comforting, threw the Doctor into such a violent Fit of passion, that he was carried off groaning and speechless to bed; where, being quite spent, he fell into a kind of slumber.[31]

[4]

It was a good satirical idea, but the Scriblerians had not invented it. The basic joke and something of its spirit was devised by a friend of theirs, the Christ Church wit Dr. William King. For years King had found amusement in mocking the follies of the virtuosi, and his delightful parodies were widely read. In January of 1713, Thomas Hearne entered his obituary into his diary. "On Christmas day last died the ingenious Dr. Wm. King of Doctors Commons and was buried . . . in Westminster Abbey. This Dr. King was a Man of excellent Natural Parts, which he employ'd in writing little trivial Things to his dying day, in so much that tho' he had a good Estate, was Student of Christ Church formerly and few Years since Judge Advocate in Ireland, yet he was so addicted to the Buffooning way that he neglected his proper Business, grew up poor, and so dyed in a sort of contemptible manner."[32] It was an accurate, if somewhat uncharitable, epitaph, but Hearne of course had little patience for any of the wits. King *was* indolent and died in poverty, despite the best that his friends, including Swift, could do for him; but he had always enjoyed a laugh at the virtuosi.

King's first target was no less than Dr. Bentley at the time of the quarrel over the letters of Phalaris.[33] More than anyone else he was the originator of the Bentlian satire, a genre that was to have a long life and contribute much to Scriblerian merriment. It all began, so King wrote afterward, when he accidentally witnessed an argument between Bentley and a bookseller. Apparently the great scholar had rudely treated a request from the editor of Phalaris, the young aristocrat Charles Boyle, to borrow a manuscript of the *Epistles* from the Royal Library. For this he was repaid by an ironic reference to his "singular humanity" in the edition of the letters which soon appeared. In turn this provoked the famous dissertation (appended to Wotton's *Reflections on the Ancients and Moderns*) in which Bentley answered Boyle and, more importantly, exposed the work as a fraud. When the Christ Church wits,

under the leadership of Francis Atterbury, gathered thereupon to defend
their young confederate, they called upon King to contribute a letter describing the event. They then mustered their resources for a combined assault on
Bentley which would both defend the authenticity of Phalaris and also
expose the rudeness and pedantry of the great scholar. (The work became
known as *Boyle against Bentley*.) On the whole, the Christ Church men
preferred to obscure the problems of Greek scholarship by an attack on
Bentley's personality and a parody of his style and manner. As Dr. Johnson
remarked, King "was one of those who tried what Wit could perform in
opposition to Learning, in a question which Learning only could decide." [34]
Not that they entirely overlooked the scholarly problem; they made a stab at
the Greek, and King did what he could to hunt out the ancient sources which
he thought might help; but they simply did not have the learning to contest
with Bentley. [35] So they concentrated instead upon the satire, which was
brilliant and original, indeed for most readers convincing, and never so able as
in the two passages which seem to have been contributed by King. In the
first, he argued that the author of the *Dissertations* could not possibly have
been Bentley, since no scholar or gentleman in his time could conceivably
have written them. He thus used Bentley's exact manner and method and
even his very words to disprove Bentley's authorship of his own work! He
also seems to have contributed the comical index (*A Short Account of Dr. B.
by way of Index*) which concluded the work and called attention under
separate headings to Bentley's pedantry, bad manners, uncouth style, and so
on. It was a device, like so many of King's, that swiftly caught on, especially
among the Scriblerians, and it certainly annoyed his victim. [36]

The quarrel with Bentley continued for the rest of King's life. When the
great scholar replied to the Christ Church work with a more elaborate set of
Dissertations on Phalaris, King set out on his own, this time with a clever
series of *Dialogues of the Dead* (1698) patterned on Lucian and ridiculing
"the Snarling Critick Bentivoglio." [37] Perhaps they were, like all King's
works, "little trivial things"; they were nevertheless effectively and engagingly
written. King's light touch concealed a profound repugnance not only for Dr.
Bentley but for all modern learning. (One of the *Dialogues* expressly mocked
Wotton's *Reflections*.) [38] His indolence and good humor kept him from the
rancor of his friend Swift, who was soon to assault the very same targets in *A
Tale of a Tub* and the *Battle of the Books*. But he was almost as ingenious,
and in his gift for parody unsurpassed. In 1712 he was still after Bentley,
assisting in the satire that followed the great man's Horace. In a way it was
too bad that he did not live to celebrate Bentley's more extravagant scholarly
adventures – his edition of Milton, for example – but the Scriblerians did not
overlook that opportunity.

Meanwhile, King had found a second target in that other characteristic modern activity, the natural science of the Royal Society. In 1700 he wrote the *Transactioneer*, which as we have seen became a problem for Dr. Woodward through its amusing parody of Dr. Sloane's *Philosophical Transactions*. King, like Dr. Arbuthnot, preferred to write everything anonymously, with the result that no one was ever sure what belonged to him. He was for example generally credited with Swift's *Tale of a Tub* until he publicly disavowed it.[39] In 1709 he returned to his theme (still anonymously) with some *Useful Transactions in Philosophy and other sorts of Learning*, a periodical which threatened to appear as long as demand required. In fact it petered out after three numbers and within the year. John Gay thought that its raillery was too much of one particular kind, so that the town grew quickly tired of it.[40] Still it was amusing while it lasted and must have pained Dr. Sloane (here Dr. Jaspar Slonenburgh, "a Learned Member of the Royal Virtuosi of Great Britain"), whose *Voyage to Jamaica* was the special butt of one whole number.[41] As always, it was the uselessness and triviality of modern learning which astonished and amused King and his friends. In other issues he ridiculed the antiquaries as well as the natural philosophers. There was thus an essay, "On the Invention of Samplers communicated by Mrs. Judith Bagford," which traced its theme from the Greeks until modern times. (Was the name meant to suggest Dr. Woodward's friend, John Bagford?) There was an account of some recent antiquarian books on ancient Greek dances, and the games of Greek schoolboys, parodying the learned reviews of the day and obviously inspiring the fifth chapter of the *Memoirs of Martin* ("A Dissertation Upon Playthings").[42] And there was another essay which must also have amused the Scriblerians, "A New Method to teach Learned Men how to write Unintelligibly, being Collections out of Softliness an Italian, Bardowlius and Bardocoxcombius . . . etc." Indeed King might despise work and so bring on his own ruin, but his enthusiasm for flogging the virtuosi never abated. About the same time that he was writing the *Useful Transactions*, he turned again to philological learning in what was probably his best work, a piece entitled *The Art of Cooking* "in imitation of Horace's Art of Poetry," inspired, according to King, by Dr. Lister's edition of Apicius Coelius, *Concerning the Soups and Sauces of the Ancients*. Some letters to Lister preceded the poem and poked fun again at the collection and criticism of manuscripts and at the triviality of philological learning.[43]

It is hard now to say what influence these casual pieces of buffoonery had upon the generation of wits who read them. The framework of satire had, after all, been set two generations earlier. King's anonymity retarded his fame both in his own time and now. But he certainly helped to furnish the literary atmosphere in which the Scriblerians worked. And in one of his earliest spoofs he probably supplied the inspiration for the specific satire on Dr. Woodward's shield in *The Memoirs of Martin*. We have met the *Journey to London* earlier as a parody of the travels of the virtuoso in search of rarities (specifically Martin Lister's *Journey to Paris*). In the course of this work, King's visitor was encouraged to inspect the various eccentric collectors in London and ran across an old gentleman who discoursed at length on ancient Roman tea and chocolate. After a time he presented his guest with "A Roman Tea Dish and a Chocolate Pot, which I take to be about Augustus's time, because it is very Rusty." By now the joke about the rust of ancient objects was probably wearing thin, but King added a fillip of his own which gave the Scriblerians their clue. The maid, so the old gentleman told his visitor, desired to scour it clean, which he of course forbade, lest it do him "an immense damage."[44] Here King's imagination stopped short and the Scriblerians took hold. They had now only to envisage what would have transpired had the maid been left to her own devices. It is not hard to see them gathered together in Dr. Arbuthnot's rooms, combining their wits in order to transform the old gentleman of the *Journey to London* into Cornelius Scriblerus, and the coffee pot into his shield, spinning out the full humor of the situation which King had first conjured up but left unfinished.

But had they read King when they wrote the *Memoirs*? I cannot truly say. The association of rust with ancient monuments was the oldest of the jokes on the antiquaries, and it was bound to occur to somebody to wipe one clean and enrich the humor. Why not to two of the wits independently? However, the Scriblerians knew King and read his works, and so might just as well have lifted it from him.[45] It is rather like a similar instance which occurred in the very next year, a coincidence which is even harder to explain. Just after the Scriblerians had devised their scheme to embarrass the virtuosi, another young London gentleman discovered Dr. Woodward's shield. Lewis Theobald set himself up in 1715 as author of a new journal, the *Censor*, patterned upon the *Spectator*. He devoted the fifth number to a familiar set piece on the "antiquary," not this time without sympathy for his achievements but in despair at one of his all-too-characteristic practices.[46] Theobald was unequivocally against anyone who sought "to impose false Wares upon the Ignorant under a Pretext of Learning and Antiquity." But this was just what Cornelius Scriblerus had done, since his shield had turned out to be nothing more than a sconce! (A sconce was apparently an embossed protective cover, resembling a shield, which was fastened to the wall with one or more arms for holding candles.) "I therefore profess," continued Theobald, "that altho' I entertain a just Veneration for the Collections of Celsus the Naturalist" (could Dr. Woodward be more obvious?), "I will no more suffer his Back of an old-fashioned Sconce to pass under the honourable Name of a Roman Shield." If the antiquary persisted in claiming its authenticity, he even threatened reprisals on the broker who sold it "and the Brazier who furbish'd it up to its present Dignity."

Had Theobald somehow seen the *Memoirs of Scriblerus*? It is unlikely; in 1715 the work remained in manuscript, and I have found nothing to suggest that Theobald knew it directly. It looks rather as though the ridicule of the wits against the virtuosi, continued now over several generations, had finally cast a veil over all antiquarian learning, leaving even its very objects in doubt. The fact that there were some notorious fakes, like those prized by the Earl of Pembroke, undoubtedly helped. Dr. Woodward was too obvious a target for the wits to ignore, and his shield, now dressed up in the Latin prose of Tom Hearne and Mr. Dodwell, too astonishing an object not to provoke their raillery. In this context, it was not inconceivable that it might indeed turn out to be nothing more than a sconce, and it is likely enough that the idea could have occurred to two or more of the satirists independently – as likely, indeed, as that other notion that its rust could have been scoured clean by a serving-maid. But whether it was a coincidence or not, Dr. Woodward's shield certainly provided its share of amusement in 1714-15. One way or another, it seems that by that time everyone in London had heard about it, and nearly everyone, whether scholar or wit, had an opinion to offer.

[5]

With the death of the Queen, the Scriblerians scattered. Arbuthnot gave up his rooms in St. James' Palace and set up practice in Chelsea. For the time being, the mirth of Martin had to be suspended. The *Memoirs* remained incomplete, although some further chapters had been roughed out that spring and others still were to be added later. (Martin's education and playthings, logic, metaphysics, medicine, and the philological criticism of Bentley were some of the topics eventually treated.) While the work seemed only temporarily suspended, Pope thought that it did not matter, "for Mankind will be playing the Fool in all weathers and affording us materials for that Life which . . . I hope to see the grand Receptacle of all the oddnesses of the world." [47] By September, however, the friends were already nostalgic for their past merriment. "It is a pleasure to us," wrote Parnell and Pope to Arbuthnot, "to recollect the satisfaction we enjoy'd in your company, when we used to meet the Dean and Gay with you; and Greatness itself [i.e., The Earl of Oxford] condescended to look in at the Door to us." [48] It is clear that Martin had become associated particularly with the Doctor. "Then it was that the immortal Scriblerus smiled upon our endeavors who now hangs his head in an obscure corner, pining for his friends that are scattering over the face of the earth." Would Dr. Arbuthnot, Parnell urged, "be pleased to let us know whether indeed he be alive." Pope too recalled their "agreeable Conferences, as well as Occasional Honours, on your account," and hoped their meetings would soon be resumed.

For his part, Arbuthnot did not deny the attribution. "I am extremely oblig'd to you," he replied, "for taking notice of a poor old distressed courtier, commonly the most despised thing in the world."[49] The recent blow had so roused Scriblerus, he continued, "that he has recovered his senses and talks like other men. From being frolicsome and gay he is turned grave and morose." Yet the work might still continue: "Martin's office is now the second door on the left hand side in Dover Street where he will be glad to see Mr. Parnell, Mr. Pope, and his old friends, to whom he can still afford a half pint of claret. It is with some pleasure that he contemplates the world still busy, and all mankind at work for him. . . . I will add no more, being in haste, only that I will never forgive you if you don't use my aforesaid house in Dover Street with the same freedom as you did that in St. James; for as our friendship was not begun upon the relation of a courtier, so I hope it will not end with it."

Dr. Arbuthnot need not have worried; friendship continued as did the Scriblerian inspiration, although the *Memoirs* languished. Here at last was a bit of luck for Dr. Woodward, who this once was spared the slight. When Pope returned finally to his notes many years later and delivered them to the world, the Doctor was dead.[50] By then, all recollection of the shield was fading, as an editorial note to the *Memoirs* makes plain.[51] In the end, despite its great fame, the shield was destined to disappear almost altogether, except perhaps for the notes of few learned commentaries on the *Memoirs*. How appalled the good Doctor would have been to discover that eventually it was the shield of Martinus Scriblerus which was destined for immortality, while his own great prize would be relegated to a footnote and pass into oblivion! But then, I think there can be little doubt that this news would have startled most of his contemporaries as well, the wits along with the scholars, and including undoubtedly the amiable Dr. Arbuthnot himself.

Part Three
CONCLUSION

XIV.
The Decline and Disappearance of Dr. Woodward's Shield

[1]

On the twenty-fifth of April 1728, Thomas Baker informed Hearne that Dr. Woodward had died. As was his custom, the Oxford antiquary immediately composed his epitaph.

Dr. John Woodward, my Friend, with whom I have corresponded many years, is very lately dead after a long Indisposition (he having been confined to his Bed for at least three Quarters of a year) and hath left his Museum to the University of Cambridge. . . . He was an ingenious learned man but undeservedly despised by many. He wrote many remarkable Things, the most remarkable of which is his Theory of the Earth. . . . It was his Misfortune to differ in his notions of Physick and Philosophy from several great and leading Men . . . which drew down great Troubles upon him, which must needs shock and vex him.[1]

Now the trials of a stormy life were over, and we are assured that he died piously and one may hope peacefully, in the familiar surroundings of his beloved College. His last years had not been happy, frustrated by illness and his failure to complete his many projects. He had even to suffer rejection by the French Academy of Inscriptions (along with Halley), despite the support of his old friend, the Abbé Bignon. However, his executors were very loyal, and though he had asked for a simple burial, they placed him in Westminster Abbey, not far from Isaac Newton, and gave him a fine monument by Peter Scheemakers.[2] There he rests silent at last, left behind by the march of history and the throngs of sightseers who scurry past without comprehension and scarcely a glance.

Now all attention was upon his things. Everyone knew about the Doctor's collections, and all eyes were riveted upon the auction catalog which announced the splendid sale. "I could not read a Page in it," wrote the antiquary William Brome, "but I had a terrible itching."[3] The executors had been very busy, canvassing the books and manuscripts, organizing the fossils and antiquities. The last were difficult and needed specialized knowledge, particularly since the Doctor had failed to complete his own inventory. A mere enumeration would not be sufficient — especially for the greatest prizes like the famous shield. For this the executors needed expert help, and they turned at length to the aging antiquary Robert Ainsworth.

John Murray, "the curious Collector of Books," told Hearne in 1723 that Ainsworth was a non-juror who "understands our English Coyns, he believes, as well, if not better than any Man in England, that he is a mighty modest Man, an excellent Scholar, and hath been about seven Years about a Dictionary in the nature of Littleton's. He was Author of the Catalogue (which is printed) of Mr. Kemp's Rarities, a thick octavo. . . . He is a married Man, and lives at Hackney, near London."[4] In short, he was the perfect man for the job, a scholar of genuine if somewhat limited accomplishment, who had spent his long life with Latin grammar, as a schoolmaster, and with Roman antiquities as a hobby. As a schoolmaster he was only moderately successful and his educational writings did not make him famous, despite one inviting title, "The Most Natural and Easie Way of Institution containing Proposals for Making a Domestic Education less chargeable to Parents and more Easie and Beneficial to Children." (His later poverty, however, was more likely to have resulted from his costly pastime and great age than from his small fees.) At his hobby he was more successful, and his contributions to the reborn Society of Antiquaries were frequent and continuous throughout these years, although largely presentations from his own collections.[5] He was finally and somewhat belatedly to make his reputation with his Latin dictionary, which replaced Littleton and long held the field. Meanwhile the Kemp Catalogue was an obvious prototype for Dr. Woodward's executors, who did not have to be reminded of the resemblance of the two collections (including the iron helmet that so resembled the Doctor's shield). In working through Kemp's antiquities, Ainsworth had had even to consider such familiar Woodwardian problems as the location of the Temple of Diana.[6]

Indeed, according to Hearne, Ainsworth had all the qualifications for the job but one; his friend Richard Rawlinson told him that he did not know how to discriminate true antiquities from counterfeits.[7] It did not help, perhaps, that Ainsworth had been hired on both occasions to assist the owners (rather than the prospective purchasers) of the antiquities. In any case, as he drew up his account of Dr. Woodward's shield for the sales catalog, he did not hesitate to defend its authenticity. He admitted that not all the arguments that had previously been used to defend it were equally valid — for example, those that had emphasized the reliability of the iconography. After all, "a modern Artificer, knowing the History of those elder Times, might conform his Workmanship to all those Circumstances." But having admitted so much, he insisted anyway that the iconography demonstrated that it must be later than the primitive Republican period and earlier than modern times. It must be later because of the several anachronisms which were otherwise unexplainable, and earlier even than the Middle Ages because of the equestrian standards which he was certain would otherwise have had crosses upon them. Above all, he believed that it was the elegance of the workmanship that was conclusive for its date. Like Dodwell – whose treatise he employed at every step, along with the whole file of Dr. Woodward's correspondence – he believed that it must be from the time of Nero, when Roman metalworking had reached its highest achievement. He was not dismayed by the iron, since the patches which were badly tarnished were later than the original which was still bright and smooth; the ancients had obviously known how to temper that metal in ways long since lost. Only in the question of its use did Ainsworth waver, content merely to summarize the arguments on both sides. His essay was thus a fair digest of previous opinion about the shield and, besides the Latin version of the catalog, it appeared also in an English abridgment and again some years later as a separate publication.[8]

Yet there is one passage in Ainsworth's argument that may have surprised or at least amused some of his readers. It was an addition to Dodwell's views, allegedly from the lips of Dr. Woodward himself, and appeared in Ainsworth's article on the shield in the *London Journal*, April 19, 1729, under the pseudonym "Atticus."

To what Mr. Dodwell had advanced I must add one Strong Argument of the Antiquity of this Monument, which is the Gold that was on it some Years since and which was scoured off by the Officiousness of a Servant in the Absence of her Master; of this I have been informed by Dr. Woodward himself. The Art of making Iron receive Gold, as this was laid on, has been long since lost.[9]

Had the Scriblerians, then, borrowed their satire from life, after all? Or had the story taken life somehow from the satire? With so many maids scouring so many monuments of their rust and gilt, it is hard now to say. Since there is nothing elsewhere to confirm the story, much less to connect it with the wits, and since the satire of Martin slumbered yet in manuscript, it is perhaps best left to coincidence. Nowhere in the whole of Dr. Woodward's large correspondence does the episode appear. And although there is one passing reference to something like this during Dr. Woodward's lifetime, it does not seem to be conclusive. In 1720 Pope prefixed to Addison's posthumous *Dialogues on Medals* some lines that he had written earlier in which the following couplet appears:

> Poor Vadius, long with learned spleen devour'd,
> Can taste no pleasure since his shield was scour'd.[10]

That this was an oblique reference to Dr. Woodward is beyond doubt. But was Pope referring to a real event or to the Scriblerian satire?

The tone of Ainsworth's essay was unmistakably defensive, and it appears that voices had been raised again to decry the shield. Certainly there were many who attended the sale and had their doubts, and in the end the shield had to be bought in for a hundred guineas by the executor, Colonel King, under a pseudonym.[11] One buyer who was present and questioned the work was Richard Rawlinson. He had just returned to England from half a dozen years abroad, bringing with him a vast collection of manuscripts, coins, and other curiosities. It was only the beginning of one of the greatest libraries of the time; at his death he left the Bodleian Library over five thousand manuscripts. Somehow he had got hold of Downes' *Stricturae* against Dr. Woodward's shield and, believing the arguments there to be entirely cogent, had them printed for the first time, waiting tactfully only until the auction was over. About the same time another buyer, James West, rehearsed Downes' arguments before the Society of Antiquaries. One difficulty was that it was easy now to read Downes but almost impossible to get hold of a copy of Dodwell.[12] Thomas Hearne was disappointed. "All the objections I have often heard in common talk before and they have been obviated and judged to be of no great moment. Indeed the Publication of these *Stricturae* is not for the credit of Mr. Downes's posthumous Reputation." Thomas Baker, on the other hand, gave his copy to Conyers Middleton, the new Woodward Professor, hoping that he would send it on to Rome, "where they would be able to form a judgment of it by comparing it with Originals." To Richard Rawlinson he wrote, "I must confess I always suspected the Age of Dr. W's famous monument, tho' having never seen it (or if I had) am not capable of judging. You and Mr. D[ownes] who have been to Italy are surely better judges of such Antiquities than Mr. Dodw. who was too fond of paradoxes and of saying something new and out of the way."[13]

Somehow the doubts were growing more insistent, and Dr. Woodward's shield, despite the vigor and sincerity of its defenders, was seriously losing credit.

[2]

Thomas Hearne was annoyed with his friend Rawlinson for publishing a criticism which he thought had been thoroughly rebutted. A year or two earlier he had been even more thoroughly exasperated when the remarks of an enemy, Walter Moyle, were similarly exhumed and published against the shield. Rawlinson was a non-juror and his opinions, even when wrong, were forgivable; Moyle was a Whig, even a republican, and thus beyond all redemption. Moreover, Moyle was a clever critic with a sharp tongue and an impressive array of critical learning at his command. "His criticisms are bold," wrote Richard Richardson to Hearne, "but often wrong."[14] Anyway, they could not be lightly regarded. Hearne was annoyed and Dr. Woodward was contemptuous; but Moyle's remarks did not enhance the reputation of his shield.

Moyle had received Dodwell's dissertation in 1719 as a gift from his friend Dr. Musgrave.[15] Although he lived far off in Cornwall in retirement and ill health, he tried to keep abreast of every development in classical scholarship. Many years before, when he was young, he had hobnobbed with the wits at Will's Coffeehouse, on friendly terms with Dryden and Congreve. For a time (1695-98) he had even served in Parliament. From the beginning Moyle displayed an omnivorous appetite for modern learning, the study of nature as well as history. He was a friend of John Ray's and collected plants and birds, especially the latter, which he kept in a well-stocked museum.[16] While still young, he wrote a tract on Roman government which allowed him to reflect on the endless cycles of change which repeated themselves throughout human history. (He leaned expressly upon Machiavelli and Harrington for his views.) His learning was not yet profound, and he accepted without hesitation the whole story of Romulus and the Alban Kings as founders of limited monarchy and the Roman constitution. It was in any case a very convenient analogue with the present. Typically, he concluded his essay with a reflection on the contest between Brennus and Camillus. " 'Tis an Observation of Machiavel that great Dangers and violent Extremities often rectify and recover a Constitution of Government tending toward Corruption; of which he gives a notable Example in the Sacking of Rome by the Gauls."[17]

In this early work, Moyle writes like a rhetorician; he uses history to draw an immediate political lesson.[18] (That legend had a purpose as agreeable in this regard as fact was of course a very old complaint.) But if his original milieu was the literary and political world of fashionable London, it was another story in Bake, Cornwall. With leisure now and a great library, Moyle had time for genuine scholarship, time for the abstruse learning of the *erudits*, philology, chronology, and the rest. "I was once setting up for a Botanist," he wrote to William Sherard, "But my ill state of Health has many years since oblig'd me to abandon that Study tho I have still a strong Relish and Inclination for it. I have chiefly apply'd my self to studies which I cou'd follow within Doors, as Criticism, Philology, etc. . . ."[19] He read widely, especially in history, and made copious extracts. Once, he sent a friend a whole sheaf of excerpts taken from Matthew Hale, "to confirm the Relation of Moses concerning the Beginning of the World and the peopling of it."[20] He did not publish anything, but he wrote some critical essays and many long

letters. No problem daunted him; he was ready to correct the editors of
Horace and the great Vossius more than once, to discourse on the Triumphal
Arch of Titus or on a fragment of Polybius published by Gronovius. He was
eager to exchange letters about Roman coins dug up at Exeter and (using
Spanheim) to show that the famous Vaillant must have got his dates wrong.
He was happy to write about the chronology of the Bactrian kings, the
Chaldean era, or any other chronological matter. Indeed, he had gotten quite
far from his early rhetoric; on one occasion he even thought to apologize for
"so long a letter and such a heap of Quotations."[21]

Moyle had an independent mind and enjoyed a disagreement. (If political
differences existed, he enjoyed it even more.) Thus he was ready to cross
swords with Henry Dodwell over the date and meaning of the Lucianic
dialogue, the *Philopatris*, in a tract he hoped to dedicate to Richard Bentley.[22] And he insisted in another place that Marcus Aurelius had persecuted
the Christians, whatever one might say about his usual clemency and good
manners. "His severities against the Christians did not flow from the cruelty
of his temper but his zeal for religion and the laws of his church." Moyle did
not fail to notice that "cruelty for the sake of religion" had been made a
virtue by most Christian sects from the time of St. Augustine. In yet another
work he argued against the famous but lately much disputed miracle of the
"thundering legion." No wonder perhaps that Hearne was ready to pronounce
him "an ingenious man, yet a man of vile Republican Principles, a great
Enemy to the Fathers of the Church, and in religion a Sceptick!"[23]

The correspondence with Musgrave centered upon their common interest
in antiquities. Musgrave was not far away in Exeter, and delighted to submit
his views to his learned neighbor. Thus he sent Moyle a copy of the famous
Bath inscription in 1709 and was full of praise for the reply, which was long
and formal. Moyle, however, declined to publish his emendation, "especially
when such able Criticks as yourself and Mr. Dodwell are preparing Dissertations at large upon it." Instead he sent to Musgrave another inscription from
Fowey, Cornwall, with an elaborate commentary. As the years passed he
helped Musgrave with further suggestions for his works (especially the *Geta*),
though he declined "to set up for an Umpire between two such Criticks as
your self and Cuperus" when they had their dispute about the golden eagle.
(This did not stop him from writing at great length on the subject.)[24] When
Musgrave's work was complete and published as the *Antiquitates Britannico-
Belgicae*, Moyle sent him a congratulatory letter. But on page 171 he found a
note mentioning Dr. Woodward's shield and Dodwell's dissertation which led
him to a characteristic comment. Apparently Moyle had not seen the original
print, as he certainly did not know the shield directly, but he had found
enough for his purpose in the engravings in Drakenborsch's *Silius Italicus* and

Clarke's *Caesar*. He decided at once that the shield could not be antique because of the stone amphitheater which appeared there amidst the ruins of smoldering Rome. Here was an anachronism, a "notorious Mistake and Incongruity," which could never have been made by any genuine Roman artist. "I suppose the shield to be a modern Invention," he asserted categorically, and with a great show of learning he proceeded to demonstrate that no stone theaters had ever existed in the time of Camillus or long afterward. In a postscript he noticed that he had since come upon Lipsius' *de Amphitheatro* and was delighted to find full confirmation there of his opinions.

Moyle's views were set out in a private letter. With the rest of his controversial remarks it was printed in 1726.[25] (He died in 1721.) When Hearne read the letter he wrote at once to Dr. Woodward. He was not surprised, somehow, since it was apparent on every page that Moyle was "a very Confident bold Writer and very many things Prophane." Woodward, of course, agreed. "As to my Clypeus Votivus, for such the best Antiquaries judge it to be, Mr. Moyle passes sentence upon it without ever having seen it from two very imperfect Sketches ... and without having consulted Mr. Dodwell's excellent Book *de Parma*." No one had claimed that the shield was from the time of Camillus, "so that Mr. Moyle in demolishing that Notion, only demolishes a mere fancy of his own. But that it was antient is argued by all the best Judges of all Nations. . . ."[26]

It was Dr. Woodward's last word on the subject and Hearne appreciated it enough to print it twice, in 1726 in his edition of John of Glastonbury's *Chronicle* and ten years later.[27] And it was a fair enough answer, although it is hard to believe that Moyle would have conceded easily. (The notorious printer Curll attempted to reply for him, but his *ad hominem* arguments probably did not help the cause.)[28] The trouble with defending the shield was that, while it was possible to answer many of the criticisms separately, when they were added together they began to look only too plausible. But so far no one had yet taken the trouble to put them all together, in public anyway. Doubt might be spreading, but the weight of the argument still lay, however precariously, with the defense.

[3]

Of all those who in 1728-29 came once again to consider Dr. Woodward's shield, perhaps the most accomplished virtuoso, certainly among the younger generation, was the Scottish gentleman, lately Baron of the Exchequer, Sir John Clerk. He is a figure strangely neglected in the story of the nascent Scottish Enlightenment, one of the most important links between the age of the Covenanters and the generation of David Hume — in the words of an

admirer, "a treasure of Learning and good Taste."[29] Ensconced at his great
house outside Edinburgh, his beloved Penicuik, he presided over a mixed
company of poets, painters, architects, and antiquaries for whom he was at
once patron and fellow worker. He had been sent as a young man to study
law at Leyden and he never forgot the lectures he had attended there, on
history by the two professors of Eloquence, Gronovius and Perizonius, and
on church history by Frederic Spanheim. "I never had half an hour to spend
in idleness," he recalled, "but diverted my mind by different successive
studies." No wonder that he developed an appetite for antiquities which he
carried with him first to Italy and then back to Scotland. In Rome he went
three times weekly to the home of one Chaprigni, "a learned Antiquary and
Phylosopher," where he met with the virtuosi and discoursed on all the new
discoveries in literature and antiquities. (When Chaprigni died he left to Clerk
a head of Cicero, a bust of Otho, and a little statue of Diana of Ephesus, all of
which found their way eventually to Penicuik.) While in Rome he studied
music with Bernardo Pasquini and Corelli, and he afterward wrote poetry and
composed music.

Altogether Clerk spent five years abroad in Holland, France, and Italy,
learned law, mathematics, and philosophy, spoke Dutch, French, and Italian,
and gained a broad education in classical studies and antiquities. In Scotland
he inherited the family estate, acquired the station of Baron, and found
himself (since the law terms were short) with the leisure and money for a life
of learning. "I had a great deal of time on my hands," he remembered, "and
this I always spent to my own satisfaction. I was naturally studious and often
laborious this way, so that, except the time I spent at my favorite diversions
of shooting and fishing in the country, all my leisure hours were spent in
Books." "Amongst my Studies," he recalled also, "the Greek and Roman
Literature chiefly delighted except now and then when phylosophical or
mathematical learning required my attention." He collected fine paintings
and Roman antiquities, but did not neglect the world of nature. To his old
friend Boerhaave he wrote describing his home at Penicuik. "There you will
find books in all departments of literature," there also "certain bronze and
marble statues, altarpieces, inscriptions, and that sort of thing." There were
paintings by Raphael, Rubens, and Veronese — rivals, he believed, to the best
of Greece and Rome. There were besides many other objects, "notable for
beauty or rarity such as bones, limbs, or skins of wild beasts, birds or fishes."

For so I would imitate Julius Caesar or Augustus (according to Suetonius)
and even if I had not the example of such great men, I should regard it as a
mean thing to build up a library of huge volumes on antiquities, and yet to
disdain the very objects which the most learned men, as Graevius, Gronovius,
and Montfaucon, have explained with such expenditure of time and toil. The

Things themselves speak and for the most part explain themselves; but descriptions, however accurate, present to the mind only confused or shadowy ideas. . . .[30]

It was thus Roman antiquities that drew his most earnest attention. When he inherited his lands, including some mines, he took a trip to the north of England to see how the work was done, and his journal is about equally divided between what he saw in the mines and what he saw of Roman remains in the rich hunting-ground of the Roman wall.[31] He hired the young Alexander Gordon with others to acquire whatever he could for him, and began to stock his house with a rich collection of altars, inscriptions, and other monuments of ancient Britain.[32] He struck up a correspondence and friendship with all the notable young antiquaries of his generation, with Roger Gale, William Stukeley, John Horsley, and Smart Lethieullier. He was elected to the Society of Antiquaries as soon as that body began to admit corresponding members.[33] It was not long before he got to know its great patrons, Lord Hertford and the Earl of Pembroke. Through all these friendships and the rich correspondence that resulted, he was able to keep thoroughly abreast of classical scholarship and antiquarian activity throughout Britain. He never doubted, as he wrote to Stukeley, that "those who have the greatest knowledge of Antiquity have *ceteris paribus* the best title to be esteemed patriots."[34] And he won for himself a reputation as a scholar and connoisseur that was richly deserved. It was at this time, for example, that John Horsley was finishing what was to be the most ambitious study of Roman Britain that had ever been attempted, and he depended upon Clerk at all stages of his work, "since there is none whose opinions the learned world (and that very justly) will pay a greater regard to." "As long as my work continues upon my hands," he wrote later, "I find your Lordship must still share in the trouble."[35] In fact, Clerk remained involved even after Horsley's death in its posthumous publication.

But it was with Roger Gale that he struck up the closest friendship. Gale was about fifty when he began corresponding with Clerk, a gentleman of considerable estate, a former member of Parliament and presently Commissioner of Excise. Gale was besides first Vice-President of the Society of Antiquaries as well as Treasurer of the Royal Society. His father, the Dean of York, had been an eminent scholar, and one of Roger's first works was the publication of the Dean's edition of the *Antonini Iter Britannicarum* (1709), with many contributions of his own. This was followed by an essay on the four main Roman roads in Britain, which was printed by Thomas Hearne in his edition of Leland. While still a very young man, Gale had translated anonymously from the French a popular work on coins, F. Jobert's *The Knowledge of Medals* (1697). To numismatics and Roman topography he

joined a passion for archaeology, traveling widely in England in search of
Roman remains and collecting whatever he could in the way of coins and
inscriptions. At his death he left an impressive collection to Trinity College.
He was thus in every respect an admirable friend for Clerk, in the Scottish
virtuoso's estimation "a person of great learning and humanity."[36]

It was Gale who read to the Society of Antiquaries Clerk's first paper, an
essay on the ancient burial practices. A remark of the Baron's on the ancient
brass instruments (about which Hearne and Thoresby had quarreled) led to a
long exchange of letters in which the two men enlarged on the theme and
displayed their antiquarian skills. Gale denied that they were Roman and with
Clerk believed them to be British, although he remained at a loss to explain
their use. He thought that Hearne's views were particularly unsatisfactory,
and it would have dismayed the Oxford scholar to hear his old friend describe
him as "an Author of a strong Imagination in all his writings, and too positive
in what he proposes, drawing very bad Conclusions from weak Premisses."[37]
How Hearne had supposed that these brass instruments could ever have been
chisels was difficult to understand, when neither their shape nor their material warranted such a guess. For once the Trajan column did not help, since it
was impossible to know what metal the soldiers had employed in their tools,
either from the prints or from the column itself. Together Gale and Clerk
explored the different uses of iron and brass in the ancient world, especially
in the making of arms, drawing upon the literary sources but also upon their
extensive acquaintance with the actual objects. It was all admirable preparation for a consideration of Dr. Woodward's shield.

It was Gale apparently who saw to Clerk's election to the Royal Society
after reading to it his "curious observations upon the fractures or dislocations
as they may be called of your Coal veins."[38] Both men were interested in
natural history, but they relished antiquities more. The Royal Society
remained active in both respects, "not refusing Dissertations upon Anti-
qityes," as Gale explained, "tho not comprehended in the plan of their
Institution." He thought that that was why its meetings were always well
attended while the Society of Antiquaries seemed relatively to languish.[39] In
the event, Clerk's next antiquarian foray, his most ambitious yet – a Latin
tract, *De Stylis*, on the ancient Roman pens – was read to both societies and
received gratefully by both.[40]

In 1727 Clerk went to London, and as was his custom kept a travel
diary.[41] For the first time he met Gale, and thenceforth they were insepara-
ble. Together they made the rounds. They visited the Earl of Pembroke, who
"was exceptionally civil to me and shewed me his pictures, Library, and gold
medals." It was the first of many happy sessions, culminating in a memorable
trip to Wilton. They traveled to Hampton Court to view the Raphael and

Mantegna cartoons. Clerk called upon the Lethieulliers, father and son, at Gray's Inn, and saw "several curious Egyptian Antiquities particularly a Mummy the first I had ever seen." At Hans Sloane's he saw "the greatest collection of things that ever I had seen in my life; not the Treasures of any Forreign prince can equal them." And at Dr. Mead's he saw books, statues, medals, pictures, and drawings, "but what exceeds everything in his possession a head of Homer in bronze." Like his host, he thought it must be the very one which was mentioned as having perished in the fire at Justinian's palace in Constantinople. Dr. Mead gave him a copper engraving to take back to Penicuik.

The next day he visited Dr. Woodward, whom he had met at the Royal Society.

The Doctor has abundance of natural curiosities and many fine antiquities particularly a piece of Mosaick work representing Hercules at the foot of a tree, I suppose in the Garden of Hesperides. He has several urns with Inscriptions, many manuscripts and a chosen Library of books. The chief things in his Cabinets are fossils by which he sufficiently proves his notion of the deluge in his printed History of the Earth. He demonstrates beyond contradiction that many of his bones and shells are really such and not *Lusus naturae* as some pretend. Most of these were found under ground.... I observed some of the shells perfectly entire and yet fill'd with solid Limestone. Others were full of solid Lead ore, the reason of which is (as the Doctor asserts) that the Deluge rendered the whole strata of the Earth soft as pap which entered all the cavities of the shells. The Doctor is a man of Learning but positive and affected in all that he says and does, in that the most accomplish'd pedant I ever saw.[42]

Clerk was obviously intrigued by the performance, for he swiftly got hold of Dr. Woodward's book and read it through on his return to Scotland. His friend Smart Lethieullier who asked his opinion, was disturbed by the "confidence" that ran through the Doctor's works, but he did not think that Woodward's conjectures could be easily dismissed.[43] Clerk sent him a long and careful reply. He had found the Gresham professor's *Essay* "a piece with himself; he aims at a certain point and right or wrong makes everything subservient to it." He objected especially to the Doctor's idea of gravity arranging the strata, since any coal miner could testify that coal (which was light and porous) was found generally under stone which was often double or triple its weight and density. Dr. Woodward's arguments on this point reminded him exactly of his opinion of his shield – his Roman shield, as he called it.

For if he had noticed that it was of Iron, he might easily have concluded that it was so far from being Roman that it could not be above 100 or 150 yeares old. His skill likeways in Mettals and rust might have demonstrated to him that had his shield been a foot in thickness it must needs have decayed since the days of the Romans and that not one of those fine figures on it had been seen.[44]

But the Baron had more to say, both about Dr. Woodward's theories and his shield. On the first score there was much to agree with, for (among other things) he had no doubt that it *was* water from the abyss that had caused the Flood and brought shells to the tops of mountains. Nevertheless, he doubted that the great work which Woodward still threatened would allay all doubts. "This Work will be a Deluge indeed as the Dr. describes it and I'll promise in his name that it shall contain a vast many things as irregular as his Strata and as heavy as Words can make them." As for the shield, the death of Dr. Woodward and the auction of his effects revived Clerk's skepticism. He was sorry, he wrote to Gale, to hear that the Doctor had died. "He was a droll sort of a philosopher, but one who had been at much pains and expense to promote natural knowledge." He thought the fossils were the most valuable things in the collection. But for the *clypeus votivus*, he could only pity the gentleman who had bought it. "Never was there any thing more absurd in my opinion than to fancy it Roman; for as it is iron it could never have lasted the fourth part of the time; for by the sculpture, if genuine, it had been as antient as the time of Herodian." Clerk thus passed quickly over the iconography and the style; like most of the critics he was content to settle upon a single argument — this time its composition. "I never saw anything of iron which was Roman except great hinges of doors and the like which had lost half their substance by rust." When Gale sent him a copy of the *Stricturae*, Clerk agreed that it contained abundant learning, but then he returned to his single track. "I may be mistaken in this opinion, but I took it to be of iron and was on the point of making this observation to Dr. Woodward, when you and I were to see his curiosities." Gale agreed entirely; "the small delicate figures could never have maintained their lineaments and beauty this long." Moreover he saw that there was a plausible alternative. "We had several good artificers here from Italy at the latter end of Harry the 7[ths] and beginning of Harry the 8[ths] reign when perhaps this was made, Shields not being then out of use, and I have seen a curious one, but of copper, said to have been born by the latter at the seige of Boloigne."[45]

It was a suggestive and damaging parallel. How many others saw it I cannot say. What is clear is that by 1729 the skeptics were coming to predominate. Perhaps the antiquaries had learned their craft better; certainly they were relying more on their experience of the objects themselves and less on the literary evidence alone. Perhaps it was the malicious gossip which had always surrounded the Doctor and his shield; clearly there was as much of it in 1729 as ever before. Certainly Dr. Freind (Woodward's old medical rival) was not alone when he referred to it then as "the top of an old andiron."[46] Anyway, Gale and Clerk were quite wrong to imagine that someone was willing to put down a hundred guineas at the auction in favor of the shield. When the sale was over and Dr. Woodward's things dispersed, the shield remained alone in the hands of the executors, and it was not long before it began that quick slide into oblivion where it has lain ever since.

[4]

Somehow the French had always appeared the most suspicious. As we have seen, there were those in the circle about the Abbé Bignon who had raised sufficient doubts to make the great patron of the arts uneasy. Whatever their feelings, it was a long time, however, before their suspicions became public. It was only on November 14, 1738, that a member of the Academy of Inscriptions delivered an address which recalled their doubts and assembled the case against Dr. Woodward's shield.

M. Anicet Melot was an antiquary of minor distinction, soon to become Keeper of Manuscripts in the Royal Library. His address was occasioned by the fundamental question of the reliability of Livy and the early Roman history, a subject of continuing interest to the Academy.[47] Specifically, Melot was disturbed by Livy's account of the Gallic sack of Rome, which he found neither flattering nor convincing. To a patriotic Frenchman it appeared obvious that Livy and the later Romans had tried to put the best face that they could on what was really a Gallic triumph by blackening the reputation of the early Celts and inventing altogether the Camillus episode. By looking about in other early sources, in Strabo for example, or Caesar, Livy's prejudiced description of the early Gauls as barbarous, primitive, and undisciplined could be corrected. And by looking at the other and earlier tradition that stemmed back to Polybius, Livy's account of the miraculous rescue of the Romans by Camillus could be seen to be an outright fiction — poetry, not sober history. Melot writes expressly to defend "la gloire de la Nation Gauloise," and his patriotism leads him to the modern conclusion: the story on Dr. Woodward's shield was evidently not history.

What then of the shield which the London antiquaries had used to authenticate the tale? Melot was ready to concede that a few literary authorities like Polybius were not able to counterbalance the evidence of monuments which were, he agreed, "the only incontestable testimony about ancient history." He knew Dodwell's work and could view Dr. Woodward's shield in its engraving. But he was not at all convinced by the arguments for its authenticity. The shield seemed to Melot obviously medieval, "une production des siècles d'ignorance et de barbarie." To make his point, he quoted at length from a letter by his distinguished contemporary, Claude Gros de Boze, recalling a reply that he had written long ago to Dr. Woodward himself. Apparently de Boze with all the other continental scholars, had been solicited for his views at the time the engraving was originally circulated, but unlike the rest had written out an argument against the shield. Whether Dr. Woodward received it and what he thought of it I cannot say; but de Boze's opinion was now set out for all to see.

De Boze was a highly respected scholar. He had been elected to the Academy of Inscriptions in 1705 and made its perpetual secretary the following year. For fifteen years thereafter he edited the *Memoirs*. It was probably at the beginning of his tenure that he learned of Dr. Woodward's shield. Coins were his specialty, and his cabinet became famous; he also built an immense library. Somehow he got hold of that rarest of antiquarian books, Dodwell's defense of Dr. Woodward's prize. [48]

What de Boze recalled telling Dr. Woodward was that it was very difficult to decide how old any monument was simply on the basis of an engraving. He had, however, compared the illustration in his hands with a dozen shields or so from the time of Francis I and Henry II. Many of them, he insisted now, were of exactly the same form and grandeur as Dr. Woodward's work, and many represented subjects from Greek and Roman history like the enslavement of Helen, the death of Lucrece, etc. Moreover – and this appeared decisive to de Boze – all of them had been made of iron and gilded. Dodwell and Hearne, it is true, had pronounced in favor of its antiquity, but Dodwell's profound knowledge of history and chronology were no help here in the absence of direct acquaintance with the monuments.

De Boze thought that it was necessary to add three other arguments. In the first place, the partisans of the shield had all placed it too late to be of any historical value. How could a monument of Imperial Rome attest to events at the beginning of her history? Secondly, it was contrary to Roman usage to make a votive shield so long after the event. (Besides, such shields were usually gold or silver, never iron.) Finally, the Rome depicted on the shield had never looked like that, especially its theaters and amphitheaters. Indeed, all the detail – the dress of the Gauls, the standards, etc. – was wrong. The shield was a modern work, like a thousand others of the same kind!

All told, it was the most formidable and complete attack on the shield yet. (How much of it he had developed thirty years before is not clear.) It was marred only perhaps by its polemical purpose. What its effect was in England, if any, I do not know. No doubt Dr. Woodward would have responded, but he probably knew no more of de Boze's criticism than what he learned from the Abbé Bignon. Outside France, Melot's essay was read and digested by at

least one learned man. We shall see in the next chapter what the Abbé
Winckelmann made of it.

[5]

Only the owners of the shield appear not to have lost confidence in it. Not
that anyone else seemed to care very much.[49] Colonel King, who was proud
to possess it, got Ainsworth to republish his defense of it a few years after Dr.
Woodward died. And John Ward, Woodward's biographer, took several paragraphs to describe and defend it in his *Lives of the Gresham Professors*
(1740). Perhaps that was not too bad, since Ward was a very capable and
respected antiquary. He became a Gresham Professor (of rhetoric) in 1720,
and a few years later even Hearne conceded that he was "an excellent
Scholar, 'tho never bred up in any University." He especially admired Ward's
long Latin tract, *de Asse*, which had appeared in Ainsworth's catalog of the
Kemp antiquities.[50] He was of great service to Horsley and saw the *Britannia
Romana* through the press. Eventually Ward became a member of both
learned societies and contributed papers regularly to each. Like so many of
his colleagues in the new generation, his special interest was in Roman
antiquities, which he knew firsthand from his own collection and from his
travels. Colonel King gave him liberal access to his manuscripts for his life of
Dr. Woodward, and he read through and summarized the correspondence he
found there about the shield. Reading over that mass of evidence and the
roll-call of famous men who had argued in favor of the shield, and remembering his own close association with Ainsworth, it is not surprising that he
found in favor of the shield, whatever the critics had said.

When Colonel King died in 1767, the shield, with the manuscripts, passed
into the hands of Dr. John Wilkinson, purchased apparently for forty
pounds.[51] Although Wilkinson became a member of both learned societies,
he was not, as far as I can tell, particularly distinguished either as an
antiquary or a physician. He did, however, publish a few occasional pieces,
including a bit of *Domesday Book*, and he appears to have traveled abroad.
He may also have invented the cork life-jacket, for there is a tract under his
name to that effect dedicated to the King. In any case, he bought the shield
at the Colonel's death and lived with it until he too died, in 1819, and
bequeathed it to the British Museum. He was deeply impressed by the "divers
Tracts and Papers" that had come with his prize, and determined to do it
justice in a new work which would comprehend them all. But he was not
apparently very persuasive; at least, he failed on one notable occasion when
he tried to sell it to the famous collector Dr. William Hunter, the founder
subsequently of the Hunterian Library. Hunter was not likely to have balked

at the price, whatever Wilkinson may have asked, for he laid out more than £20,000 for ancient coins alone in the years between 1770 and 1783. It is much more likely that he rejected it as inauthentic despite the engraving that Wilkinson sent him and his efforts to justify the piece.[52]

There is not much point in rehearsing Wilkinson's arguments here; they were largely drawn from the papers before him, from Woodward himself, from Cuper, Batteley, and Dodwell.[53] He met the chief objections that had been raised against the shield easily enough. Thus he found the iron no problem; on the contrary, had the shield been anything more precious, it would even more likely have perished. Besides, the ancients had known how to temper iron to make it last. The analogy of sixteenth-century shields was easier still. He simply recalled Dr. Woodward's own opinion on the matter, drawn from the authority of the Earl of Pembroke. "Hans Holbein was no Scholar," he had written, "Not one Piece of good Work in his Times." And again, "we at this day have lost the Art of Embossing Iron."[54] More to the point, Wilkinson seems to have seen in the museum at Naples some Roman objects, including a helmet and fine Roman "casque" found near Capua, which he thought very similar. He remembered too a *tabula Isaica* in the King of Sardinia's library at Turin, "made of Iron and covered with Silver." Here were analogies much more useful than anything later. Finally, he found the arguments of Downes about the anachronisms of the shield even easier to meet. Everyone knew that the literary works of Antiquity, even the histories, were filled with them. Perhaps he remembered Dr. Woodward's own note on the subject. "Everything in the Shield," Woodward had admitted, "is not designed justly, anymore than in the other Works of Antiquity in which there frequently occur much greater Oversights some set forth Tritons, Gryphins and Chimaeras. Even the very Inscriptions are sometimes wrong, false spell'd with false Grammar." Strangely, however, Wilkinson supplied only Renaissance examples – Michelangelo, Raphael, and the Caracci. In general, and like his sources, he found the iconography (despite its mistakes) too accurate for the moderns, though not contemporary with the event.[55]

Wilkinson's essay was apparently the last time that the shield was ever defended for its antiquity; and if his arguments were not entirely original, at least they were complete and virtually all that could be said in its favor. The essay was, however, unfinished and in manuscript, and there it remained, unkown for all practical purposes to the world. Whether it would have had any success in print is hard to say. It added little that had not been said before – and while it attempted directly to rebut the critics, it was their cumulative suspicions and not the merits of their individual arguments that had begun to tell against it. Like Dr. Woodward's theory of the earth, the theory of the shield was at least as defensible in the state of existing knowledge, and the critics had not really been more successful in drawing up critical arguments against it than Dr. Woodward and his friends in defending it. On each point there was a plausible reply. It is true that there was something disturbing about the cumbersome efforts that were required to keep patching it against the blows that were beginning to rain down upon it from all directions, but what was still missing in both cases — what was required before absolute conviction could be attained — was a realistic alternative. If Dr. Woodward's shield was not Roman, what was it, and why had so many great men been taken in by it?

[6]

Even while Wilkinson was scribbling away defending the shield, a much better scholar was reviewing the arguments and coming to opposite conclusions. Richard Gough was one of those prodigies for whom all the circumstances of life combined to provide the ideal antiquarian career.[56] Already at twelve and a half he had enough learning to publish a translation from the French, *The History of the Bible* (1747). At Cambridge he got all encouragement in the college (Corpus Christi) which had always fostered antiquarian study. And it was not long before he could claim "leisure, reputation, and an unencumbered income." He used it to travel, collect, and observe, for the many ambitious treatises he intended. "When I quitted College," he recalled, "I by little and little every year made excursions over the greatest part of England and Scotland with only a servant." It was preparation for his most laborious achievements, among them a new greatly enlarged edition of Camden (1789) and his own *Anecdotes of British Topography* (1768, 1780). At one time he thought to republish Horsley's *Britannia*, and he wrote many occasional pieces and contributed to others still. He became a fellow of the Society of Antiquaries in 1767 and its director for twenty-five years (1771-96); he was also its first historian. And he built a fine library, purchased many antiquities and manuscripts — including a number by Dr. Woodward — and inevitably he encountered the Roman shield.[57]

Gough was thus very well equipped to review Dr. Woodward's antiquarian career and the debate about the shield. In the second edition of the *British Topography* he devoted some considerable space to Woodward's London contributions.[58] For it, it seems that he even looked at the manuscripts of John Conyers in the Sloane collection. And now, for the first time since Ward, he decided to rewrite the story of the shield. For this he had a transcript made of some of the Woodward correspondence[59] and he read through the foreign critics carefully, especially Cuper. He pointed out some errors of detail in the accounts by Hearne, Ward, and Gronovius. On the main point, he knew Moyle's objection to the antiquity of the shield and he did not think that Ward (whom he admired very much as a scholar) had removed it. As far as he could tell, both Dr. Woodward's shield and the Kemp helmet that matched it were probably made in Italy in much later times. "No antient artist could be so ignorant as to ascribe such buildings to that period." And he suggested that other later shields resembled it more closely than anything in antiquity, e.g., one depicting the labors of Hercules from the time of the Spanish Armada and another in the Duke of Norfolk's possession engraved by Vertue.[60]

Of the two, the first shield apparently provided the climactic monument in Gough's time for the visitor who went to look at the collections in the Tower of London. "The last thing they shew," according to a popular guide, "is the Spanish General's Shield, not worn by, but carried before him as an Ensign of Honour. On it are depicted in most curious Workmanship, the Labours of Hercules, and other expressive Allegories which seem to throw a Shade upon the boasted Skill of modern Artists. The Date is 1376, near 100 Years before the Art of Printing was known in England."[61] The second was described by Wilkinson, with an enigmatic note in his manuscript. "I believe the Beautiful Target painted by Julio Romano now in the Norfolk family to have been one of K. Henry's Collection." When it was displayed before the Royal Society in 1767, it was noticed that it had been given early in the sixteenth century by the Duke of Savoy to Norfolk's ancestor, the Earl of Surrey. It was just about the size of Dr. Woodward's shield, though made of plaster and painted in gold. Everyone thought that it was executed in a "most masterly manner, great Spirit, and a fine Stile." It was James West who proposed that it might be the work of Julio Romano. The shield had two representations on each side, including an impressive battle scene and yet another story from Livy, the tale of Mucius Scaevola and the Etruscan King, Porsena. Discussion was so lively that almost the whole meeting was given over to it and no other business could be transacted.[61]

Fortunately, no one seems to have thought to compare the Norfolk shield with Dr. Woodward's on that occasion and Wilkinson does not seem to have been moved very much by their similarity either. Nor does it appear that he ever responded to Gough's criticisms. In short, the debate was not renewed. Although Gough's remarks were published, Wilkinson's notes lay silent in manuscript; the shield had no more defenders. And when at last in 1819 it came into the possession of the British Museum, it seems unlikely that the curator was still under any illusions about its ancient provenance. Somehow, without anybody's really noticing it and without any new arguments being offered, the weight of opinion had tipped against the shield, and it was no longer possible to believe that it was Roman.[63]

XV.

Fakes and the Progress of Modern Scholarship

[1]

Dr. Woodward's shield was a fake — and like most such objects, once this was understood, it immediately lost value for the collector and interest for the scholar. In a way this was only fitting, for both its value and interest had certainly been inflated by the false claims that were advanced for it by the Doctor and his friends. And yet fakes and forgeries have their continuing interest, especially to the student of method. For perhaps nothing reveals so well the condition of critical scholarship as how successfully it can distinguish between the true and the false object.

The Augustans were certainly aware that their cabinets were open to suspicion. The Earl of Pembroke, we have seen, actually collected counterfeit coins in order to display them and show off his skills as a connoisseur. The standard manuals on the subject all gave advice to the collector to help him to detect forgeries. Everyone knew the history of the subject. "Jean Cavin who liv'd in the last Century," went one popular account, "had the art to counterfeit those Pieces which Antiquaries call'd Paduan. Alexander Bassien, his Companion, imitated his Cheat. Lawrence Parmesan rendered himself Famous by the same Cheat [and] Cratero, a Fleming by birth, falsified a great Number of Gold Medals. These four Impostors fill'd the World with false Pieces. . . ." Nor did this exhaust the possibilities, "Valerius Bellus of Vicenza knew how to put false Reverses to Greek and Roman Medals"; and there were still other kinds of fraud.[1] As the demand grew, so the supply was found to meet it. Thomas Hearne recorded a conversation with Aaron, an old Jew of Oxford, who was shown a silver shekel belonging to Captain Dove. "I being present with Mr. Pryse and others, when the Captain insisted it was genuine, Aaron, said if so, it was worth five hundred pounds, insinuating, what I said at first sight, that it was spurious. And yesterday passing by Aaron in High Street, I mentioned this shekel to him. *Foo, foo*, says Aaron, *they make such every day*."[2] It was certainly a hazard to the collector and to the scholar. Coins had been used more and more to shore up the authenticity of historical fact against the pyrrhonists – yet even they, it seems, were subject to doubt and error. "All this must certainly harass the Antiquaries and please the Sceptics in matters of History, who have nothing so strong against them as Arguments drawn from Medals."[3]

But outright forgeries were only one concern; fakes and inadvertent misrepresentations were even more hazardous and harder to detect. The Augustans were not innocent here either, and urged caution on the scholar and collector. Moreover, they understood that the representations on ancient objects could also be misleading or false, like the coin of Pope Urban VII which showed his coronation, although he had died before it could take place.[4] In the end, an antiquity, like a literary document, had to be examined and criticized as well as authenticated, before it could be put to historical use.

Gradually the lesson was digested by the scholars, who grew steadily more sophisticated about the possibilities of error and the skills required for detection. When Alexander Gordon (the friend of Baron Clerk) translated Scipione Maffei's *Complete History of the Ancient Amphitheaters* (1730), he saw the value of direct personal observation, of skillful excavation, and of continuous criticism of the received accounts. Only the monuments themselves, he argued, as well as genuine coins and inscriptions, could furnish reliable testimony. Maffei himself was particularly alive to the danger of using engravings rather than the real thing, and to the peril of forgeries. He tells a story about a rare medal he discovered which he believed to be authentic and helpful. As soon as it was found, an artist in the neighborhood took a genuine coin of Constans, with the same reverse, erased the original inscription, and inserted the word *Verona*. Here was a subtler fraud than outright invention! But it did not fool Maffei for a moment (so he says), and he bought it anyway as a proof to confirm the authenticity of his original. It was obvious, he insists, even to someone with only "a middling Skill in such Things," that the forged inscription differed from the original in its characters, in the size, clarity, form, and position of the letters. But he had a more ingenious proof: "It was impossible to make any new Inscriptions in the Contour of the Medal so long but that the very Weight of the Coin would shew the cheat." A pair of gold scales and a comparison of coins was thus sufficient to demonstrate the forgery. Even so, Maffei drew the proper conclusion: "Not to found an Opinion or a new Discovery on the Faith of one Medal, tho none should oppose it," so many and so surprising were the "falsifiers of medals."[5]

Thus the Augustans were on the right track; they understood that criticism depends upon comparison; that critical scholarship and history were therefore, to a considerable degree, collaborative and progressive enterprises. (Here lay the "modern" claim to superiority over the ancients.) By collecting, comparing, and classifying the various objects of their scrutiny, they were furnishing an indispensable groundwork for advance. Only through the systematic accumulation of evidence, scrupulously arranged and compared, could the fake and the fraud, the misrepresentation and the error, be finally revealed. And only through the discrimination of the evidence could the legends that passed for history be distinguished finally from the truth. If simple comparison was not enough by itself to reconstruct the past, it was at least a preliminary condition for success. But everything depended upon the availability of useful objects; and this was why the science of numismatics advanced, with all its hazards, more rapidly than the rest. Still, the problem of detecting a false coin, or extracting its truth, was not essentially different from any other kind of antiquity – a shield, for example, or a fossil.

[2]

Apart from the fact that there were some who thought that all fossils were a kind of fraud perpetrated by the Creator, there were one or two cases of fossil fakes that exactly resembled those of the antiquaries. At least two celebrated instances occurred in Dr. Woodward's lifetime, though on each occasion he seems somehow to have been spared the news.[6] The first involved a German geologist named Beringer and was a case of pure and deliberate fraud. The second involved Dr. Woodward's great friend J. J. Scheuchzer, and was simply a case of mistaken identity.

Johann Bartholemew Adam Beringer was, like so many of our subjects, a physician with a craving for natural curiosities and a theory about the Flood.[7] One day in 1725 he was presented with three stones by some local youths and was elated to discover some remarkable images engraved upon them. He sought out more and was soon rewarded by a great store of even more startling objects, all dug from the same hillside near Wurzburg. Within a year he proudly announced his discoveries in his *Lithographiae Wirceburgensis* (1726) and had them engraved. Here were to be found stones with almost every kind of insect, fish, flower, and bird illustrated upon them, as well as the sun, moon, stars, and comets (complete with fiery tails), and finally, "as the supreme prodigy," some magnificent tablets engraved in Latin, Arabic, and Hebrew characters. Beringer's enthusiasm carried him beyond even the hesitations of his suppliers.

What transpired next is more difficult to say.[8] In the traditional story, not long after the publication of his work Beringer found, among the rocks at Wurzburg – most wonderful of all – a stone bearing upon it in unmistakable letters his own name! At last Beringer realized that he had been duped by his students; he tried vainly to buy back all copies of his book, but ruined himself in the attempt and died not long afterward in poverty and mortification. Some legal documents recently come to light do not alter the story much, although they suggest a more sinister hoax, with Beringer duped rather by jealous colleagues at the university than by playful students. Whether he received any satisfaction in the suit that followed remains unclear, but he did live on until 1740, leaving behind him two more books as well as the recollection of his folly. Later generations remembered the moral: Beringer's exposure "served not only to render his contemporaries less liable to imposition, but also more cautious in identifying an unsupported hypothesis."[9]

Indeed, the discovery that such a hoax was possible was edifying, but the whole episode might have been vastly more instructive had Beringer's work been allowed to run its course and meet the refutation of the learned world. (One would have liked to hear Dr. Woodward on the subject!) As it was, Beringer withdrew his treatise almost before its lesson could be learned, and even managed thereby to gain a modicum of credit for his honesty.[10]

It was a very different case with J. J. Scheuchzer, whose discovery was altogether more significant and more obstinate, although almost as astonishing.[11] We have met the Swiss geologist before as Dr. Woodward's lifelong friend and the ardent exponent of his ideas. In work after work he argued the case for the Flood and the reality of fossils as once-living creatures. Perhaps the most amazing of his many publications was the brief polemic which he dedicated to Dr. Woodward and entitled *Piscium Querelae et Vindiciae, The Complaints and Claims of Fishes* (1708). In this piece the fossil fish, abused throughout the centuries, speak out at last to defend themselves against the sneers of the men who have persistently disbelieved them. They are led by a pike ("of the class of non-spiny fishes with a single dorsal fin") drawn in the first plate and discovered in the Oeningen Quarry near Constance. Since no water had ever been found there, it was clear that the pike (which had only to be compared with a living fish to be identified) was incontrovertible evidence of the Flood. "Judge whether this Pike," Scheuchzer urged his readers, "which we submit for the first course of this philosophical feast, does not deserve to be designated as an authentic witness, more than equal to any objection, of that Noachic Cataclysm . . . ?" "Be not swayed from your assent," he continued, "by Plautus's maxim *None but a fresh fish is good*, for this is a bona fide witness, having been actually present at the Catastrophe. Old indeed he is — 4,000 years not counting his life-span before the Flood; no tasty morsel to the gullet [he], to win the plaudits of Plautus; no welcome guest to the frying pan, but welcome indeed to Museums that sate the mind with his mute utterance. . . ."[12]

For all his conviction and Dr. Woodward's, there was still lacking one piece of evidence which was required to complete the proof of the Deluge. Despite all the plenty and variety of fossil remains so far come to light, no one had yet discovered any sign of antediluvial man. If the Flood had indeed encapsulated the shells and bones of countless creatures that had once walked the earth and turned them to stone, what had happened to the many men who must also have been destroyed in that event? It is true that Dr. Woodward had indicated that the bones of mammals were generally rare, and even proposed an explanation for this, but he had raised no theoretical objection to discovering a human fossil, and there was obviously every incentive for doing so.

It was not surprising then that on the 25th of December, 1725, Scheuchzer wrote to the Royal Society to announce the discovery of two petrified human beings in that same remarkable quarry at Oeningen near Constance. *Homo Diluvii testis*, antediluvian man, had at last been found! In no time the *Philosophical Transactions,* the *Journal des Scavans,* the *New Memoirs of Literature,* and other periodicals were all proclaiming the event.[13] Within a year Scheuchzer drew his conclusions together in a short Latin tract which he published with a large fold-out plate at the end.[14] And he continued to write about his discovery until his death in 1733.

It was some time before doubt was expressed, and even longer before the error could be demonstrated beyond question. Perhaps refutation was only thoroughly accomplished by Georges Cuvier in the fifth volume of his epochal work, *Recherches sur Ossemens Fossiles,* in a chapter entitled "Sur le prétendu Homme Fossile des carrières d'Oeningen."[15] Scheuchzer's *homo diluvii,* it turned out, was no more nor less than a giant fossil salamander!

What had gone wrong? Obviously, Scheuchzer in his eagerness to demonstrate his theory had leaped from a superficial resemblance to a dogmatic conclusion. "Nothing less than total blindness, on the scientific level," Cuvier wrote, "can explain how a man of Scheuchzer's rank, a man who was a physician and must have seen human skeletons, could embrace such a gross self-deception."[16] It seemed impossible that even a cursory examination would not have shown the difference. But Cuvier had forgotten the labor of a hundred years, not least his own, in accumulating fossil remains and establishing the whole new science of comparative anatomy.

No need to trace that development here. Suffice it to say that in Scheuchzer's time animal anatomy barely existed, and comparison was hardly possible. It was Cuvier himself who brought the science to maturity.[17] In his first work, published in 1799, he dealt with elephants. Never before had a competent anatomist with sufficient information studied these fossil animals. By comparing the shape and surface of the teeth, the shape and position of the tusks, the geometrical proportions of the crania, and so on, Cuvier demonstrated the differences between mammoth and mastodon, between extinct and living species.[18] In passing, he showed that the resemblance of elephant to human bones had been the cause of many giant tales, still common among scientists of Dr. Woodward's day. In the meanwhile, among many other activities, Cuvier built a great collection of fossils, discovered many new extinct species, and learned how, with his comparative method, to develop a picture of the whole animal from the merest fragment. He could, he boasted, detect a genus and distinguish a species from only a single fragment of a bone.[19] For a man with his technique and experience among the fossil vertebrates, Scheuchzer's *homo diluvii* must appear immediately ludicrous, and Cuvier had no trouble identifying it correctly, even though it was an unknown and extinct animal, by comparing it with other lizards of its class.

The difference between Scheuchzer and Cuvier, between failure and success, was thus not the result of theory or predilection. It was not Scheuchzer's diluvial fantasies so much as his ignorance of animal anatomy that sustained his error. Cuvier did not abandon the Flood; rather, he extended the systematic comparison of animal structures beyond anything known in Scheuchzer's time. Despite his belief in extinction (itself simply the result of greater comparative knowledge), Cuvier clung stubbornly to the notion of the fixity of the species and the short duration of the world.[20] Exactly like Dr. Woodward, he believed that natural science demonstrated the reliability of Scripture, and when he composed his own *Theory of the Earth* (for so it was entitled in the English translation of 1813)[21] he too insisted that Moses must have been inspired, since he knew the conclusions of modern minerology although he did not know its method. And, again like our hero, he believed that the ultimate goal of his researches was to restore the ancient history of the globe. "As an antiquary of the new order," he wrote, "I have been obliged to learn the art of deciphering and restoring these remains, of discovering and bringing together in their primitive arrangement, the scattered and mutilated fragments of which they are composed, of reproducing in all their original proportions and characteristics the animals to which these fragments belonged, and then of comparing them with those animals which still live . . . an art which is almost unknown." The end was the recovery of the successive epochs of the history of the earth, the setting for universal history.[22]

Thus Cuvier was merely using and improving the method of natural history which Dr. Woodward and his fellows in the Royal Society had already begun to employ.[23] If he was extending it to new territory with wonderful results, it was nevertheless without altering any of their underlying assumptions about the nature of the world. He was still developing a "Theory of the earth," still convinced of the reliability of the Scriptural account of nature, still persuaded that the earth was young and the species fixed, still disputing even with the proponents of the *lusus naturae*.[24] By collecting, observing, comparing, and by putting together the findings of his predecessors and contemporaries, he was able to settle many of the vexed questions that had troubled Dr. Woodward and his generation, not only about elephants and giants, but also about the Irish deer and other such matters.[25] But there was hardly a sentiment or a proof in the whole of Cuvier's essay which Dr. Woodward would not have endorsed, excepting only the extinction of the species – and perhaps he might have been won over even to that by Cuvier's obviously greater knowledge. No doubt Dr. Woodward would have approved also the elimination of Scheuchzer's fossil man which he himself had looked for in vain; but then even the Swiss might well have conceded. "Should someone contend that these are Ichthyospondyls," he had written of his fossil discovery, "I would decline to join issue with him, were he first to show me that they are more congruously associated with the back of a fish than with that of a man."[26] He had the right idea, only the wrong comparison.

[3]

And so it was with Dr. Woodward's shield. The method of the antiquaries had been right; only their comparisons were too circumscribed. The evidence was not yet in, at least in sufficient order and quantity to demonstrate the truth.

What evidence was required? Obviously, before anything else, knowledge of classical antiquities, shields especially, but also all other objects sufficiently related or analogous. Much evidence was accumulating in the cabinets and publications of the learned; much more was being unearthed daily, especially in the new excavations at Herculaneum and elsewhere. What was still missing was a Cuvier to assemble it, someone with sufficient experience of the things themselves, able to compare and classify it systematically and to extract the appropriate generalization. It was Cuvier who had complained of his predecessors (a little unfairly) that they were all either "cabinet naturalists" or "mere mineralogists."[27] And something of the same handicap dogged the antiquaries, with their division between collectors and philologists. What was needed was a comparative method for antiquities to match the comparative anatomy of biological science.

But of course it had been there all along, in embryo anyway, and when it was satisfactorily assembled, it antedated Cuvier by at least a generation. (From the beginning, the classification of antiquities had advanced ahead of natural curiosities.) Dr. Woodward himself understood what must be done for antiquities as for fossils, but he failed with all his business to do it. When at length a Cuvier for classical antiquities did appear, he was none other than the famous Abbé Winckelmann — though to be sure, since the French scientist followed after, it would be fairer to reverse the compliment and entitle Cuvier the 'Winckelmann of paleontology.' (In his own terms, he was "the antiquary of a new order.") But it was not alone in the method and emphasis upon classification that these two great men resembled each other; it was in their invention of principles of generalization which they extracted from the study of their materials and which they could then apply back to objects or parts of objects unknown. For Cuvier this involved discovering the invariable "structures" of animal anatomy and their adaptation as systems to different purposes; for Winckelmann this meant the discovery of artistic "style." Here was a method of classification based ultimately upon simple comparison, but more sophisticated and successful than anything yet attempted. Like Cuvier, Winckelmann learned to place even the merest fragment in a meaningful whole and thus startle his contemporaries by his genius in authenticating objects and detecting fraud even when only a limb or broken torso survived.[28]

To consider Winckelmann only as an antiquary is of course to overlook his greater reputation and importance as a connoisseur and "philosophical teacher of the beautiful," as the leader in an aesthetic revolution and the fountainhead of the notorious "tyranny of Greece over Germany."[29] But from our perspective, and whatever else might be claimed for him, Winckelmann was an antiquary and a scholar in a continuing tradition, and he would probably not have objected to the title. (He was so pleased with his election to the English Society of Antiquaries in 1761 that he had the list of its members framed and hung in his room.)[30] He was self-taught, it is true, but he had trained himself systematically first in classical literature and philology, then in the antiquities themselves, coins, gems, and especially the statuary of ancient Greece and Rome. He learned from artist friends the mysteries of restoration and fabrication, and from the excavations at Herculanium, which he inspected several times for himself, the problems of practical archaeology. (His reports and criticisms have established for him a claim as founder of the modern science.)[31] He annotated a collection of several thousand antique gems, the greatest of its kind, an exercise in identification and classification paralleled only by the achievements of the numismatists and perhaps even more demanding.[32] Above all, he looked more intently than anyone else both at the ruins of the Eternal City and through the vast collections of ancient monuments that had been assembled there. Eventually he was appointed Superintendent of Antiquities in Rome and became famous throughout Europe for his guided tours of the city. The *éloge* upon his death by the famous classical scholar Heyne is a paean to the extraordinary confluence of skills and opportunities that produced the greatest antiquary of his time.[33]

Heyne's eulogy did not overlook the parallel in method between the naturalist and the antiquary. Just as the former, he says, must have arranged in his memory in a methodical fashion all the objects of Nature, so too the latter must have a perfect knowledge of the monuments of Antiquity, each of which he must classify and explicate, determining the art, the century, the authenticity, and the merit of each. The essence of the task is *examiner, comparer et juger.* Winckelmann's triumph here, his most characteristic antiquarian work, was the *Monumenti Inediti* compiled at the end of his career. (The Abbé had a copy sent to the Society of Antiquaries.)[34] Although less well known now than his theoretical tracts and aesthetic theories, it was the major work of his last years, an effort to explicate a series of classical monuments, either unknown or mistaken, according to the standards of critical scholarship. "Before doing any research," Heyne had emphasized, "it is necessary to begin by knowing which monuments are really ancient, or which parts are ancient and which modern. That manner of proceding has not yet been adopted by everyone, least of all the Italian antiquaries. . . . M. Winckelmann has shown us by his work the new means that one can follow faithfully."

The new method was not so much new as it was more systematic. Winckelmann began his famous *History of Art* with a long introduction to the point. First, he noticed that all the collections and learned works of his contemporaries were crowded with errors. At Wilton House, for example, among the statues of the Earl of Pembroke, there were said to be four by the ancient artist Cleomenes. Winckelmann was particularly astounded that one of them was asserted to be the very one that Polybius had brought with him from Corinth to Rome. Pembroke, he thought, might just as well have sent the artist directly to Wilton![35] But everywhere it was the same. Montfaucon, who "saw with the eyes of others," was led into innumerable errors, like attributing the Statue of Hercules and Antaeus at the Pitti Palace to Polyclitus himself, "a statue of inferior rank of which more than one half is of modern restoration." Spon had presented as genuine antiquities a round marble from the Giustiniani villa and a vase on which he wrote a whole essay, despite the fact they were both completely modern. Moreover scholars had naturally drawn wrong conclusions from these errors. "From a relievo in the Mattei palace, which represents a hunt by the Emperor Gallienus, Fabretti wished to prove that horseshoes, nailed upon the foot in the mode of the present day, had already come into use at that time; he did not know that the leg of the horse had been repaired by a sculptor not versed in such matters."

In Cuper's dissertation on the *Apotheosis of Homer*, the draughtsman had considered Tragedy to be a male figure and overlooked a buskin clearly depicted in the marble; he had placed a scroll instead of a plectum in the hands of a Muse and entirely misunderstood a sacred tripod depicted there. There was not a scholar in Europe, apparently, who had avoided these mistakes, and Winckelmann piled example upon example to make his point.[36]

One trouble, he explained, was that they did not have sufficient firsthand experience with the monuments themselves. "It is difficult, indeed almost impossible, to write in a thorough manner of ancient art, and of unknown antiquities anywhere but in Rome."[37] Even a residence of two years was insufficient, as he himself discovered. Another difficulty was ignorance of restoration; for since few ancient statues survived intact, modern additions were normal and usually mistaken for genuine. "I soon discovered the general rule, that the detached parts of statues, especially the hands and arms, are for the most part to be looked upon as new."[38] A critical examination of art (as of literature) depended upon distinguishing the modern from the ancient, the genuine original from the additions; but there was no substitute for practical experience in the workshops of the restorers. Reliance on engravings or drawings was necessarily misleading, however honestly they were intended.

But the difficulty lay deeper still; the scholars of the past had expended all their energies on the attempt to ascribe particular works to particular artists. (It was exactly like the effort of antiquaries to ascribe the foundation of cities or the invention of the arts, or anything else of consequence in the past, to specific nameable individuals.) In doing this they had underestimated the general characteristics that marked off the work of one culture from another, or one period from another; they had failed to develop a really useful scheme of classification. In a word, they had lacked an understanding of the historicity of artistic style.[39] Winckelmann boasted that he was able to define the special characteristics that defined Egyptian and Etruscan art, and even more confidently, the several periods in the history of Greek art. He had not meant to compile another chronicle of artists; he was creating instead a new kind of art history, "intended to show the origin, progress, change, and downfall of art, together with the different styles of nations, periods, and artists, and to prove the whole as far as it is possible, from the ancient monuments now in existence."[40]

How much this was the result of preconception, how much the fruit of Winckelmann's study, I leave to others to decide. Viewed simply from the point of view of scholarship, Winckelmann's contribution was to make clear the necessity of prolonged saturation in the objects of antiquarian study, coupled with the evidence of ancient literature. Whatever the value of his generalizations about Greek style (and they have been variously received), his method made him more conscious of the problems of authenticity than anyone had been before, and thus more capable of detecting a fake. Of course he was far from infallible, and his erstwhile friend, the painter Mengs from whom he had learned so much, palmed off on him a forgery of his own which appeared in the *History of Art* and for a time destroyed their friendship! The fact that he had never visited Greece and knew classical Greek art largely through Hellenistic works and Roman copies was a serious handicap. (Thus he placed the famous Laokoon three centuries too early and misrepresented its intention.) But his judgments on his predecessors were largely correct, and he advanced the state of the discipline enormously by his acute and learned criticism.

Nor did Winckelmann overlook Dr. Woodward's shield. His knowledge of classical antiquities was too complete to pass by that well-known work. He may have met it first in Ward's *Lives of the Gresham Professors*, which he read and excerpted.[41] Eventually he seems to have come upon Henry Dodwell's treatise and the engraving there. Certainly he read Melot and de Boze in the *Memoirs of the Academy of Inscriptions*. It was enough; he did not hesitate to form an opinion which he printed in the preface to his celebrated treatise on the ancient monuments.[42] For Winckelmann, it was but another example of the mistakes of his predecessors: a modern work masquerading as an antiquity. Inevitably he compared it to Spon's shield, but with the surprising submission that that work too represented an error – though of a different kind. Despite Dodwell's arguments, he had no doubt that the great scholar was mistaken about Dr. Woodward's shield; and despite the affirmations of Spon and a host of others, he had no hesitation in suggesting that they had completely misunderstood that work. The representation on its face, he urged, was not Scipio Africanus but must be rather the return of Briseis to Achilles; it was therefore not a Roman but a Homeric representation, though late and decadent in style. Although he did not elaborate, and offered not a word to justify his judgment on Dr. Woodward's shield, it would have been hard to deny the authority of his verdict. He had seen too many of the real objects and knew too much about classical style – and Roman art – to be taken in by a sixteenth-century imitation.

[4]

The first difficulty, therefore, for the scholars who undertook to evaluate Dr. Woodward's shield, was that they were unable to compare it with enough undoubted works of antiquity. Unfortunately they knew no more – and in some ways even less – than the artist who had created the shield in the first place and tried to make it look genuine. There was as yet little classical sculpture in England, and what there was was at least as dubious as Dr. Woodward's work and likely only to compound the difficulty. As a result, they were forced back upon the faulty and inadequate evidence of engravings and medals.

Take the Earl of Pembroke's collection, for example. Here beyond doubt was the finest accumulation of classical objects — books, manuscripts, coins, and sculpture — to be found in England. The Earl had worked hard to assemble it, alert to every opportunity to purchase. He had bought from the collections of the Earl of Arundel, Cardinal Mazarin, and at the great sale of the Giustiniani, assisted as always by his close friend Sir Andrew Fountaine. "He resolved," we are told, "not to run into all sorts of curiosities but to buy such as were illustrative of ancient history and literature," especially busts and statuary. He confined his choice always "to the best Ages."[43] His collection was appreciated, indeed celebrated, by all the scholars of his generation. When Humfrey Wanley was invited to dinner, he described the coins he was shown as "perhaps the most Curious and Gentlemanly Collection in Europe. And my Lord gives us the best account of them of any man I know."[44] It was the universal opinion, echoed by Nicolson and Thoresby, by Heneage Finch, by Joseph Wasse, and indeed by Dr. Woodward himself, who contributed to the Earl's collection. There was not a prince in Europe, wrote Alexander Gordon, with a nobler treasure of medals and coins, statues, busts, and bas-reliefs. And when Baron Clerk visited Wilton, he thought he had looked upon "the greatest collection of Greek and Roman statues that ever I saw in any palace abroad." His friend Roger Gale agreed.[45]

Nor was Pembroke merely a dilettante collector. Despite much business and high office, he constantly fussed over his collections and impressed everyone with his learning. "He is very busy from morning to night in marshalling his old fashioned Babys as Sir Isaac Newton calls 'em; he is distributing them in proper classes, such as Busto's, Inscriptions, basso relievos, Statues, etc. These he is placing in distinct Rooms, his Egyptian, Syrian, Lydian, Thracian, Greek, Roman Marbles. Then he is mustering by themselves the old Greek Persons, the learned Persons, the Consular, the Emperors, Empresses, the Divinitys, etc. So that you may suppose that he generally walks 10 miles a day in his own house and sometimes in his Slippers and sometimes is so busy that unless Ladys come to visit him he will fobb off his beard (as he calls it) for two or three days."[44]

Yet it was this collection which, as Winckelmann suggested, was riddled with forgeries and mistaken attributions, many of them caused by the bad judgment of the Earl himself. Once or twice this was suspected by contemporaries, though no one seems to have understood the extent of the problem. On one occasion Thomas Hearne wrote to his friend, the collector James West, to question an ancient Greek inscription in the Earl's collection. He thought that the inscription must have been added after the object came to Pembroke, since it had been seen ten years before without it. If true, "'twould make many other things in that Collection suspected." In any case, he confided to his diary that Pembroke "was not a man of that deep Penetration, nor of that profound Learning he is taken to be."[47] Even Pembroke's friends, Clerk and Gale, agreed that his judgments sometimes appeared arbitrary and dogmatic. Nevertheless, these two men, both of them considerable scholars and practical archaeologists, both of them critics of Dr. Woodward's shield, accepted most of the Earl's judgments and were entirely taken in by his collection. Indeed, Gale admired the sculpture so much that he urged Pembroke to have it engraved and published. "I believe nothing on this side of the Alps exceeds his Collection," he wrote to Clerk, "and nothing consequently could do more honour to him and the nation among the Learned than such a Description of his Curiosityes." Clerk agreed. "The Antient Statues and Bustoes here make a noble collection and such as wou'd befit the house of a great King."[48]

When, however, Pembroke had them actually engraved by one Cary Creed, the result was disappointing. Gale hoped they would be published in a "scientific" manner; upon seeing a specimen, he thought the drawing poor and doubted that Creed would be allowed to finish. When the work was completed, no one, except perhaps the Earl himself, was satisfied. Thomas Hearne and his friend West agreed that it was "ill-executed." Not only were the drawings poor, but the descriptions (in English) were "miserably spelt, and the Syntax horrid."[49]

But bad grammar was the least of it. It was through Pembroke's descriptions that the Greek sculptor Cleomenes was assigned several pieces, and many of the rest were mistakenly identified or attributed to other classical artists. "The Bustoes," Clerk had been told at Wilton, "comprehend the heads of most of the Antients as far as can be guessed from Coins or antient heads extant in other collections."[50] Pembroke simply could not bear an anonymous or unidentified piece of antiquity. ("No unknown Heads nor Fragments," it is said, were admitted into the Collection.)[51] The identifications were essential to their purpose. "The great Number of Persons," wrote a contemporary catologer of Pembrokes's many busts, "is not only an Ornamental Furniture but more Instructive and Pleasant to those who have read of them and an Incitement even to young Folks (as my Lord observ'd among his children), by seeing the Persons, to learn something of them. . . ."[52] The descriptions on the published plates were his own idea; Clerk recognized the style at once. "It seems that he has there set down his notion of each piece, and has obliged the etcher or engraver to make it, as he wrote it, part of the copper-plate."[53] This explained why some things were so "dogmatically" expressed there. Pembroke, however, was not satisfied with simply identifying his statues, he often had the forged names chiseled into the monuments themselves. One incinerary urn, perhaps the same one previously owned by Dr. Woodward, was thus attributed directly to Horace.[54]

But if neither Clerk nor Gale was entirely pleased with Cary Creed's work,[55] they neither, so far as I can see, raised any objections to Pembroke's attributions. Like everyone else, they accepted the marbles as genuine antiques and submitted to the Earl's judgment. Indeed, when Clerk visited Wilton, he picked out for special praise the same works that Pembroke most admired, including his greatest favorite, the relief of "Curtius precipitating himself into the gulf." This was the very piece that Pembroke had attributed to Cleomenes, "the finest Work by that Greek Sculptor," brought from Corinth to Rome by Polybius. The Earl described it as "among the best pieces of Sculpture relating to the Romans," and Clerk agreed that it was "very fine."[56]

There is a wonderful irony in the thought of Clerk standing reverentially in front of the Earl of Pembroke's forgery and admiring it (like his host) even above the genuine statuary in the collection. The story of Curtius, like the story of Camillus, could be found dramatically told by Livy and with just about as much veracity. Here again was false testimony to a legendary event, although one does not know whether Clerk or Pembroke employed it that way. (Even Livy seems to have had his doubts.) For the scholar today – as for Winckelmann – "the relief is entirely modern" and Pembroke's effusion "highly amusing." The great work was a fraud, no more nor less than Dr. Woodward's shield, although for the moment it passed entirely without question.[57]

No wonder then that the Earl accepted Dr. Woodward's shield as genuine. (I do not know what Dr. Woodward thought of his Cleomenes – or if he ever visited Wilton.) Pembroke's whole collection, despite some undoubted works, had been spoiled by "the large number of spurious pieces, the abominable restorations, and the absurd nomenclature." To compare one Renaissance imitation with another was bound only to confirm one mistake by another. In either case, the predisposition to believe was not easy to dispel, except by a knowledge of original works more intimate and more extensive than could possibly be achieved in England or on a casual tour of the Continent. It was only surprising that Clerk had been able to raise any doubts about Dr. Woodward's shield. But these rested after all not so much on grounds of style as on disbelief in its material; and if Dr. Woodward's sculpture had been made of marble rather than metal, it would probably have passed the scrutiny even of the suspicious Baron. In art, as in literature, the neoclassical sensibility triumphed in large part for lack of scholarship, and Cleomenes, Camillus, and the rest all took on an eighteenth-century air, despite the best efforts of classical erudition.

[5]

But if the Earl of Pembroke's marbles illustrate the general problem, the shield of Dr. Spon shows it even more exactly. Here, after all, was the single most obvious comparison that could be imagined for the newfound monument of Dr. Woodward, an object of apparently identical purpose and pedigree. Over and over again it was illustrated and explicated during the period, and there was not a single commentator on the Doctor's shield who did not notice it.[58] What a shock then to discover that that piece also was completely misunderstood and therefore entirely misleading!

Once again it was the Abbé Winckelmann who pointed the way. He insisted that the picture on its face could not possibly be Scipio but must be Agamemnon returning Briseis to Achilles; moreover, the art was not of the classical period but belonged to a later and more degenerate time.[59] A generation or so afterward another famous scholar, the curator of the Bibliothèque Nationale (the successor to the Cabinet du Roi, where the shield had finally come to rest), Aubin-Louis Millin, confirmed Winckelmann's judgment with a more elaborate argument.[60] He showed that Spon's iconography simply did not match the literary evidence; that he was wrong in his identification of almost every detail from the costume and the beards on the figures to the architecture in the background. Worse yet — and most disconcerting — Millin strongly doubted that the work was even a shield! Using Spon's own evidence, the representations on ancient coins, he suggested that nothing so elaborate had even been hung in a temple. This was no *clypeus votivus*. It was clear upon closer inspection that the object was really a great dish, an example of what the Greeks called *diskoi* and the Romans *lances* or *tympana*. (In fact, the suggestion had already been made in Dr. Woodward's day, though it was not generally accepted.)[61] He placed it on grounds of style in the Empire, perhaps in the reign of Septimius Severus. His identification was accepted and has remained the official description.[62]

But it is unnecessary to belabor the point. Wherever the antiquary of Dr. Woodward's day turned, he was beset by difficulties. On the whole there were simply not enough comparable objects yet to be seen. Where there were some, they had to be examined largely through the medium of engravings, which were likely to introduce the very inaccuracies and anachronisms that were being hunted out. Finally, on those unusual occasions when an exactly comparable object did appear, it was just as likely itself to be a fake as Dr. Woodward's shield. Winckelmann and Millin were right; the shield of Dr. Spon was not a shield, and it did not illustrate Livy. It has even been suggested recently that it too was a Renaissance copy![63] In any case, it was no better for the purpose of authenticating Dr. Woodward's shield than *Curtius leaping into the gulf.* Before antiquaries could be certain about what they had before them, when for example they read Dodwell's dissertation or looked at Dr. Woodward's cabinet, they had to know a great deal more about what still lay hidden beneath the earth. When museums at last became stuffed with the fruits of Herculaneum, Pompeii, and a thousand lesser sites, when Millin resumed the study of Dr. Spon's shield, many of those difficulties began to disappear and the work of comparison became relatively simple. (For the study of shields, it was the Naples Museum, with the spoils of the new archaeology, that became most important.)[64] Not that discrimination is ever perfect; fakes continued to abound in the museums of the nineteenth century, and without doubt some still evade their more sophisticated custodians today.[65] The detection of forgery is but one aspect of the continuous evolution of modern scholarship. Meanwhile, even as Dr. Woodward had seen, piece by piece, each new discovery had to be fit into its class so that gradually the material contours of the ancient world could take more precise shape. Along the way, some of the mistakes had to drop out; sometimes, like Scheuchzer's *homo diluvii* or like Dr. Woodward's shield, they could no longer be made to fit.

[6]

I have left for last the other side of the dilemma. It was one thing to discover what Dr. Woodward's shield *was not,* and for this a knowledge of classical antiquities was required. It was another thing to learn just what Dr. Woodward's shield *was.* For this – since it was a sixteenth-century object – knowledge of Renaissance art was necessary. But here too the antiquary was beset by difficulty. It is true that much of the classical art of Renaissance Italy had begun to flood into England, though more of painting and drawing than of sculpture. But of modern armor, very little seems to have been known. The fact is that in 1700 almost no interest seems to have been taken in the many decorative shields that had been created in abundance in the sixteenth century and which were the real counterparts of Dr. Woodward's weapon. It seems incredible that no one should have suggested the comparison which alone would have immediately identified the Doctor's object, excepting only Cuper's Dutch friend Vilenbroek, who happened to own one, and the admirable de Boze.

So backward was the general state of knowledge about arms that when a visitor was shown the collections in the Tower he was directed among other things to a musket employed by William the Conqueror! A sixteenth-century suit and great sword were also attributed to the Normans, along with the armor of his jester.[66] Medieval armor was not studied at all; the "age of the tournaments" invoked by the Paris antiquaries to oppose Dr. Woodward's shield was regarded generally as Gothic and barbarous, and banished on the whole to that limbo to which the Renaissance humanists had consigned it. It was Woodward's friend Harris who characteristically defined "Gothick" syle in the *Lexicon Technicum* as a manner of building imposed upon the "civilized world" by a rude people who "demolish'd what they could of the Ancient Greek and Roman Architecture, and instead of those Admirable and Regular Orders and Manners of Building, introduc'd a licentious and fantastical Manner . . . without any of that August Beauty, and just Symmetry, which the Fabricks of the Ancients entertain us with." Just so had Addison described the Cathedral at Siena.[67] If classical archaeology was relatively young, medieval archaeology had to await the Gothic revival and the nineteenth century to awaken any interest at all. With the Renaissance it was different, but it is doubtful whether the eighteenth century was much aware that there was a distinction between Gothic and Renaissance armor. Certainly no one in England seems to have thought seriously of collecting or studying knightly armor until the very end of the eighteenth century.

There are one or two exceptions, perhaps the best-known belonging by an odd chance to Dr. Woodward's old enemy Dr. Richard Mead. When or how he obtained his shield I do not know, but it was a splendid work indeed. In the words of the Wallace Collection Catalogue (where it may now be seen), it is "boldly embossed, damascened and plated with gold, with a composition representing Scipio Africanus receiving the keys of Carthage after the battle of Zama, B.C. 202."[68] There are several figures, a background landscape of Carthage and the sea, a "winged genius bearing the crown of Victory," and a border festooned with fruits, flowers, and vases. In size, metal, and splendor, it far outshone its rival. It seems first to have come to scholarly attention in Francis Grose's *Treatise on Ancient Armour* (1786), a work memorable for being the first to treat its subject. Grose printed it as his frontispiece and explained it only as some conquering general receiving the keys to some unidentified ancient city. He saw that the shield was not ancient, however, and suggested that it might be from the fifteenth century. "It was probably used as one of the insignias of dignity commonly borne before the generals in chief of that time."[69] Apparently he was unable to read the cartouche at the top which contained the entwined initials of Henry II and Diane de Poitiers. Its date is thus about 1550, and it may be that the shield commemorates the capitulation of Calais in 1558 – although an Englishman might perhaps be forgiven for overlooking that unfortunate event.

In short, there was at least one contemporary shield in England in Dr. Woodward's day which might have helped to place his work. To a modern expert its grand effect and richness of design make it "almost Giorgionesque in its simplicity."[70] But one has the uneasy feeling that Dr. Mead might well have purchased it as an antique, and that he could anyway have palmed it off as one if he liked. In Grose's day this was harder to do; now of course it would be impossible. We know too many other shields of the same vintage, possibly even by the same artist. On grounds of style and date it has been tentatively assigned to the Milanese, Lucio Piccinino.[71]

Were there any others? Probably, but it is unlikely that there were very many. It would have been helpful if the sixteenth-century wooden buckler now in the Tower which is painted with scenes from the *Life of Camillus* had been known to Dr. Woodward. It might certainly have awakened suspicion – but it only came to the Tower at the Bernal sale in 1855.[72] In 1737 Baron Clerk learned of a similar work. "One of our physicians at Glasgow," he wrote to Gale, "has lately got a present from a friend of his in Spain, which is a shield of the same size and metal with that which Dr. Woodward had. It has a good deal of raised work upon it, representing a concert in a field, by 7 or 8 figures, all females with different instruments." He would not say that it was definitely Roman, no doubt thinking of the Doctor's shield, though it was "certainly of great antiquity." One obvious problem was an organ with pipes among the instruments as well as "two figures much like Violins, which I never observed on any Roman monument."[73]

So the antiquaries were thwarted here too; there was simply not enough to go on, although there was enough surely to suggest doubt. The obscure provenance of Dr. Woodward's shield certainly did not help his cause. Yet it was a shrewd argument to place its transmission back in the reign of Henry VIII. The greatest obstacle to the study of Renaissance shields in England was that the real center of their manufacture had been on the Continent, and the one occasion for their transmission Henry's love of pageantry.[74] This accounted for their scarcity in England. And so with no classical shields and few modern ones, it was clearly not possible to do much with the whole subject except by analogy.[75] Under those circumstances – and always bearing in mind their strong inclination to believe in any classical object – it was really quite remarkable how clever and sophisticated the defense of Dr. Woodward's shield was. Unfortunately, as with all his other opinions, Dr. Woodward had attempted a conclusion before a positive determination was possible. The critics however had as little to go on and did not do much better. But they had the great advantage of being right. Dr. Woodward's shield, we are told now, is without doubt a sixteenth-century shield, although its exact origin remains even yet a mystery.[76]

[7]

And so our story ends in anticlimax. Yet perhaps it is not the less instructive for that. At the outset, I warned that the interest of the tale of Dr. Woodward's shield lay less in the narrative than in its underlying intentions: to lay bare a neglected portion of the Augustan intellectual landscape and to place the scholarly achievement of the time in the sweep of modern learning. For the first, I hoped that the story might be left largely to speak for itself and that in resurrecting it, in all of its obscure detail, something of the significance of Augustan learning to Augustan culture might be the better appreciated. For the second, it is time now perhaps to draw some conclusions, if only provisionally, so that we may close our story with an appropriate moral or two.

My first suggestion is that history, or rather antiquarian study, despite a frequently misleading appearance, already possessed in Dr. Woodward's day a "scientific" dimension — that is to say that it employed a method that could be used to accumulate reliable information and to distinguish between truth and falsehood, "fact" and "fiction." In this respect it exactly resembled those other scientific disciplines so dear to Dr. Woodward: geology, paleontology, and the rest of "natural history." If it was not always, nor even usually, successful, it was on the right track, committed to a general intention and to a set of procedures that could in the end accomplish the job. If we have at our disposal now a mass of dependable information in all these fields to which we are constantly adding and amending and from which we try to mold our general views of the world and its history, we are, it seems to me, building still upon the foundations laid by Dr. Woodward and his contemporaries. And if we no longer accept the validity of the *Epistles of Phalaris,* the *homo diluvii* of J. J. Scheuchzer, or the Roman origin of Dr. Woodward's shield, if each of these has been finally assigned to its correct place in the world of history or nature, that is because we have inherited both the method of the antiquary and its assumptions. Thus it was Dr. Woodward's activity, as the "ancients" well understood, that made him modern, not his "hypotheses" or conclusions which, in the way of all science, had perpetually to give way to something better. And this must be my excuse for describing a "failure" rather than a triumph; for in the end it is more illuminating of the real evolution of a discipline to concentrate upon its techniques rather than its results, and nothing perhaps is so revealing in this regard as a mistake. It was after all really no change in method, but rather the same set of procedures applied persistently and cumulatively to the same problems, that gradually turned Dr. Woodward's shield from a Roman object giving evidence about Roman history to a Renaissance object of no use whatever to understanding ancient Rome.

So we may conclude that history had, and retains still, a critical or "scientific" dimension. Whether that entitles it to be called fully a science or not I leave to the philosophers, who have for a long time now been much exercised by the question. But the story of Dr. Woodward's shield is a reminder, in any case, that historically the discipline of history, with all its ancillary techniques, did in fact take its rise in close relation to natural science (or some part of it), and that many of its problems, procedures, and proponents were often the same, or at least analogous. This is not to claim that the one was the cause of the other; nor is it to say that the achievement of the antiquaries furnished either the whole or necessarily the most important ingredient to modern historiography. History may in some sense be a science, but if so it is a science *sui generis* and the ancients, with their ideas about narrative and causation and about the connections between history and literature, had much to offer also. Moreover, the whole problem of "historical explanation" so dear to modern philosophy (and so central to the contemporary debate about history as a science) was of little interest to the antiquaries and falls outside our purview. But it *is* to say, even to plead, that no consideration of the nature of history and its possibilities can or should entirely dispense with an estimate of the value and character of the kind of thought that transformed Dr. Woodward's shield from one sort of object to another, the quality that I have for convenience labeled "scientific."

There is a second and related conclusion that seems to me to arise from our story. It is that the great transformations of thought in natural science and in history that followed upon Dr. Woodward's work depended at least as much upon the slow and gradual accretion of information, upon the normal work of the disciplines, as upon any sudden revelations or any large theories brought from outside. Thus the world of the dinosaurs and the world of prehistoric men, both of which required a vastly extended duration of time, were both conjured into being only after, and largely as a result of, the discovery, fossil by fossil and implement by implement, of those lost worlds. It was not I think possible to anticipate those discoveries (nor, if anyone could, then to make them plausible), and it was equally impossible *after* enough evidence had accrued to withstand for very long their cumulative effect. It seems to me quite certain that Dr. Woodward, despite his stubbornness and his tenacity, would have accepted (like almost everyone else) the new worlds which his own discoveries helped to bring into being but which his own theories denied.

And so perhaps the story of Dr. Woodward's shield was fated from the first to anticlimax. To the extent that the history of ideas is the story of evolutions and not revolutions, something of the drama must be lost.[77] While there are problems in human thought that may be solved definitively and at a single go, some of the more intricate and many of the more intractable are only puzzled through gradually, cooperatively, sometimes even inadvertently, in the course of other things. Thus without anyone really attending to the matter very directly, Dr. Woodward's shield was transformed from one kind of an object into another. How had it happened? As we have seen, the advance of the antiquarian discipline, the immense accumulation of archaeological detail, slowly but steadily altered the whole configuration of classical antiquity and the Renaissance until they appeared very different to the scholars who were attempting to piece them together from their remains. Where once the shield had fit nicely into one configuration of objects, a century later it quite clearly belonged to another. The very gradual character of this process meant that no single point was decisive, and it is hard now to know just when the proofs became conclusive. But in a hundred years everything had changed, not only Dr. Woodward's shield, but – like the whole idea of the prehistory of the world – those other great constructions of the historical imagination, the ideas of Antiquity and of the Renaissance as well. Our story is thus a good example of that peculiar circularity of reasoning, so well known to the philologist,[78] whereby the parts may be shown to determine the whole and the whole the parts – not statically, but in some progressive relationship that takes us gradually closer and closer to the truth. That Dr. Woodward understood something of this process is undeniable, but he was temperamentally too dogmatic and too impatient to rest content with so little. Here at last the wits found a genuine mark for their satire; and here, perhaps with echoes of their laughter to egg us on, we may draw the final moral from our tale.

Appendix: A Newly Discovered Woodward Manuscript

Dr. Woodward's books and manuscripts were largely dispersed at his death, and it has not been possible to recover all of them. One would give a lot to find the missing file of letters that we know he kept and that would almost certainly have contained correspondence from Newton, Locke, and Leibniz, among many others! Recently I learned that Dr. Woodward's own interleaved copy of the *Essay toward a Natural History of the Earth* had been slumbering all this while in the Geology Library of Pomona College in California. There in a clear and unmistakable hand was further evidence that Woodward never gave up his ambition to defend and elaborate his geological theories.

The notes are copious but a little disappointing. The great majority consist of quotations and citations in several languages to a wide range of authorities in support of his original views. The latest seems to be a note on the weather in 1718. Occasionally Woodward records information from his own experience and from his unpublished works. So, for example, to support his view that the flood could raise metals, he reports an experiment he made with a solution of copper in water (p. 208). He refers also to a manuscript tract, *Propositions concerning Bodies Supported in Fluids*, where he tries to show that the resistance of a fluid to a solid immersed in it is proportional to the extent of the surface of the solid (p. 206). Elsewhere he recalls the fossil river shells that he had found himself and in the collection of the apothecary John Conyers (p. 255), and information from Hugh Howard about incrustations in the aqueducts of Rome (p. 199). At the end of the book, he offers a "Catalogue of the Bodies mentioned in the Nat. Hist. Earth," a four-page list of fossils. Once in a while his irascible personality shines through. It appears that Woodward had contributed to John Harris's friendly *Remarks* a review of past geological opinions (see above, p. 39). "It had not been worse for the Discourse," Woodward complained, "if this ingenious Gentleman had every where kept my *Words*, and spared some Interpolations of his own that are inserted here and there in it" (*Essay*, p. 39).

Throughout, Woodward remains inflexible. He amasses new evidence to prove that the flood must have occurred in the spring (p. 165). And in a long note (p. 80), he defends his problematical view of the formation of strata. But most of the time he is content simply to reinforce his old views with new citations.

Abbreviations in the Notes

A. Manuscripts

The bulk of the manuscripts employed in this work may be found in one of three great libraries: the British Museum (now the British Library); The Bodleian Library, Oxford; and the Cambridge University Library. Each of these libraries defines its manuscripts by collection as well as by volume, and I have abbreviated in the obvious and customary ways, e.g., B.M. MS. Add. 6127 means British Museum, Additional Manuscript no. 6127. For the many other libraries and collections consulted I have used the following abbreviations, preferring otherwise to give the full citation in the notes:

B.M.	British Museum, London
Bodl.	Bodleian Library, Oxford
C.C.C.C.	Corpus Christi College, Cambridge
Camb.	Cambridge University Library
Cuper Corr.	The correspondence of Gisbert Cuper in the Koninklijke Bibliotheek, The Hague
Fitz. Mus.	Fitzwilliam Museum, Cambridge
Nicolson Diary	The diary of William Nicolson at Tullie House, Carlisle
Penicuik	The Penicuik MSS. of Sir John Clerk on deposit in the Scottish Record Office
R.S.	The Royal Society, London
Scheuchzer Corr.	The correspondence of J. J. Scheuchzer in the Zentralbibliotheek, Zurich

Soc. Antiq.	Society of Antiquaries, London
Sperling Corr.	The correspondence of Otto Sperling in Det Kongelige Bibliotek, Copenhagen
Yorks. Arch. Soc.	Yorkshire Archaeological Society, Leeds

B. Published Sources: Primary and Secondary

A.O.	Anthony Wood, *Athenae Oxoniensis,* ed. Philip Bliss (4 vol., London, 1820).
Aitkin, *Arbuthnot*	George A. Aitkin, *The Life and Works of Dr. Arbuthnot* (Oxford, 1892).
Beattie, *Arbuthnot*	Lester M. Beattie, *John Arbuthnot, Mathematician and Scientist* (1935; reprinted New York 1967).
Bentley Corr.	*The Correspondence of Richard Bentley,* ed. J. and C. Wordsworth (2 vol., London, 1842).
Bibl. Top. Brit.	*Bibliographia Topographica Britannica,* III (London, 1790).
Biog. Brit.	*Biographia Britannia* (6 vol., London, 1747-66).
Birch	Thomas Birch, *The History of the Royal Society* (4 vol., London, 1756-57).
Britannia	The *Britannia,* ed. Edmund Gibson (London, 1695).
Bull. Inst. Hist. Sci.	*Bulletin of the Institute of the History of Science.*
Chauffeppié	Jacques Georges de Chauffeppié, *Nouveau dictionnaire historique et critique* (4 vol., The Hague, 1750-56).
Cuper, *Lettres*	Gisbert Cuper, *Lettres de critique, d'histoire, de litterature, etc.* (Amsterdam, 1742).
DNB	*Dictionary of National Biography.*
Eng. Hist. Rev.	*English Historical Review.*
Gent. Mag.	*The Gentlemen's Magazine.*

Harris, *Remarks*

John Harris, *Remarks on Some Late Papers Relating to the Universal Deluge* (London, 1697).

Hearne, Leland, *Coll.*

John Leland's *Collectanea*, ed. Thomas Hearne (6 vol., Oxford, 1715).

Hearne, Leland, *Itin.*

John Leland's *Itineraries*, ed. Thomas Hearne (9 vol., Oxford, 1711-12).

Hearne, *Remarks*

Thomas Hearne, *Remarks and Collections*, ed. C. E. Doble, D. W. Rannie, and H. E. Salter (11 vol., Oxford, 1885-1921).

Hooke, *Posth. Wks.*

The Posthumous Works of Robert Hooke, ed. R. Waller (London, 1705).

Huddesford

[William Huddesford], *The Lives of those Eminent Antiquaries, John Leland, Thomas Hearne and Anthony a Wood* (2 vol., Oxford, 1772).

JEGP

Journal of English and Germanic Philology.

JWCI

Journal of the Warburg and Courtauld Institutes.

J. Soc. Bibliog. Nat. Hist.

Journal of the Society for the Bibliography of Natural History.

J. Hist. Med.

Journal of the History of Medicine.

J. Soc. Bibliog. Sci.

Journal of the Society for the Bibliography of Science.

Johnson, *Lives*

Samuel Johnson, *Lives of the English Poets*, ed. G. B. Hill (3 vol., Oxford, 1905).

Kippis, *Biog. Brit.*

Andrew Kippis, *Biographia Britannica*, 2nd ed. (5 vol., London, 1778-92).

Lhwyd Letters

Life and Letters of Edward Lhwyd, ed. R. T. Gunther (*Early Science in Oxford*, vol. 14, 1945).

Lit. Anec.

John Nichols, *Literary Anecdotes of the Eighteenth Century* (9 vol., London, 1812-15).

Lit. Ill.   John Nichols, *Illustrations of the Literary History of the Eighteenth Century* (8 vol., London, 1817-58).

MP   *Modern Philology.*

Memoirs   *Memoirs of Martinus Scriblerus,* ed. Charles Kerby-Miller (New Haven, 1950; reprinted New York, 1966).

Monk, *Bentley*   Samuel Monk, *Life of Richard Bentley,* 2nd ed. (2 vol., London, 1833).

NQ   *Notes and Queries.*

N.R.   *Notes and Records of the Royal Society.*

Nicéron   Jean Pierre Nicéron, *Mémoires pour servir à l'Histoire des Hommes illustre dans la République des Lettres* (43 vol., Paris, 1729-45).

Nicolson, *Letters*   William Nicolson, *Letters on Various Subjects,* ed. John Nichols (2 vol., London, 1809).

PMLA   *Publications of the Modern Language Association.*

PQ   *Philological Quarterly.*

Phil. Trans.   *Philosophical Transactions of the Royal Society.*

Pope, *Corr.*   *Correspondence of Alexander Pope,* ed. George Sherburn (5 vol., Oxford, 1956).

Pope, *Works*   *The Poems of Alexander Pope,* Twickenham edition, ed. John Butt (10 vol., London and New Haven, 1939-67).

Portland MSS.   *The Manuscripts of the Duke of Portland at Welbeck Abbey,* Historical Manuscripts Commission Reports (10 vol., 1891-1931).

Proc. Amer. Antiq. Soc.   *Proceedings of the American Antiquarian Society.*

Proc. Geol. Assn.   *Proceedings of the Geological Association.*

Proc. Mass. Hist. Soc.   *Proceedings of the Massachusetts Historical Society.*

RES

Ray, *Corr.*

Ray, *Further Corr.*

Richardson, *Corr.*

Scot. Hist. Soc.

Spence, *Anecdotes*

Swift, *Corr.*

Thoresby, *Diary*

Thoresby, *Letters* (1832)

Thoresby, *Letters* (1912)

Trans. Card. Nat. Soc.

VCH

Ward, *Lives*

Woodward, *An Attempt*

Woodward, *Cat.*

Woodward, "Egyptians"

Woodward, *Essay*

Review of English Studies.

The Correspondence of John Ray, ed. Edwin Lankester (Ray Society, 1848).

The Further Correspondence of John Ray, ed. W. T. Gunther (Ray Society, 1928).

Extracts from the Literary and Scientific Correspondence of Richard Richardson (Yarmouth, 1835.)

Scottish Historical Society.

Joseph Spence, *Observations, Anecdotes, and Characters of Books and Men,* ed James, M. Osborn (2 vol., Oxford, 1966).

The Correspondence of Jonathan Swift, ed. Harold Williams (5 vol., Oxford, 1963).

The Diary of Ralph Thoresby, ed. Joseph Hunter (2 vol., London, 1830).

Letters of Eminent Men Addressed to Ralph Thoresby (2 vol., London, 1832).

Letters Addressed to Ralph Thoresby, ed. W. T. Lancaster (Thoresby Society, 1912).

Transactions of the Cardiff Naturalists Society.

Victoria County History.

John Ward, *Lives of the Professors of Gresham College* (London, 1740).

John Woodward, *An Attempt Towards a Natural History of Fossils* (London, 1729).

Catalogue of the Library, Antiquities, etc. of the late Dr. Woodward (London, 1728).

John Woodward, "Of the Wisdom of the Antient Egyptians;" *Archaeologia,* 4 (1777), pp. 212-311.

John Woodward, *An Essay Toward a Natural History of the Earth* (London, 1695).

Woodward, *Fossils*   John Woodward, *Fossils of all Kinds Digested into a Method* (London, 1728).

Woodward, *Nat. Hist.*   John Woodward, *The Natural History of the Earth,* ed. and trans. Benjamin Holloway (London, 1726).

Woodward, *Select Cases*   John Woodward, *Select Cases and Consultations in Physick,* ed. Peter Templeman (London, 1759).

Woodward, *State of Physick*   John Woodward, *State of Physick and of Diseases* (London, 1718).

Wren Soc. Pub.   *Wren Society Publications.*

Notes

Chapter 1: Doctor of Physick

1. *Hippocratis de Morbis Popularibus* (London, 1717). The book consisted of a text and translation of the first and third books of Hippocrates with nine commentaries by Freind appended, the seventh and largest of which was devoted to purging as a cure for smallpox. It received an English translation in 1729, *The Benefit of Purging in the Confluent Smallpox*, trans. J. Sparrow.

2. See Woodward's *Select Cases*, the case of Mr. Cosin, p. 185ff.

3. Besides his dedication to the Hippocrates, see also Freind's *History of Physick*, pt. 1 (London, 1725), p. 303ff.

4. From John Ward's account, quoted by Phyllis Allen, "Medical Education in 17th-Century England," *J. Hist. Med.*, 1 (1946), p. 127. When Sir Richard Blackmore asked Sydenham what to study for his practice, he was told "Read Don Quixote, it is a very good book, I read it still" – so low an opinion, concluded Blackmore, "had this celebrated man of the learning collected out of the authors, his predecessors," Blackmore, *Treatise upon the Smallpox* (London, 1723), preface, p. 11.

5. "Instead of poring over a multitude of Books, he [had] read Men, and benefited more by a profitable and free Acquaintance with the Living, than any one there of the same Profession, by making Comments in the bulky Writings of the Dead." William Pittis, *Some Memoirs of the Life of John Radcliffe* (London, 1715), pp. 5-6. See also Steele in *The Tatler*, no. 44, July 21, 1709, on "Aesculapius."

6. Woodward to J. J. Scheuchzer, April 25, 1718. (For this correspondence, see chap. 2 below, n. 92.) See also Maurice Emmett to Woodward, July 20, 1717, Camb. MS. Add. 7647, no. 124.

7. Woodward to Scheuchzer, Jan. 2 and Feb. 13, 1719.

8. The best account is in Beattie, Arbuthnot, p. 242ff.

9. Woodward to Scheuchzer, April 28, 1719, and April 5, 1720. Scheuchzer eventually translated Woodward's *State of Physick* into Latin, *Medicinae et Morborum Status* (Tiguri, 1720).

10. Woodward, *State of Physick*, pp. 52-57.

11. *Ibid.*, p. 34.

12. *Select Cases*.

13. *State of Physick*, p. 57.

14. The Gulstonian Pathological Lectures, read Jan. 27-31, 1711, "Of the Body of Man; Of Health, of Diseases, and of Remedies; Three Physico-Mechanical Lectures read in the Theater of the College of Physitians." There is a digest, largely reduced to headings only, in B.M. MS. Sloane 2039, ff. 115-29^V. I have not been able to discover much about Woodward's other services to the College except that he was elected censor in 1714-15 and that he attended meetings of the Council regularly. See the College of Physicians MS. Annals, vols. 1695-1710 and 1711-22. There is also a letter from Charles Goodall thanking Woodward for some unspecified service to the College in Camb. MS. Add. 7647, Sept. 23, 1709.

15. "Observations on the Dissection of Daniel Coates, who was executed at Tyburn, 1710," B.M. MS. Sloane 2039, ff. 130-35^v. Woodward's design here was to illustrate the Gulstonian Lectures on the bile. These observations appear to have been read again in 1714 and 1719.

16. Tom Brown, *Amusements Serious and Comical*, ed. Arthur L. Hayward (London, 1927), "Physic," p. 79.

17. Now in the Bodleian Library, 8°D55 Jur, seven tracts with attributions in pen and pencil. See George A. Aitkin, *The Life of Richard Steele* (2 vol., London, 1889), II, p. 202n.

18. The controversy, Dr. Woodward wrote to a friend, was "raised by Dr. Mead, Dr. Friend, and Dr. Cade, against whom I have wrote in that Work pretty smartly. They are in great Vogue with the People here and not having Reason or Argument to offer in their own Defence, in great Rage, they had recourse to Reproach and Slander. . . ." Among his defenders he was proud to enumerate Dr. Sewell, Dr. Harris, and Dr. Burnet. Fitz. Mus. MS. Spenser-Percival Packet B, Oct. 4, 1720. For Mead, see Matthew Maty, *Authentic Memoirs of the Life of Richard Mead* (London, 1755).

19. The tract was reprinted in Wagstaffe's *Miscellaneous Works* (London, 1726), pp. 351-414. Sir Charles Dilke preferred to attribute it to Swift in *NQ*, 3rd ser., I, pp. 381-84, and II, pp. 253-54, 396, but was answered by James Crossley, who insisted on Arbuthnot; *ibid.*, II, pp. 131-33. Another possibility could have been Dr. Henry Levitt, Wagstaffe's editor and also an enemy of Dr. Woodward's.

20. Also for Dr. Woodward was *Dr. Friend's Epistle to Dr. Mead render'd faithfully into English* (London, 1719). Woodward believed that the annotation which was hostile to his enemies was by Dr. George Sewell. He attributed the favorable *Appeal to Common Sense* to Burnet. For other titles, see Beattie, *Arbuthnot*, pp. 242-62.

21. Arbuthnot, *The Miscellaneous Works* (Glasgow, 1751), I, pp. 172-91. I am not convinced by Beattie's doubts about Arbuthnot's authorship.

22. "The Antidote," in a letter to *The Freethinker*, June 1719, in Steele's *Tracts*

and Pamphlets, ed. Rae Blanchard (Baltimore, 1944), pp. 505-06. *The Freethinker*, no. 126, dealt with the same subject. Woodward reported his successful cure of Steele in 1715 in his *Select Cases*, pp. 369-71. In 1720 the cure was repeated; see Aitkin, *Steele* (n. 17 above), p. 201.

23. William Munk, *The Roll of the College of Physicians*, 2nd ed. (3 vol.; London, 1878), II, p. 8. The two men were clearly not friendly; see Dr. John Allen to Woodward, Jan. 29, 1722, Camb. MS. Add. 7647, no. 6.

24. Among many instances of appreciation, see Camb. MS. Add. 7647, nos. 77-81, 86, 103-13, 135, 136, 144, 146. See also John Postlethwayt to his cousin, C.C.C.C. MS. 587, f. 34.

25. Dobson, *Eighteenth Century Vignettes*, 3rd ser. (London, 1923), p. 42. See also Charles R. Weld, *A History of the Royal Society* (2 vol., London, 1848), I, p. 337n.

26. Some of these tracts are quoted in Steele's "Antidote" (n. 22 above), pp. 513-14. See also *The Weekly Journal*, June 20, 1719, in *Lit. Anec.*, VI, pp. 641-42; *Mist's Journal* (*The Weekly Journal or Saturday Post*), June 13, 1719, with two different accounts; and Defoe's *Mercurius Politicus*, Vol. 7, June 1719, p. 355.

27. "The Antidote," pp. 518-19; *The Weekly Journal*, June 20, and *Mercurius Politicus*, pp. 356-58, all printed it, as well as the *Flying Post or Post-Master* for June 13-16, 1719.

28. "Their Pride and Vanity," Woodward wrote to Scheuchzer (April 5, 1720) about his critics, "had made them very clamorous and angry but they have been effectually silenced by these Defences which are wrote by Dr. Sewell, Dr. Burnet, Dr. Harris, Sir Richard Steele and some other the greatest and most polite Pens in England." There was profit in it too. "The unfair Opposition of my Adversaryes hath much raised my Honour and carryd me into a far greater Business in Physick than ever before!" See also the letter cited in n. 18 above.

29. One respectful correspondent was "unquestionably Europe's greatest physician," Hermann Boerhaave. There are seven letters from Boerhaave to Woodward written between 1716 and 1726 in Camb. MS.

Add. 7647, nos. 64-70. They were overlooked by G. A. Lindeboom in his edition of *Boerhaave's Correspondence* (*Analecta Boerhaaviana*, III, Leyden, 1962). For the quote above, see Lester S. King, "The Background of Herman Boerhaave's Doctrines," *Boerhaave-Voodrachten van de Ryksuniversiteit te Leiden*, I (1965), p. 3.

Chapter 2: The Natural History of the Earth

1. In 1752 two doctors killed each other in a repetition of the duel over fevers! See *Essays in the Bilious Fever: Containing the Different Opinions of those Eminent Physicians John Williams and Parker Bennet of Jamaica which was the Cause of a Duel and Terminated in the Death of Both* (Jamaica; reprinted London, 1752). Mead was chastised for his behavior in "Dr. Andrew Tripe's" *Dr. Woodward's Ghost* (London, 1748). For the rest, see Beattie, *Arbuthnot*, pp. 242-62.

2. The source for the date 1665, which appears in all subsequent accounts, seems to be John Ward, who wrote Woodward's biography in his *Lives* pp. 283-301. Ward knew Woodward and made scupulous collections for his life, some of which remain in the British Museum, e.g., B.M. MS. Add. 6209-10, 6194, 6271. Ward also gives Woodward's epitaph from his monument in Westminster Abbey with the same date. The epitaph was probably composed by Woodward's old friend and executor, Col. Richard King. However, in a letter to Scheuchzer dated Jan. 4, 1705 (New Style?), in which Woodward reviewed his medical history for his fellow physician, he wrote categorically, "I am now 37 years old and have been thin and lean all my life." This would put his birth in 1667 (or possibly 1668) and make him several years younger than has been thought, and thus even more of a prodigy. Does a man going on forty, a man of uncommon precision about details, mistake his birthdate? The advantage of the later date is that it exactly explains an otherwise mysterious gap in Woodward's life, which worried Clark and Hughes, between his apprenticeship and his residence with Dr. Barwick. The most useful

accounts of Woodward's life after Ward are John Willis Clark and T. M. Hughes, *The Life and Letters of Adam Sedgwick* (Cambridge, 1890), I, chap. 5, pp. 166-87; the article in the *DNB* by B. B. Woodward; and recently, Dr. V. A. Eyles, "John Woodward, Physician and Geologist," *Nature,* 206 (1965), pp. 868-70.

3. Hearne, *Remarks*, II, p. 13.

4. Bodl. MS. Tanner 25, ff. 370-92. Ward made a copy, B.M. MS. Add. 6194, pp. 14-18^V, and used some of the information in his life of Woodward. Woodward later repaid Barwick with a Latin life now lost; see Woodward to Postlethwayt, June 14, 1720, C.C.C. MS. 587, f. 232. Barwick says in 1692 that he had known Woodward for eight years, and Ward says that Woodward left school at sixteen, thus leaving an unexplained gap of three years.

5. For the following, see my article (with references), "Ancients, Moderns and History," in *Studies in Change and Revolution*, ed. Paul Korshin (Scolar Press, 1972), pp. 43-75.

6. No doubt Barwick helped. He often told Woodward that his best moment had been his defense of Harvey's theory about the circulation of the blood, one of the outstanding "modern" achievements. *Vita Johannes Barwick* (London, 1721), preface, sig. a; *Biog. Brit.*, "Barwick," note C.

7. See the letters of Dr. John Allen and John Fisher to Woodward in Camb. MS. Add. 7647, nos. 4, 126. Natural history, Woodward wrote to Scheuchzer, "can never be delivered with too great Plainness and Simplicity." Scheuchzer Corr., March 26, 1708.

8. Woodward to Postelthwayt (n.d.), C.C.C. MS. 587, ff. 230-31.

9. Perhaps this is too strong. The *Biog. Brit.* remarks on his "fluency in writing and speaking Latin for which he is deservedly famous." "Barwick," note C.

10. Woodward to Edward Lhwyd, Nov. 28, 1691, Bodl. MS. Ashmole 1817b, f. 347.

11. Woodward to Lhwyd, Jan. 19, 1692, Bodl. MS. Eng. Hist. C 11, ff. 104-05. The story seems to have come from John Aubrey. The naturalist John Ray found it equally implausible; Ray to Lhwyd, Jan. 18, 1691, Bodl. MS. Ashmole 1817a, f. 216.

12. Woodward to Lhwyd, March 2, 1692, Bodl. MS. Eng. Hist. C 11, f. 107.
13. Lhwyd to Lister, July 1, 1690, Bodl. MS. Lister 36, ff. 11-12.
14. Woodward, *An Attempt*, I, pt. 2, preface, p. 1.
15. See the postscripts of Barwick's letters of 1690 to Hans Sloane, B.M. MS. Sloane 4036, ff. 74, 90, 96.
16. Woodward, *An Attempt*, p. 1. His Derbyshire background may have helped to inspire his geological interest; cf., for example, the reminiscences of John Whitehurst, "An Inquiry into the Original State and Formation of the Earth," *Works* (London, 1792), preface. For Sherborne, see Sir Robert Atkyns, *The Ancient and Present State of Gloucestershire* (London, 1712), pp. 643-45; *VCH: Gloucester*, VI (1965), pp. 120-27; *Memorials of the Duttons* (London and Chester, 1901), pp. 98-102. It appears that Sir Ralph was of no help whatever in intellectual matters, but his house was wonderfully situated with respect to the quarries.
17. Plot, *Oxfordshire* (Oxford, 1677), p. 111, Plot and Lister both recommended Woodward for the Gresham post, and Hooke was a fellow professor.
18. B.M. MS. Add. 28104, no. 16. See also Burnet to Southwell, July 10, 1688, B.M. MS. Add. 10, 039, ff. 63-64. Burnet's book appeared in two parts and went through three Latin editions and eight English ones between 1681 and 1759; it was reprinted as late as 1826, and there were Continental versions as well. See V. A. Eyles, "Bibliography and the History of Science." *J. Soc. Bibliog. Sci.*, 3 (1955), p. 64.
19. Burnet, *The Sacred Theory of the Earth*, 2nd ed. (London, 1691; reprinted Carbondale, Ill., 1965), p. 25.
20. Basil Willey, *The Eighteenth Century Background* (1940, reprinted Boston, 1962), pp. 27-39; Ernest Lee Tuveson, *Millennium and Utopia* (1949, reprinted New York, 1964), pp. 113-26; Marjorie Hope Nicolson, *Mountain Gloom and Mountain Glory* (1959, reprinted New York, 1963), pp. 184-224.
21. Ray to Lhwyd, March 22, 1692, *Further Corr.*, pp. 234-37. For the controversy in general, see Nicolson, *Mountain Gloom*, pp. 225-70.

22. Woodward, *Essay*, p. 242ff.
23. Erasmus Warren, *Geologia or a Discourse Concerning the Earth before the Deluge* (London, 1690), pp. 26, 34, 41; Hubert Croft, Bishop of Hereford, *Some Animadversions upon a Book Intitled a Theory of the Earth* (London, 1685), p. 41.
24. Locke to James Tyrrell, quoted by Tyrrell in a letter to Robert Boyle, May 25, 1687. Boyle, *Works* (London, 1772), VI, pp. 619-20.
25. Ray to Lhwyd, Nov. 7, 1690, in Ray, *Further Corr.*, pp. 209-12. See Lhwyd's return, Nov. 25, 1690, *Lhwyd Letters*, pp. 110-12.
26. "He told me that Mr. Woodward, a Londoner, shewed him very good Draughts of the common female Fern, naturally formed in Cole, which himself found in Mendip Hills," Lhwyd to Lister, *Lhwyd Letters*, pp. 106-07. Ray's initial thoughts on the problem were published in his *Observations made in a Journey through Part of the Low Countries* (London, 1673), pp. 113-31). See also Ray to Tancred Robinson, Oct. 22, 1684, in Ray, *Corr.* pp. 151-56; and in general Edwin Lankester, *Memorials of John Ray* (London, 1846), and Charles E. Raven, *John Ray, Naturalist* (London, 1950).
27. Ray to Lhwyd, Oct. 8, 1695, in Ray, *Further Corr.*. pp. 259-62.
28. R. T. Gunther, *The Life and Work of Robert Hooke*, pt. 2 (*Early Science in Oxford*, vol. 7, 1930), pp. 697-99, 700-01, 712, 760-68. Hooke had argued in favor of fossils as once-living creatures in his *Micrographia* (1665), obs. 17, and had quarreled with Lister on the matter in 1683 before the Royal Society. See Birch, IV, pp. 237-38. For Hooke's views in general, see the series of papers assembled in *Posth. Wks.*, pp. 210-450. There is some brief commentary in Gordon L. Davies, "Robert Hooke and the Conception of Earth History," *Proc. Geol. Assn.*, 75 (1964), pp. 493-98, and in Albert V. Carozzi, "Robert Hooke, Rudolf Erich Raspé, and the Concept of Earthquakes," *Isis*, 61 (1970), pp. 85-91.
29. I know of only one, John Aubrey, who had worked with Hooke on his experiments in the early days of the Royal Society. See Aubrey's *Natural History of*

Wiltshire, ed. John Britton (London, 1848), pp. 46-47 (summary only); Ray to Aubrey, Sept. 22, 1691, *Further Corr.*, p. 171; Lhwyd to Aubrey, April 3, 1692, *Lhwyd Letters*, pp. 161-62.

30. Lhwyd to Lister, July 19, 1689, *Lhwyd Letters*, pp. 88-90. The only biography of Lhwyd to date is by Richard Ellis in *ibid.*, pp. 1-51.

31. The intermediary was the virtuoso William Charleton or Courten; see Charleton to Lhwyd, April 29, 1690, Bodl. MS. Ashmole 1814, f. 308, and Lhwyd's return, May 23, B.M. MS. Sloane 4062, f. 234.

32. Hearne, *Remarks*, I, p. 244; Ray, *Further Corr.*, pp. 228-30.

33. Woodward to Lhwyd, Oct. 20, 1692, Bodl. MS. Ashmole 1817b, f. 355; Lhwyd to Woodward, Oct. 23, Bodl. MS. Ashmole 1816, f. 17. Lhwyd had earlier complained about their relationship; March 24, 1692, Bodl. MS. Eng. Hist. C 11, f. 109. But see below.

34. Woodward to Lhwyd, Dec. 31, 1692, Bodl. MS. Ashmole 1817b, f. 356; Lhwyd to Lister, n.d., Bodl. MS. Lister 3, f. 132.

35. Lhwyd to David Lloyd, Jan. 9, 1686, *Lhwyd Letters*, pp. 76-80; Lhwyd to Ray, Nov. 25, 1690, *ibid.*, pp. 110-12.

36. Lhwyd to Ray, Nov. 25, 1690, *ibid.*

37. Ray to Lhwyd, Bodl. MS. Eng. Hist. C 11, f. 47; Lhwyd to Lister, n.d., *Lhwyd Letters*, pp. 282-83.

38. Woodward to John Morton, March 19, 1696, Bodl. MS. Eng. Hist. C 11, ff. 116-17. Woodward's support for Lhwyd is evident in the letters of Bodl. MS. Ashmole 1817b, e.g., ff. 107, 350.

39. Lhwyd to Lister, n.d., *Lhwyd Letters*, pp. 173-74.

40. Woodward to Lhwyd, June 21, 1693, Bodl. MS. Ashmole 1817b, f. 360.

41. *Ibid.*, Nov. 7, 1693, f. 363. The lecture appears to have been given on May 18, 1693; Woodward, *An Attempt*, I, pt. 2, p. 19; Hooke Diary in Gunther, *Life* (n. 28 above), p. 244. "I have already read 7 lectures," Woodward wrote proudly to the Bishop of Norwich (John Moore), "having had an Auditory every day the last Terms that were Lecture days, both for Number and Quality as far beyond my

Expectation as beyond my Merit." Bodl. MS. Tanner 25, ff. 229-30. His early lectures seem to have combined medical and geological interests. "I have engaged myself to read upon some of the Druggs of our Pharmacopeia. The Natural History of some of them being very imperfect," Woodward to Lhwyd, July 13 and Nov. 7, 1693, Bodl. MS. Ashmole 1817b, ff. 362-64.

42. Hooke diary in Gunther, *Life*, p. 242. "The seven Professors have Fifty Pounds per annum each, and handsome Apartments in the College, which is a large Piazza of above an Acre of Ground Square, but much decayed through Age." *A Journey Through England* (3 vol., London, 1714), I, pp. 164-68. "I have contrived," Woodward continued in his report to the Bishop of Norwich (n. 41 above), "so that I let as much as I have 30 pounds a year for and have reserved a very convenient apartment for myself besides." When some difficulties arose in 1703, Woodward composed a petition to Parliament in which he noticed that the Physick Professor got the largest house, "that besides his dwelling house, he might have room for a library, repository, and other conveniences requisite for a physician." See B.M. MS. Add. 6193 and 6271, ff. 17-18. In 1720 it was reported that Woodward had a coach house and stable on the premises; B.M. Add. 6194, f. 64. At some point Dr. Woodward composed an account of the College which looks like it was prepared for publication, "A Letter to John Bembde Esq. Concerning the Original and the Present State of Gresham College." B.M. MS. Add. 6209, ff. 330-40.

43. R.S. MS. Jour. Bk. 1690-96, pp. 203, 207.

44. *Lhwyd Letters*, pp. 210-11, 218-20.

45. Nicolson to Lhwyd, Feb. 16, 1694, Bodl. MS. Ashmole 1816, f. 468.

46. Oct. 27, 1691, Bodl. MS. Ashmole 1817b, f. 348.

47. "Some Thoughts and Experiments Concerning Vegetation," *Phil. Trans.*, 21 (1699), pp. 193-227. We are properly reminded, however, that Woodward did not really understand the part played by water in the formation of carbohydrates. But then neither did anyone else for a

century. See H. Hamshaw Thomas, "Experimental Plant Biology in Pre-Linnaean Times," *Bull. Inst. Hist. Sci.*, 2 (1955), pp. 20-21. In 1812 it was reported that "Dr. Woodward's experiments have acquired great celebrity and are constantly referred to." Thomas Thompson, *History of the Royal Society* (London, 1812), p. 58.
    48. May 7, 1692, Bodl. MS. Ashmole 1817b, f. 350. He mentions here also "a discourse about the cause of the Gravity of Natural Bodyes which I am not yet resolved to publish but intend as soon as I am able to finish it."
    49. B.M. MS. Add. 6209, p. 19.
    50. Woodward to Lhwyd, Bodl. MS. Ashmole 1817b, f. 373; Lhwyd to Lister, Dec. 5, 1695, *Lhwyd Letters*, pp. 295-96; Ray to Lhwyd, Sept. 9, 1694, in Ray, *Further Corr.*, pp. 253-55.
    51. See Melvin E. Jahn, "A Bibliographical History of John Woodward's *An Essay Toward a Natural History of the Earth*," *J. Soc. Bibliog. Nat. Hist.* 6 (1972), pp. 181-213.
    52. Hooke, *Posth. Wks.*, pp. 410, 321.
    53. Gideon A. Mantell, *The Medals of Creation* (London, 1844), I, p. 17. Mantell seems to have borrowed the term from the Swedish geologist Tobern Bergman, perhaps from the work translated by William Withering as the *Outlines of Minerology* (Birmingham, 1783), app. 2, p. 123. Mantell was an intimate of Charles Lyell, who began his epoch-making *Principles of Geology* with a similar comparison. The parallel is explicit in most writers on the subject, e.g., Buffon, *Oeuvres Complètes*, ed. Pierre Flourens (Paris, 1853-54), IX, p. 455; James Parkinson, *Organic Remains of a Former World* (London, 1804), I, pp. 7-8 (quoting Bergman); William Whewell, *History of the Inductive Sciences* (London, 1837), III, pp. 482-83. It is noticed briefly by Cecil Schneer, "The Rise of Historical Geology in the 17th Century," *Isis*, 45 (1954), pp. 263-67.
    54. Woodward, *Nat. Hist.* pp. 155-58.
    55. See for example the praise by John Challinor, *The History of British Geology: A Bibliographical Study* (Newton Abbot, 1971), p. 63; Eyles, "Woodward" (n. 2 above), pp. 403-04. In this connection one should notice also Woodward's "Directions for making Observa-

tions on Corals, Corallin . . . and other like Bodies," drawn up for the use of his assistants and published at the end of his life in *Fossils*, pp. 88-119.
    56. Woodward, *Essay*, pp. 63-64, 70, 39-40.
    57. *Ibid.*, sig. A-A^V.
    58. *Works of the Learned*, 4 (Dec. 1702), pp. 748-51 (reviewing the 2nd ed. of Woodward's *Essay*).
    59. Bodl. MS. Lister 36, f. 118.
    60. June 29, 1695, Bodl. MS. Ashmole 1829, f. 23.
    61. *Phil. Trans.*, 19 (Oct. 1695), pp. 115-23.
    62. Ray, *Further Corr.*, pp. 250-53, 253-54, 270-71. See also Ray to Aubrey, May 7, 1695, *ibid.*, pp. 182-83, and a series of lost letters to Tancred Robinson summarized by Derham and listed in *ibid.*, pp. 301-02.
    63. Nicolson to Lhwyd, Bodl. MS. Ashmole 1816, f. 474; Ray to Lhwyd, June 8, 1696, in Ray, *Further Corr.*, pp. 265-66; Ray to Robinson in a lost letter of March 24, 1696, where he "Objects largely against Dr. Woodward's Essay and haughty imposing his Hypoth of Origin of Springs," *ibid.*, p. 301.
    64. Bodl. MS. Lister 36, f. 118; Archer to Lhwyd, April 22, 1695, Bodl. MS. Ashmole 1829, f. 16.
    65. Ray preferred forty; see Ray to Lhwyd, June 8, 1696, in Ray, *Further Corr.*, pp. 265-67. Both numbers are based on Genesis 8, but Ray's figure seems more to the point and a fairer reading.
    66. June 16, 1695, Bodl. MS. Ashmole 1816, f. 117.
    67. March 19, 1696, Bodl. MS. Eng. Hist. C 11, f. 117.
    68. Ward says that it was generally thought that Lister was the author; *Lives*, p. 286. According to Robinson, he answered Dr. Woodward in three chapters of a Latin tract; see Robinson to Lhwyd in Bodl. MS. Ashmole 1817a, f. 332 and Lister to Lhwyd, Bodl. MS. Ashmole 1816, f. 119.
    69. Bodl. MS. Tanner 25, f. 386.
    70. Robinson, "A Letter sent to William Wotton . . . concerning some Late Remarks etc. written by Mr. Harris," n.d., a single-sheet broadside. See also Robinson to Lhwyd in Bodl. MS. Ashmole 1817a. f.

328. Lhwyd told Lister he did not know who the author was, except that he was not at Oxford. "It seemed to me only that some good naturalist had supply'd another with materials." *Lhwyd Letters*, pp. 276-78.

71. John Morton wrote to Ray that Robinson was "Dr. Woodward's professed adversary, particularly one of Christ Church in Cambridge informed me that he was publicly reflecting upon him in one of the London Coffee-houses and therefore I may be suspicious of the fairness of that account." March 25, 1696, Bodl. MS. Ashmole 1816, ff. 406-07. Robinson did admit in his broadsheet that he had helped Thomas Robinson of Ousby to write his criticism of Dr. Woodward. This was in reply to Harris, who had smoked him out in his apology for Dr. Woodward (treated below). See the section "A Word of Advice to Mr. Robinson ... Author of a Late Book call'd New Observations," in Harris, *Remarks*. Robinson had been corresponding with Ray (who he thought was the greatest naturalist of his time) from 1683. Unfortunately, most of these letters have perished, except for the summaries made by William Derham for his biography of Ray and now in the Botany Library of the British Museum, South Kensington. They contain many tantalizing references to Dr. Woodward, as in March 1695, "Character of Dr. Woodward and his Essay, he ignorant, paradoxical." Still another attack on Woodward, apparently unrelated, appeared in the *Miscellaneous Letters giving an Account of the Works of the Learned*, I, no. 14 (April 1695), pp. 322-31, and II, no. 2 (Feb. 1696), pp. 49-57. It was apparently by Dr. Ashe, the Irish bishop and friend of Swift. See also Lhwyd to Lister, April 23, 1696, Bodl. MS. Lister 36, f. 154.

72. Lhwyd to Lister, n.d., *Lhwyd Letters*, pp. 277-78; Ray to Lhwyd, Feb. 3, 1695, in Ray, *Further Corr.*, pp. 263-65.

73. Morton to Lhwyd, March 25, 1696, Bodl. MS. Ashmole 1816, f. 413; Lhwyd to Lister, Feb. 9, 1696, *Lhwyd Letters*, pp. 298-300.

74. Lister to Lhwyd, Dec. 3, 1695, Bodl. MS. Ashmole 1816, f. 121; Woodward to Morton, March 19, 1696, Bodl.

MS. Eng. Hist. C11, f. 117. The phrase "implacable enemy" appears in Archer to Lhwyd, March 21, 1696, Bodl. MS. Ashmole 1829, f. 14, and Lister to Lhwyd, Feb. 9, 1696, *Lhwyd Letters*, pp. 298-300; "mortal enemy" in Ray to Lhwyd, Feb. 3, 1695, Ray, *Further Corr.*, pp. 263-65. It looks suspiciously like Tancred Robinson's work; see the extract of his letter quoted by Ray.

75. Richardson, *Corr.*, pp. 37-40.

76. William Whiston, *New Theory of the Earth* (London, 1696). Whiston admits his indebtedness to Woodward in the preface to his *Vindication of the New Theory of the Earth* (London, 1698). "This Book," he also recalled, "was shewed in MS to Dr. Bentley and to Sir Christopher Wren, but chiefly before Sir Isacc Newton himself, on whose Principles it depended, and who well approved of it." *Memoirs of the Life and Writings of William Whiston* (3 vol., London, 1749), I, p. 43.

77. See Petty to Southwell, Oct. 4, 1686, *The Petty-Southwell Correspondence*, ed. Marquis of Lansdowne (London, 1928), pp. 235-36. In 1694 Halley delivered a paper on the subject, "Some Considerations about the Cause of the Universal Deluge," *Phil. Trans.*, 33 (1724), pp. 118-23, 123-25.

78. Nicolson to Lhwyd, Jan. 31, 1698, in Nicolson, *Letters*, I, pp. 103-05.

79. Harris, *Remarks*.

80. For the story of the long confusion about the fossil, see Clifford M. Nelson, "Ammonites: Ammon's Horns into Cephalopods," *J. Soc. Bibliog. Nat. Hist.*, 5 (1968), pp. 1-18.

81. Harris, *Remarks*, pp. 164-65.

82. *Ibid.*, pp. 34-36, 253.

83. *Ibid.*, p. 66.

84. Harris was rewarded afterward by Dr. Woodward, who got him the coveted Boyle Lectures as well as support for his *History of Kent*. See Woodward to Hearne, May 15, 1710, Bodl. MS. Rawl. Lett. 12, f. 428; Woodward to Evelyn, Sept. 25 and Oct. 5, 1697, Christ Church Oxford MS. Upcott Coll. III, nos. 99, 100; Gibson to Lhwyd, Nov. 4, 1697, Bodl. MS. Ashmole 1829, f. 154. Unfortunately, not much is known about Harris, whose papers disappeared in the 18th century,

including a lively diary from which only
an extract remains at Trinity College,
Oxford, much of it printed in the appropriate volume of the College Histories by
H. E. B. Clakiston (London, 1898), pp.
172-76. Harris was a man of considerable
talent, as will appear; see *Lit. Anec.*, IX,
pp. 769-75. The Trinity MS. has a covering
letter of April 1761 lamenting the disap-
pearance of Harris' papers; I am grateful to
Mr. J. P. Cooper for allowing me to look
at it.
    85. Evelyn to the Archbishop of Can-
terbury, Sept. 26, 1697, Christ Church
Oxford MS. Letter Book II, no. 794. (He
recommended it anyway.)
    86. Review of Scilla in *Phil. Trans.*, 19
(1696), pp. 181-95; letter to Dr. Arbuthnot "Concerning an Abstract of Agostina
Scilla's Book," appended to Arbuthnot's
Examination (cited in n. 88 below), pp.
65-84.
    87. Lhwyd to Morton, Feb. 23, 1698,
B.M. MS. Sloane 4062, f. 277.
    88. Arbuthnot, *An Examination of
Dr. Woodward's Account of the Deluge*
(London, 1697), pp. 14, 29-30.
    89. Nicolson, *Letters*, I, pp. 103-05.
Ray naturally thought Arbuthnot's arguments conclusive, Ray to Lhwyd, in Ray,
Further Corr., pp. 270-71.
    90. Nicolson to Wotton, Jan. 13,
1698, Nicolson, *Letters*, I, pp. 96-100. See
also his notes, "Dr Arb agt Dr Woodward," meant apparently for Dr. Woodward, Oct. 3, 1698, Bodl. MS. Add. C 217,
ff. 50-51. Nicolson was never entirely won
over by Dr. Woodward, but he never gave
up hope, either, that the doctor would
answer his critics. Thus his letter to
Lhwyd: "We shall have no occasion to set
the whole Globe to Leaven, in an Universal Deluge, meerly for the convenience
of sprinkling over Shells like fruit in a
Bagg-Pudding." March 18, 1697, Bodl.
MS. Ashmole 1816, ff. 488-89.
    91. Ch. Kinnaird to Sloane, Oct. 10,
1705, B.M. MS. Sloane 4040, ff. 77-78.
    92. The correspondence is largely in
the Zentralbibliothek Zurich and consists
of 74 letters of Woodward to Scheuchzer,
MS. H 150 a, b, c, and 50 draft replies by
Scheuchzer, MS. H 293-95. Through the
courtesy of the Zurich library and Dr. J. P.
Bodmer I was furnished with a Xerox

copy. Unfortunately, it was impossible to
determine which manuscript any given
letter was from and so I have limited my
citations to dates. There are excerpts from
a few of these letters in Melvin E. Jahn,
"Some Notes on Dr. Scheuchzer and on
Homo Diluvii Testis," *Towards a History
of Geology*, ed. Cecil J. Schneer (Cambridge, Mass., 1969), pp. 158-83. The
Scheuchzer manuscripts are described in
Rudolf Steiger, *Verzeichnis des wissen-
schaftlichen Nachlasses von J. J. Scheuchzer* (Zurich, 1933). See also Rudolf Wolf,
*Biographien zur Kulturgeschichte der
Schweiz* (Zurich, 1858), I, pp. 181-281.
    93. Woodward to Scheuchzer, Jan. 25
and May 11, 1703.
    94. See Katherine B. Collier, *Cos-
mogonies of our Fathers* (New York,
1934), p. 133, citing Jehan, *Dict. de
cosmologie et de paleontologie* (Paris,
1854), col. 629.
    95. Scottish interest may be pursued
in the correspondence of Robert Wodrow,
some of it with Lhwyd, much of it with
William Nicolson, part of it reproduced in
the *Early Letters of Robert Wodrow*, ed.
L. W. Sharp (Scot. Hist. Soc., ser. 3, vol.
24, 1937), esp. pp. 24-26, 27-28, 117-19,
124-26, 187-89, 234-38, 263-64. Other
letters of interest, including one from
James Sutherland, appear in the *Analecta
Scotica* (2 vol., Edinburgh, 1834-37), I,
pp. 339-41, 359-61. Some unpublished
correspondence between Wodrow and
Nicolson remains in the National Library
of Scotland in the Wodrow Correspondence; see esp. Quarto Letters, I, ff. 56,
66, 97. Finally, there is a letter of Wodrow
to Cotton Mather expressing their common interest in Dr. Woodward in the
*Correspondence of the Rev. Robert Wod-
row* (3 vol., Wodrow Soc., 1842-43), II,
pp. 359-63. Among the English critics of
Dr. Woodward should be added Charles
Leigh, *The Natural History of Lancashire,
Cheshire and the Peak* (Oxford, 1700), pp.
114-30. Two of the doctor's friends,
Nicolson and Jabez Cay, came to the
rescue; see Thoresby, *Letters* (1912), pp.
81-82, 96-97. For the Germans, see below.
    96. Woodward, *An Attempt*, I, p. 242.
    97. Woodward, *Fossils*, p. 57.
    98. He made his opinion plain enough,
however. "Woodward saith he [i.e., Ar-

buthnot] neither understands mathematics nor philosophy and that his Books shewd him self to be an ignorant and scurrilous fellow and accuses him of impudency in making a collection of shells and mineralls but he never heard that Arbuthnet ever layd out 5 shillings." C. Hatton to Lhwyd, Dec. 31, 1699, Bodl. MS. Lister 36, f. 244. Woodward did write a rebuttal to his critics, but it remained in manuscript and has since disappeared.

99. John Hutchinson, *Observations made by J. H. mostly in the Year 1706*, reprinted in *The Philosophical and Theological Works* (12 vol., London, 1749), XII, pp. 259-359. There are a number of letters of that year reporting his discoveries to Dr. Woodward in Bodl. MS. Gough Wales 8, ff. 4-25. "He has made me a vast return," Woodward admitted to Scheuchzer, Sept. 10, 1706. According to Robert Spearman, Hutchinson's editor, the marginal notes in the 1706 tract were by Woodward himself; *A Supplement to the Works of John Hutchinson* (London, 1776), p. iii. See also Woodward, *Fossils*, p. 55.

100. Spearman, *Supplement*, pp. iv-v.

101. Hutchinson, *Moses Principia* (1724) in *The Philosophical Works*, I, p. 97.

102. B.M. MS. Add. 5860, ff. 155^V-56 (extracts from the correspondence of Dr. Taylor and Col. King after Woodward's death relating to the dispute). For the Hutchinsonians, see Leslie Stephen, *History of English Thought in the Eighteenth Century* (2 vol., reprinted New York 1962), I, pp. 330-33. A characteristic later work related to both Woodward and Hutchinson is Alexander Catcott, *Remarks on the Creation and the Deluge* (London, 1756-61).

103. "I assure you," he wrote to Lhwyd, he was . . . pleas'd to shew me several parts of it, of a good bulk (and full grown) size." Feb. 12, 1706, Bodl. MS. Ashmole 1816, f. 546.

104. Woodward to Scheuchzer, Scheuchzer Corr., Oct. 17, 1704; June 25, 1708; Sept. 28, 1711; March 6, 1721.

105. Woodward, *Naturalis Historia Telluris* (London, 1714).

106. *Phil. Trans.*, 32 (1723), pp. 419-20. This correspondence is in Bodl.

MS. Gough Wales 8, f. 29ff. and in the Royal Society MS. Letter Book H-3, ff. 94, 111. For Wotton's aid with the translation, see Bodl. MS. Gough Wales 8, ff. 38, 40-41.

107. Woodward, *Nat. Hist.* In some copies the title is given as *A Supplement and Continuation of the Essay towards a Natural History of the Earth*. Holloway began the translation about 1715, closely supervised by Woodward.

108. Woodward, *Essay*, p. 53; *Nat. Hist.*, pp. 154-69.

109. Southwell to Woodward, Jan. 13, 1702, B.M. MS. Stowe 747, f. 152. Southwell died later that year. Long before, he had written a dialogue criticizing Thomas Burnet's theories; see B.M. MS. Add. 10, 039, ff. 65-72.

110. More likely it was Scheuchzer's intercession. The retraction may be found in a paper "de Arena Conchifera" in *Academicae Caesario-Leopoldianae Carolinae Natura Curiosorum Ephemerides*, centuriae V-VI (Norbergae, 1717). See G. R. de Beer, "Johann Gaspar Scheuchzer FRS" *N.R.*, 6 (1949), pp. 59-60.

111. Woodward, *Nat. Hist.*, p. 106; *Fossils*, pp. 43-44; Harris, *Remarks*, p. 66. Woodward believed that the Americas had been colonized by the early peoples.

112. They are dispersed in the preface thus: pp. 9-11, 12-36, 39-49, 50-66, 67-104.

113. Woodward, *Nat. Hist.*, pp. 43-44.

114. *Ibid.*, pp. 72-94.

115. Among the missing manuscripts, Holloway mentions a *Physiological Treatise on the Structure of the Parts of Animals* and a more elaborate work on the *Idea of the Nature of Man*; also *Notes on the first Chapter of Genesis*; Woodward, *Nat. Hist.*, pp. 100n. and 104-05.

Chapter 3: Science and Religion

1. "The first practical systematic work on conchology," Guy L. Wilkins, "Notes on the *Historia Conchyliorum* of Martin Lister," *J. Soc. Bibliog. Nat. Hist.*, 3 (1957), pp. 196-205. A second edition appeared in 1697; Charleton, Sloane, Lhwyd, and Petiver all contributed. For Lister in general, see R. P. Stearns, intro.,

Lister's *Journey to Paris* (Urbana, Ill., 1967).

2. See E. A. N. Arber, "A Sketch of the History of Palaeobotany," *Studies in the History and Method of Science*, II (Oxford, 1921), p. 473.

3. Though Lister found more and more similarities with living animals, he refused to concede; see *Phil. Trans.*, 20 (1698), pp. 279-80. I am delighted to discover a view similar to mine in Martin Rudwick, *The Meaning of Fossils* (London, 1972), pp. 61-63. Lister, incidentally, had been writing about the problem since 1671. The Italian Malpighi, told Tancred Robinson in 1684 that he could *prove* Lister wrong about the fossils and Robinson sent the information on to Ray. See the letter in the Ray Correspondence, Botany Library, British Museum, South Kensington, f. 92.

4. See, for example, Ray's absurdly low estimates in *The Wisdom of God* (London, 1691), pp. 5-7.

5. Woodward, *An Attempt*, I, pt. 2, p. 7; Ray, *Miscellaneous Discourses* (London, 1692), p. 120.

6. Andrew C. Ramsay, *Passages in the History of Geology* (London, 1848), p. 21.

7. Harris, *Remarks*, p. 84.

8. Woodward, *Essay*, p. 23; in the 3rd ed. (London, 1723) he enlarged upon it further, p. 71.

9. Lhwyd to Ray, Oct. 7, 1692, *Lhwyd Letters*, pp. 167-68; B.M. MS. Sloane 3341 (Minutes of the Royal Society), May 26, 1697, f. 44; Hooke, *Posth. Wks.*, pp. 438-39. See also "Chartham News: Or a Brief Relation of some Strange Bones lately digg'd up . . . ," *Phil. Trans.*, 20 (1701), pp. 882-93, 1030-38.

10. Nov. 17, 1697, R.S. MS. Jour. Bk., IX, p. 72.

11. B.M. MS. Sloane 3341, f. 39; Thomas Molyneux, "A Discourse Concerning the Large Horns frequently found underground in Ireland," *Phil. Trans.* 19 (1697), pp. 489-512; *A Natural History of Ireland by Several Hands* (Dublin, 1726), pp. 137-49. Molyneux was a physician, frequent visitor to London, and contributor to the Royal Society; see "Sir Thomas Molyneux Bart.," *Dublin University Magazine*, 18 (1841), pp. 305-27; Capel

Molyneux, *An Account of the Family and Descendents of Sir Thomas Molyneux* (Evesham, 1820). Molyneux was early in touch with Lhwyd; see Bodl. MS. Ashmole 1816, f. 359ff.

12. "That is a supposition which Philosophers hitherto have been unwilling to admit, esteeming the extinction of any one Species to be a dismembering the Universe and rendering it imperfect, whereas they think the Divine Providence is especially concerned to preserve and secure all the Works of the Creation. Yet granting that some few Species might be lost, it is very unlikely that so many should and still more unlikely that such as were so diffused all over Europe and found in so many distinct and remote places, . . . should be so utterly extinct and gone . . . That there should not one in an age be found." Ray, *Observations* (ch. 2, n. 26 above), p. 127. See also Lhwyd to Ray, Oct. 7, 1692, *Lhwyd Letters*, pp. 167-68.

13. Nicolson to Woodward, July 29, 1697, in Nicolson, *Letters* p. 69. See also Nicolson's letter of Oct. 2, 1696, *ibid.*, p. 53.

14. I quote from a copy of a letter written by Molyneux on file in the Sedgwick Museum, by courtesy of the Curator, Dr. C. L. Forbes; I have not been able to trace the original. For Josselyn's "imperfect" description, see *Phil. Trans.*, 31 (1721), pp. 165-68; for Mather's contribution read before the Royal Society Oct. 28, 1714, see George L. Kittridge, "Cotton Mather's Scientific Communications to the Royal Society," *Proc. Amer. Antiq. Soc.* (1916 offprint), pp. 15-16. William Stukeley put the problem well; it was impossible, "for want of a Natural History of Sceletons," to determine what kind of animal any of these were. *Phil. Trans.*, 30 (1719), p. 967.

15. Woodward, *Nat. Hist.*, p. 143.

16. Woodward, *Essay*, p. 46ff.; *Nat. Hist.*, pp. 58-59.

17. See Frederick E. Zeuner, *Dating the Past*, 4th ed. (1958; reprinted London, 1970).

18. Edward Halley, "A Short Account of the Causes of the Saltness of the Ocean," *Phil. Trans.*, 29 (1715), pp. 296-300. In 1898 Joly calculated that

80-90 million years were required, and that was apparently too few! See Aswit K. Biswas, "Edmund Halley, Hydrologist Extraordinary," *N.R.*, 25 (1970), p. 55.

19. Lhwyd to John Lloyd, July 6, 1693, *Lhwyd Letters*, pp. 194-95; Lhwyd to Ray, Feb. 30, 1692, *ibid.*, pp. 156-60.

20. *Barr's Buffon: Buffon's Natural History* (10 vol., London, 1797), I, p. 132.

21. Ramsay, *Passages* (n. 6 above), p. 24; Charles Lyell, *Principles of Geology* (London, 1830), pp. 36-37. See also Lyell in the *Quarterly Review*, 36 (1827), pp. 481-82.

22. Richard J. Chorley, Antony Dunn, and Robert J. Beckinsale, *The History of the Study of Landforms* (London, 1964), p. 11. See also Gordon L. Davies, *The Earth in Decay* (London, 1969), pp. 10, 44, 74-83. The view is standard in the older histories – e.g., Archibald Geikie, *The Founders of Geology* 2nd ed. (London, 1905), p. 67; Horace B. Woodward, *History of Geology* (London, 1911), pp. 10-12. The contemporary exception is Rudwick, *Meaning* (n. 3 above), preface and pp. 87,155, 266.

23. *The Correspondence of Isaac Newton*, ed. H. W. Turnbull (Cambridge, 1960), II, pp. 321-35; also John Keill, *An Examination of Dr. Burnet's Theory of the Earth* (Oxford, 1698), p. 170.

24. Letter to an anonymous foreigner written after 1726 in Woodward, *Fossils*, p. 21. Rudwick thinks "it did provide a moderately convincing explanation of organic resemblances in fossils . . . moderately consonant with both Scripture and reason." *Meaning*, p. 87.

25. See for example Davies, *Earth in Decay* (n. 22 above), for some of Woodward's followers in the 18th century – or as he calls them, "bizarre works in the tradition of Woodward." Ch. 5, p. 129ff.

26. Mark Pattison, "Tendencies of Religious Thought in England, 1688-1750," *Essays* (2 vol., Oxford, 1889, reprinted London, [1909]), II, p.4.

27. *Ibid.*, p. 5

28. Boyle, *The Christian Virtuoso* (London, 1690), p. 2.

29. Charles Blount, *The Oracles of Reason* (1693), reprinted in *The Miscellaneous Works* (London, 1695), pp. 2-3, 87. Blount's book arrived at Cambridge in

closed parcels, "for those that were sent then durst not be none and because that they were Atheisticall the Vice-Chancellor sent the bedells to demand them all from the booksellers and counseld them to be burnt." Diary of Abraham de la Pryme, Camb. MS. Add. 7519, Oct. 29, 1693; printed in the Surtees Soc. Publications, vol. 54 (1870), p. 30.

30. Woodward to the Bishop of Norwich, n.d., Bodl. MS. Tanner 25, ff. 229-30.

31. The work appeared anonymously as *Reflections on Learning wherein is shewn the Insufficiency thereof in its several Particulars in Order to evince the Usefulness and Necessity of Revelation* (London, 1699).

32. "Mr. Baker's book hath been held in great esteem and for many years after its publication, it was even commended as a standard of fine writing. In this view it was formerly put into our hands by a tutor." "Baker" in Kippis, *Biog. Brit.*, I, p. 519. See also Cole to Walpole, May 28, 1778, *Horace Walpole's Correspondence with the Rev. William Cole*, ed. W. S. Lewis (London, 1937), I, p. 342. The *Reflections* passed through five editions by 1714; eight altogether.

33. *Reflections*, pp. 82-85.

34. Camb. MS. Add. 7647, no. 26.

35. *Ibid.*, no. 27.

36. *Ibid.*, nos. 28-29.

37. Nov. 13, 1699, *ibid.*, no. 34.

38. April 15, 1700, *ibid.*, no. 35.

39. *Ibid.*, nos. 34, 36 (April 23, 1700).

40. *Ibid.*, nos. 38-43. There is a letter in the University of Illinois Library from Woodward to Collins, May 30, 1701, that also seems to refer to this controversy and to threaten publication. I am grateful to Prof. George W. White of that university for sending me a copy.

Chapter 4: The Early History of Mankind

1. Jan. 2, 1667, *The Petty-Southwell Correspondence*, ed. Marquis of Lansdowne (London, 1928), pp. 11-12.

2. Southwell to Hale's grandson Matthew, Sept. 26, 1689, in the flyleaf of the British Museum copy of the German translation. Southwell seems to have re-

ceived a copy of the unpublished portion of the manuscript in 1696; see B.M. MS. Add. 9001.

3. Gilbert Burnet, _The Life and Death of Sir Matthew Hale_ (London, 1682), pp. 81-85.

4. Matthew Hale, _The Primitive Origination of Mankind Considered and Examined According to the Light of Nature_ (London, 1677), p. 5.

5. _Phil. Trans._, 9 (1674), pp. 78-83.

6. _Origination_, p. 176.

7. _Ibid._, p. 182.

8. Isaac de la Peyrère, _Praeadamitae_ (Amsterdam, 1655), trans. _Men before Adam_ (London, 1656).

9. _Origination_, p. 192ff.

10. "Edwards" in Kippis, _Biog. Brit._; Baker to Rawlinson, April 18, 1716, Bodl. MS. Rawl. Lett. 42, f. 267. "I am of opinion that he is one of the best Writers of the present age . . . a great Critick, an acute Philosopher, and a perfect Divine." _Miscellaneous Letters_, I, no. 18 (1695), pp. 463-64.

11. The Bishop of Bath and Wells, _A Commentary on the Five Books of Moses_ (2 vol., London, 1694); the Bishop of Ely, _A Commentary upon the First Book of Moses called Genesis_ (London, 1695). According to Maurice Cranston, Edwards was Locke's most formidable critic and most distressing; _John Locke_ (1957, reprinted London, 1968), pp. 390-92, 429-33. For a summary of the controversy, see H. O. Christophersen, _A Bibliographical Introduction to the Study of John Locke_ (1930; reprinted New York, 1968), pp. 57-66.

12. John Edwards, _Some Thoughts Concerning the Several Causes and Occasions of Atheism_ (London, 1695), pp. 82-83. Edwards specifically excepts Dr. Woodward here.

13. Edwards, _A Demonstration of the Existence and Providence of God from the Contemplation of the Visible Structure of the Greater and the Lesser World_ (London, 1696), p. 161.

14. Edwards, _A Discourse concerning the Authority Stile and Perfection of the Books of the Old and New Testament_ (London, 1695), III, pp. 46-47. Compare

the two Bishops above (n. 11): the Bishop of Bath, pp. xiv-xv, and the Bishop of Ely, preface, sig. A^V.

15. Edwards, _Some Thoughts_, pp. 82-83.

16. Edwards, _A Demonstration_ pp. 150-52.

17. _Ibid._, pp. 148, 150, 257. John Morton commended the first of these thoughts to Lhwyd as an example; March 25, 1696, Bodl. MS. Ashmole 1816, ff. 406-07.

18. Edwards, _Brief Remarks upon Mr. Whiston's New Theory of the Earth_ (London, 1697), ded. to Dr. Woodward, sig. A2^V.

19. Edwards to Woodward, Feb. 4, 1697, Camb. MS. Add. 7647, no. 114.

20. So Edwards to Woodward, May 16, 1699, _ibid._, no. 115. Woodward's letter, dated May 2, is now in the possession of Dr. V. A. Eyles, who has kindly furnished me with a copy. While the bulk of the correspondence is at Cambridge, Dr. Eyles possesses, besides this letter, a summary of the exchange that was drawn up by Dr. Woodward.

21. The following is based upon Camb. MS. Add. 7647, no. 116 and the Eyles letter above (n. 20).

22. June 15, 1699, Camb. MS. Add. 7647, no. 117.

23. June 26, 1699, _ibid._, no. 118.

24. Aug. 23, 1699, _ibid._, no. 119.

25. Woodward, _Nat. Hist._, p. 49ff.

26. J. Estlin Carpenter, _The Composition of the Hexateuch_ (London, 1902), pp. 36-43, etc. For the state of the controversy then over the authorship of the Mosaic narrative, see the _Journal des Scavans_, suppl., Sept. 1708, quoted in J. S. Spink, _French Free-Thought from Gassendi to Voltaire_ (London, 1960), p. 291n.

27. _Proc. Mass. Hist. Soc._, 1873-75, p. 110.

28. "Memoir of Inigo Jones," John Webb, _The Most Notable Antiquity of Great Britain_ (London, 1655), n.p. See also W. Jerome Harrison, "A Bibliography of the Great Stone Monuments of Wiltshire," _The Wiltshire Magazine_, Dec. 1901, pp. 1-170; Richard J. Atkinson, "Stone-

henge and the History of Antiquarian Thought," *Stonehenge* (London, 1956), pp. 181-204.

29. Ray to Lhwyd, May 27, 1691, Bodl. MS. Eng. Hist. C 11, f. 46. A few years later Samuel Gale wrote of it in his travel diary that "the literati are still in suspense as to the origin." "Reliquiae Galeana," *Bibl. Top. Brit.*, III, p. 25.

30. Woodward to Lhwyd, Dec. 12, 1693, and Jan. 23, 1694, Bodl. MS. Ashmole 1817b, ff. 365, 367-68. There are some notes from a later period dealing with the subject in Bodl. MS. Top. gen. C 23, f. 2ff. Here as elsewhere Woodward does not seem to have altered his original opinions.

31. Apparently it was still being argued seriously in the *Quarterly Review* of 1860 (p. 209) that Stonehenge and Avebury were post-Roman monuments; see John Lubbock, *Pre-Historic Times* (London, 1865), p. 53ff.

32. See Lubbock. pp. 6-7; Glyn Daniel, *The Idea of Prehistory* (London, 1962). Lhwyd stuck to the Druids in his contribution, "Pembrokeshire," to Camden's *Britannia* (1695); Halley, on the other hand, told Hearne that Stonehenge was "as old, at least almost as old, as Noah's Flood." Hearne, *Remarks*, VII, p. 350.

33. Woodward to Hoskins, in Woodward, *Fossils*, app., pp. 24-25.

34. Woodward to Lhwyd, March 24, 1692, Bodl. MS. Eng. Hist. C 11, f. 109.

35. Woodward commented on Greaves' *Pyramidographia* before the Royal Society, June 24, 1702, R.S. MS. Jour. Bk., IX, p. 317. For Greaves' achievement, see Fred G. Bratton, *A History of Egyptian Archaeology* (New York, 1968), p. 58. In general, see Karl H. Dannenfeldt, "Egypt and Egyptians in the Renaissance," *Studies in the Renaissance*, 6 (1959), pp. 7-27; John D. Wortham, *The Genesis of British Egyptology* (Norman, 1971); Leslie Greener, *The Discovery of Egypt* (London, 1966), p. 42ff.

36. "Of the Wisdom of the Antient Egyptians; A Discourse concerning their Arts, their Sciences, and their Learning; their Laws, their Government, and their

Religion. With Occasional Reflections upon the State of Learning among the Jews; and some other Nations," *Archaeologia*, 4(1777), pp. 212-311. The work, "seemingly prepared for the Press," was transmitted to Michael Lort by a Mr. Herbert and read by Lort at five sessions of the Society: Dec.-Jan. 1775-76, Soc. Antiq. MS. Min. Bk., XV, pp. 275-78, 280-83, 285-87, 289-93, 295-99.

37. Spencer's works were entitled *De legibus Hebraeorum* (London, 1685) and *Dissertatio de Urim et Thummim* (London, 1699). Both books created controversy and drew many replies including, inevitably, John Edwards. See "Spencer" in the *DNB*.

38. Woodward, "Egyptians," p. 215. Burnet had written that "A Moral or Philosophic History of the World well written would certainly be a very useful work, to observe and relate how the scenes of Humane life have chang'd in several Ages, the Modes and Forms of Living, in what simplicity Men begun at first, and by what degrees they came out of that way . . . what new forms and modifications were superadded by the Invention of Arts, what by Religion, what by Superstition. This would be a view of things more instructive and more satisfactory, then to know what Kings reign'd in such an Age, and what Battles were fought; which common History teacheth, and teacheth little more." Quoted in Tuveson, *Millenium and Utopia* (ch. 2 above, n. 20), p. 158.

39. Woodward, *Nat. Hist.* pp. 106-07.

40. Southwell to Boyle, May 30, 1661, in Boyle, *Works* (London, 1772), VI, p. 299.

41. E. W. Budge, *The Mummy*, 2nd ed. (reprinted New York, 1964), p. 124. For a more favorable estimate, see John E. Fletcher, "Astronomy in the Life of Athanasius Kircher," *Isis*, 61 (1970), pp. 52-53; "Athanasius Kircher and the Distribution of his Books," *The Library*, 5th ser., 23 (1969), pp. 108-17.

42. There is a fragment by Woodward to this effect also, in Camb. MS. Add. 7647, which begins, "All who have set forth the Hieroglyphs as any other than

meer History-Paintings and Sculptures . . .
are very wrong."
43. Woodward, "Egyptians," p. 227n.
44. *Ibid.*, p. 232ff.
45. *Ibid.*, p. 238ff.
46. *Ibid.*, p. 280ff.
47. Robertson Smith, *The Religion of the Semites* (London, 1894), p. vi.

Chapter 5: The Embattled Philosopher

1. "Flintshire," in *Britannia*; Morton to Lhwyd, March 25, 1696, Bodl. MS. Ashmole 1816, ff. 406-07.
2. Robinson to Lhwyd, March 5 and April 8, 1698, Bodl. MS. Ashmole 1817a, ff. 341 and 342; Sept. 28, 1697, Bodl. MS. Eng. Hist. C11, f. 85. Robinson's rebuttal was entitled "A Letter Sent to Mr. William Wotton," n.d. In it he admitted that he had helped L.P. with the passages dealing with America, "touching" them and supplying information.
3. Cole to Lhwyd, Oct. 31, 1697, Bodl. MS. Ashmole 1829, f. 146; Robinson to Lhwyd, Sept. 28, 1697, Bodl. MS. Eng. Hist. C 11, f. 85.
4. Edited by William Derham from manuscript additions made in 1703-04; see pp. 165-67, 203, 211-12, 218-20. Robinson told Ray in 1704 that Woodward's influence with the booksellers was making it difficult to publish it; Lhwyd to Richardson, B.M. MS. Sloane 4064, f. 15.
5. Lhwyd to Lister, Dec. 5, 1695, Bodl. MS. Lister 36, f. 138; *Lhwyd Letters*, pp. 295-96.
6. Woodward to Lhwyd, March 5, 1695, Bodl. MS. Eng. Hist. C 11, f. 115; Woodward to Hearne, May 3, 1711, Bodl. MS. Rawl. Lett. 12, ff. 444-45; Woodward to Scheuchzer, Scheuchzer Corr., July 22, 1709.
7. The expression "sympathizing Virtuosos" occurs also in a letter of Robinson to Lhwyd, Feb. 12, 1698, Bodl. MS. Eng. Hist. C 11, f. 95. See also April 8, 1695, Bodl. MS. Ashmole 1817a, f. 342. The manuscript is in a commonplace book, B.M. MS. Add. 15067. Lhwyd made similar annotations in his copy of Woodward's *Essay* now in the Bodleian Library, shelfmark 8° Rawl. 704. Robinson insisted that everyone believed that Woodward was the true author, though he

thought that Lhwyd's comparison was too detailed and tedious for publication. See Robinson to Lhwyd, Jan. 20 and March 17, 1698, Bodl. MS. Ashmole 1817a, f. 261, and Bodl. MS. Eng. Hist. C 11, f. 94. For Lhwyd's contribution to Woodward's critics, see Robinson to Lhwyd, April 8, 1898, Bodl. MS. Ashmole 1817a, f. 342.
8. Lhwyd to Lister, Feb. 9 and April 23, 1696, *Lhwyd Letters*, pp. 298-300, 302-04.
9. See Melvin E. Jahn, "A Note on the Editions of Edward Lhwyd's *Lithophylacii Britannici Ichnographia*," *J. Soc. Bibliog. Nat. Hist.*, 6 (1971), pp. 86-97.
10. Lhwyd to Lister, Bodl. MS. Lister 3, f. 164; *Lhwyd Letters*, pp. 320-21.
11. *Lithophylacii*, pp. 128-39, translated by Lhwyd himself in Bodl. MS. Ashmole 1820a, f. 28ff., and printed in *Lhwyd Letters*, pp. 381-96.
12. He was apprised of this by Richard Mostyn and replied on Dec. 27, 1696, *Lhwyd Letters*, p. 318. He wrote at once to both Robinson and Ray, only to be reassured by each. See Robinson to Lhwyd, April 8, 1698, Bodl. MS. Ashmole 1817a, f. 342; Lhwyd to Morton (including part of a copy of a letter to Ray), Feb. 23, 1698, *ibid.*, ff. 277-78; and a reply from Ray (extract only), March 12, 1698, *ibid.*, f. 277. For Hale, see above ch. 4, Sect. 1.
13. Lhwyd to Richardson, July 17, 1702, *Lit. Ill.*, I, p. 318.
14. "I hear some of the letters have already distasted our old friend D W and some others," Lhwyd to Lister, Feb. 12, 1699, *Lhwyd Letters*, pp. 413-14.
15. Lhwyd to Richardson, April 18, 1699, B.M. MS. Sloane 4062, f. 300; Richardson to Sloane, Nov. 29, 1712, B.M. MS. Sloane 4063, f. 114.
16. Woodward to Scheuchzer, Scheuchzer Corr., Oct. 2, 1703. The manuscript was Woodward's "Two Discourses: An Account of Ores and Metals . . . The Art of Assaying," B.M. MS. Add. 25095, f. 6^V. He insisted (correctly) that few had been willing to go into the field to seek out information.
17. Lhwyd to Morton, Dec. 26, 1704, B.M. MS. Sloane 4064, f. 48.
18. Woodward's contributions are re-

corded in the Journal Books of the Royal Society and in the Catalogue of Donations, R.S. MS. D, no. 416. Among those whom Woodward nominated for membership were Samuel Garth, Otto Sperling, Sir Philip Sydenham, Dr. Woodford, Dr. Wolfius, Joshua Barnes, William Nicolson, and John Morton. Typical was the experience of Emanuel Swedenborg: "I have been at Woodward's who was so polite as to introduce me to some of the learned and to members of the Royal Society." Swedenborg to Eric Benezelius, Aug. 15, 1712, Documents Concerning the Life and Character of Emanuel Swedenborg, ed. R. L. Tafel (2 vol., London, 1875), I, p. 223.

19. Robinson to Lhwyd, Feb. 4, 1699, Bodl. MS. Ashmole 1817a, f. 343. Other disagreements appear in the minutes for 1696-97, B.M. MS. Sloane 3341, ff. 3, 6^V, and for 1699-1700, R.S. MS. Jour. Bk., IX, pp. 176, 219.

20. "Dr. Woodward said that having bought several of the late Dr. Hookes Bookes he had found several loose Manuscripts of his own hand in them which he was pleased to offer the Sight of to the Council of the Royal Society if they may be found usefull to Explane any of his Writings." May 5, 1703, R.S. MS. Jour. Bk. of the Council, II, p. 171.

21. Robinson to Lhwyd, Jan. 20, 1698, Bodl. MS. Eng. Hist. C 11, f. 94.

22. See Sloane's sketch of John Beaumont, a critic of Burnet's, sent to the Abbé Bignon in 1740, the only place I have found where Sloane publicly declared himself; in Jean Jacquot, "Sir Hans Sloane and French Men of Science," N.R., 10 (1953), pp. 94-96 (from a manuscript in the Bibliothèque Nationale).

23. Petiver to Hugh Jones (draft), Nov. 1, 1697, B.M. MS. Sloane 3033, ff. 91^V-93. When Sloane showed a large piece of amber the size of a man's fist to the Royal Society, Woodward typically said that he had seen a piece four times the size! R.S. MS. Jour . Bk., IX, p. 200.

24. Anon., The Transactioneer with some of his Philosophical Fancies (London, 1700), p. 74.

25. Southwell to Cole, Nov. 9, 1699, B.M. MS. Add. 18599, ff. 89-90.

26. Harris to Sir John Hoskins, Pres.

of the Royal Society, denying his authorship, B.M. MS. Sloane 4026, f. 253. Nevertheless Jabez Cay, a friend of Woodward's, thought that the "transactioneer" was "no witch at rallying, and some of his jests that are the most tolerable are not his own but taken from Mr. Harris's answer to Parson Robinson." Cay to Thoresby, August 29, 1700, in Thoresby, Letters (1912), pp. 86-87.

27. The Transactioneer was reprinted in The Original Works of King, ed. John Nichols (3 vol., London, 1776), III, pp. 1-56. For King, see below ch. 13, sect. 4.

28. Feb. 28, 1700, B.M. MS. Sloane 3334 (Petiver copy), ff. 58-59. See Charles R. Weld, A History of the Royal Society (2 vol., London, 1848), I, pp. 353-55.

29. R.S. MS. Jour. Bk. of the Council, Feb. 28, 1700, p. 154.

30. Thoresby, Letters (1912), I, pp. 115-116.

31. Harris to Hoskins, B.M. MS. Sloane 4026, f. 253.

32. R.S. MS. Jour. Bk., IX, p. 297.

33. Woodward to Sloane, May 14, 1703, B.M. MS. Sloane 4039, f. 128; Sir Godfrey Copley to Thomas Kirke, June 17, 1703, B.M. MS. Stowe 748, ff. 9-10.

34. Richardson to Thoresby, May 11, 1702, Richardson Corr., pp. 40-43; Thoresby, Letters (1912), I, pp. 409-10. Richardson had been extravagantly praised by Ray in a letter to Lhwyd, Feb. 1, 1698; Ray, Further Corr., pp. 274-76.

35. Woodward to Thoresby, May 16, 1702, Fitz. Mus. MS. Spenser-Percival Bequest, Packet B; Richardson to Sloane, May 23, 1702, B.M. MS. Sloane 4028, f. 348; Sloane to Thoresby, May 26, 1702, in Thoresby, Letters (1912), I, pp. 414-15.

36. Richardson to Thoresby, July 1702, in Richardson Corr. pp. 65-68. For other letters relating to this affair, see ibid., pp. 35-36. 37-40, 41-42, 43-46, 53-55; Thoresby, Letters (1912), I, 409-10, 410-11, 420-21.

37. July 17, 1703, Thoresby, Letters (1912), II, pp. 26-29. Thoresby described himself accurately enough as one who liked "to pass calmly and silently thro' this wrangling world." Camb. MS. Baumgartner Papers, IV, pt. iii, no. 277.

38. Thoresby to Sloane, May 29,

1703, B.M. MS. Sloane 4039, f. 136; Thoresby to Richardson, Feb. 22 and March 1, 1709, Bodl. MS. Radcliffe Trust C 3, ff. 52-53, 54. In 1712 Woodward reprimanded Thoresby for endorsing Richardson for the Royal Society; *ibid.*, f. 82.

39. Minutes of the Council, B.M. MS. Sloane 3342, f. 32.

40. Woodward to Scheuchzer, in Scheuchzer Corr., Sept. 10 and March 8, 1706.

41. Petiver to Scheuchzer, April 19, 1706, B.M. MS. Sloane 3335, ff. 53-54 (draft); Thorpe to Petiver, n.d., B.M. MS. Sloane 4067, f. 167. Thorpe eventually edited it for the Royal Society.

42. Woodward to Scheuchzer, in Scheuchzer Corr., March 26, 1708; John Chamberlayne to Sloane, Oct. 19, 1713, B.M. MS. Sloane 4043, f. 195.

43. "Nothing will restore it but those Means that first rais'd it to so great Reputation; the setting of a Correspondence and the making useful Experiments and Observations. . . . 'Tis judged his [Sloane's] Consciousness of his own Incapacity to joyn in that Way, and his Hopes to defeat it, are the true Motives to all this Pother, whatever else may be pretended." From the letter of an anonymous supporter of Woodward to Newton, March 28, 1710, King's College, Cambridge MS. 151.

44. As the letter above (n. 43) continues, speaking of Dr. Woodward: "He has in Readyness Materials, for Entertainment of you and the Society, that are proper and suitable to our Institution and not such as have now for some years past, wont to be produced. There is likewise a Disposition to settle a general Correspondence, to make Experiments and Observations and set on foot such Methods as may serve both to promote the Design and advance the Honour of the Society."

45. Flamsteed to Sharp, Oct. 25, 1709, in Francis Baily, *An Account of the Reverend John Flamsteed* (1835; reprinted London, 1966), pp. 271-72.

46. See the anonymous letter cited above in n. 43.

47. Lhwyd to Richardson, Oct. 8,

1708, *Lhwyd Letters*, pp. 546-47; Robinson to Lhwyd, Aug. 12 and Sept. 25, 1708, Bodl. MS. Ashmole 1817a, ff. 360-61, 363; Thorpe to Lhwyd, Oct. 23, 1708, Bodl. MS. Ashmole 1817b, f. 154.

48. Lhwyd to Thorpe, May 20, 1708, B.M. MS. Sloane 3336, ff. 66^V-67 (copy); *Phil. Trans.*, 26 (1708), pp. 157-58.

49. Richardson to Thoresby, Feb. 28, 1708, Yorks. Arch. Soc. MS. 8; Thoresby, *Letters* (1912), II, p. 195.

50. Thorpe to Lhwyd, Oct. 25, 1708, Bodl. MS. Ashmole 1817b, f. 157.

51. Quoted in Richardson to Thoresby, Feb. 28, 1708, Yorks. Arch. Soc. MS. 8, n. 49. See Langius to Lhwyd, June 30, 1706, B.M. MS. Sloane 4064, ff. 113-14.

52. Lhwyd to Sloane, Feb. 16, 1709, B.M. MS. Sloane 4041, f. 280; Robinson to Lister, June 23, 1709, Bodl. MS. Lister 37, f. 151.

53. Nov. 25, 1708. Bodl. MS. Ashmole 1817b, f. 157.

54. Smith to Lister, July 16, 1709, Bodl. MS. Lister 37, f. 154. Dr. Woodward was also, as a result, dropped temporarily from the Council; Thorpe to Lhwyd, Nov. 30, 1708, Bodl. MS. Ashmole 1817b, f. 160. Woodward tried also to oppose Lhwyd for the job as Beadle at Oxford; Woodward to Hearne, Nov. 29, 1708, and May 24, 1709, Bodl. MS. Rawl. Lett. 12, ff. 422, 424.

55. Robinson to Lister, Nov. 18, 1708, Bodl. MS. Lister 37, f. 132; Thomas Smith to Robinson, July 16, 1709, *ibid.*, f. 154; *Phil. Trans.*, 26 (Jan.-Feb. 1709), p. 252.

56. Lhwyd to Woodward, Jan. 17, 1709, an incomplete draft reply to an "expostulating" letter (now lost), *Lhwyd Letters*, pp. 548-50.

57. See above n. 6. Woodward had written a postscript to Lhwyd just after publishing the *Essay*, apologizing for the oversight; March 5, 1695, Bodl. MS. Eng. Hist. C 11, f. 115.

58. See the anonymous letter cited above in n. 43.

59. Woodward to Scheuchzer, in Scheuchzer Corr., Dec. 6, 1709; Woodward to Thoresby, Jan. 10, 1710, in Thoresby, *Letters* (1912), II, pp. 216-17;

Harris to Leibniz, June 18, 1710, from a
microfilm copy on deposit in the London
University Library; see Eduard Bodemann,
*Der Briefwechsel des Gottfried Wilhelm
Leibniz in der Koniglichen Offentlichen
Bibliothek zu Hannover* (1895; reprinted
Hildesheim, 1966), no. 370.

60. Woodward to Scheuchzer, in
Scheuchzer Corr., July 3, 1710; Flamsteed
to Sharp, July 14, 1710, in Baily, *Flamsteed* (n. 45 above), pp. 276-77.

61. This account is based on the
King's College letter (n. 43 above) and
Sloane's report in B.M. MS. Sloane 4026,
ff. 266-67, 296; on the Minutes of the
Royal Society in B.M. MS. Sloane 3342, f.
68ff.; and on the R.S. MS. Jour. Bk. of the
Council, II, pp. 113-18. It alters and
enlarges the traditional story as found in
David Brewster, *Memoirs of Sir Isaac
Newton* (2 vol., Edinburgh, 1855), pp.
244-47, and repeated by, among others,
Louis Trenchard More, *Isaac Newton*
(New York, 1934), pp. 534-38.

62. Woodward to Newton, May 30,
1710, Camb. MS. Add. 3965, f. 288[V].
There is a suggestion in the minutes that
the two men were already in disagreement
over the bezoars in Feb. 1701, R.S. MS.
Jour. Bk. IX, pp. 245, 247.

63. B.M. MS. Sloane 3342, ff. 77-77[V].

64. Halley to Sloane, May 14, 1710,
B.M. MS. Sloane 4042, f. 131.

65. So the King's College letter (n.
43), which gains some support from the
Minutes; see B.M. MS. Sloane 3342, f. 75.
Woodward seems to have thought that the
King's College letter or something like it
would be published; he promised Hearne a
copy. Aug. 19, 1710, Bodl. MS. Rawl.
Lett. 12, f. 432.

66. Woodward to Scheuchzer, in
Scheuchzer Corr., Oct. 26, 1710; B.M. MS.
Sloane 3342, f. 83. For an account by one
of the opposition, see *An Account of the
Late Proceedings in the Council of the
Royal Society in order to remove from
Gresham College to Crane Court* (London,
1710). In appears to be by a Mr. Low-
throp; see B.M. MS. Add. 6194, f. 55. For
the defeat of Harris by a vote of 70-14, see
Thorpe to Hearne, Dec. 7, 1710, Bodl.
MS. Rawl. Lett. 10, ff. 400-01.

Chapter 6: The Collector

1. Woodward, *An Attempt*, I, pt. i, p. 45.

2. *Ibid.*, p. 242; preface, p. xiv.

3. Woodward, *Fossils*, p. 93ff. He sent
Thomas Lower to Cornwall, John Groom
and Richard Meales to the North.

4. Nicolson to Lhwyd, Nov. 7, 1700,
Bodl. MS. Ashmole 1816, f. 511; Nicolson
to Thoresby, Aug. 19, 1700, Bodl. MS.
Don d 89, f. 205-06; Woodward to
Thoresby, July 18, 1702, in Thoresby,
Letters (1832), I, pp. 420-21; Lhwyd to
Richardson, Mar. 28, 1698, B.M. MS.
Sloane 4062, ff. 279-80, and Sept. 2,
1701, B.M. MS. Stowe 747, f. 146;
Richardson to Lhwyd, Dec. 20, 1700, and
June 25, 1701, Richardson, *Corr.*, pp.
31-32, 35-36; Richardson to Sloane, May
23, 1702, B.M. MS. Sloane 4038, f. 348.

5. Dr. Shaw complained that a fossil
discovered in Palestine "was borrowed of
him by Dr. Woodward who was never
prevailed upon to return it." To which
someone else noted that "It was thus the
Dr. served Mr. A— also." W. N. Edwards,
The Early History of Paleontology (London, 1967), p. 45.

6. Woodward to Lhwyd, Jan. 19,
1692, Bodl. MS. Eng. Hist. C 11, ff.
104-05.

7. Woodward to Lhwyd, March 24,
1691, *ibid.*, f. 109. Lhwyd had accused
Woodward of running down his gifts to
him; Woodward denied it.

8. "I find our Whitney Lithoscopist
betray'd us not only instructing Mr.
Morton how to find of the Siliquestra
Fish-teeth and Vertebrae but also selling
to him several curious Stones which of
right belong'd to us." Lhwyd to Lister,
April 1, 1694, Bodl. MS. Lister 36, f. 93.
For Morton, see below, sect. 3.

9. Lhwyd to Lister (?), April 18,
1693, *Lhwyd Letters*, pp. 177-78; Lhwyd
to Morton, June 25, 1695, B.M. MS.
Sloane 4062, ff. 259; Lhwyd to Richardson, Aug. 28, 1704, B.M. MS. Sloane
4064, f. 35.

10. The use of the Woodward Museum
and collections may be traced in part by
the memoranda and inspector's reports

(required under Dr. Woodward's will) that
are still in the Sedgwick Museum. Thus the
1798 report refers to the collection characteristically as "a precious deposit of
Minerological Knowledge." Lhwyd's fossils in the Ashmolean Museum were
already neglected by 1758; see *Lit. Ill.*, IV,
pp. 457-59.

11. Woodward to Arthur Charlett,
Aug. 18, 1720 (copy), Bodl. MS. Ballard
24, f. 85. Woodward did furnish some
fossils to the Ashmolean Museum in
1716-17; see the Book of Benefactors,
Bodl. MS. Ashmole 2, p. 20^V and MS.
Ballard 24, ff. 81, 83. He had been
thwarted earlier somehow by his enemy
John Thorpe; see Woodward to Hearne,
Feb. 10, 1711, Bodl. MS. Rawl. Lett. 12,
f. 429. For his conversion to Cambridge
by a friend, Richard Graham, see Clark
and Hughes ch. 2 above, n. 2), p. 189.

12. I use the copy in the Baker papers,
Camb. MS. Mm 44. pp. 139-47. There is
another in the Sedgwick Museum with an
indenture added, July 29, 1731, showing
how it was implemented. (The original is
dated Oct. 1, 1727.) And another with
related documents in Camb. Univ. Reg.
39-17-1.

13. William Whewell, *History of the
Inductive Sciences* (London, 1837), III, p.
496.

14. *An Attempt*, Cat. of Foreign Fossils, p. 47. "In your letter," he wrote to
Scheuchzer, "you have given Names to
each but I wish you had rather mentioned
the Place where each was found . . . For I
do not so much please my self with Names
and Words as with Things and the Uses of
them." July 21, 1702, Scheuchzer Corr.
Woodward's care in these matters is
compared with the modern curator's by
F. J. North, "From Giraldus Cambrensis
to the Geological Map," *Transactions of
the Cardiff Naturalists Society*, 64 (1931),
p. 62.

15. Title page of the manuscript of *An
Attempt*, Cat. of Extraneous Foreign
Fossils, July 10, 1725, in the Sedgwick
Museum. It was corrected and prepared
for publication by A. Taylor and M. Duke-
son.

16. *An Attempt*, Cat. of English Fossils, pref. p. 2.

17. *Ibid.*, Cat. of Extraneous Foreign
Fossils, pp. 6-7.

18. *Ibid.*, pp. 31, 52, etc.; "Dampier,"
in *DNB*. There is a letter from Flamsteed
to Woodward, July 24, 1701, asking the
doctor to get information from Dampier
for him. It suggests a fairly close relationship between these two difficult men, at
least at this time; see Christ Church,
Oxford MS. Upcott Coll., I, no. 106.

19. Francis Maximilian Misson, *A New
Voyage to Italy* (2 vol., London, 1699), I,
pp. 167-71.

20. Woodward, *An Attempt*, Catalogue of Foreign Fossils, second add., pp.
8-9.

21. "Inspector's Report, 1798," in the
Sedgwick Museum.

22. Jacob von Melle, *Epistola de
Echinitis Wagricis* (1718), and A. D. Leo-
pold, *De Itinere Suecico*; see Woodward,
Fossils, pp. 54-55.

23. See G. R. de Beer, "Johann Hein-
rich Hottinger's Description of the Mountains of Switzerland, 1703," *Annals of
Science*, 6 (1950), pp. 327-60.

24. Louis Bourguet, *Lettres Philoso-
phiques* (Amsterdam, 1729), p. 177ff.;
Traité des Petrifactions (Paris, 1742), pp.
59-64, 81ff.; the letters are in Camb. MS.
Add. 7647 and in the possession of Dr.
V. A. Eyles.

25. In the summer of 1697, for exam-
ple, Dr. Woodward received a large cargo
from America including many "Shells,
Bones and Teeth of Fishes." Woodward to
Evelyn, Christ Church, Oxford MS. Upcott
Coll., I, no. 100.

26. In the possession of V. A. Eyles;
Aug. 14, 1697. The quotation is from
Louis B. Wright and Marion Tinling, *The
Secret Diary of William Byrd* (Richmond,
Va., 1941), p. vi. See also Maude H.
Woodfin, "William Byrd and the Royal
Society," *Virginia Magazine of History
and Biography*, 40 (1932), pp. 23-34,
111-23.

27. Woodward to Evelyn (n. 25
above).

28. Byrd to Sloane, April 20, 1706,
B.M. MS. Sloane 3335, ff. 77-79^V. See
"Letters of William Byrd II and Sir Hans
Sloane," *William and Mary Quarterly*, ser.
2, 1(1921), p. 186.

29. Lhwyd to Morton, Feb. 23 and April 16, 1698, B.M. MS. Sloane 4062, ff. 277, 282; Petiver to Jones (?), Nov. 1, 1697, and March 10, 1698, B.M. MS. Sloane 3333, ff. 91^V-93, 119^V. In general, see Raymond P. Stearns, "James Petiver, Promotor of Natural Science," *Proc. Amer. Antiq. Soc.*, 1952. pp. 243-365.

30. Vernon to Sloane, July 24, 1698, B.M. MS. Sloane 4037, f. 103; Stearns, "Petiver," pp. 269-71. See also Lhwyd to Richardson, July 17, 1702, *Lit. Ill.*, I, p. 318.

31. Frank Brydges, Aug. 20, 1699, Camb. MS. Add. 7647.

32. Woodward, "Of Providence and Conduct of it in the Government of the Natural World"; the manuscript was listed by Col. King among Woodward's effects at his death. King to Windsor, May 15, 1728, B.M. Add. 5860, ff. 154-55^V; Ward, *Lives*, p. 300.

33. Cotton Mather, *The Christian Philosopher*, facs. ed. Josephine K. Piercy (Gainesville, Fla., 1968), p. 1.

34. April 3, 1721, *Proc. Mass. Hist. Soc.* 1873-75, pp. 110-11. Extracts of a number of letters appear in *Selected Letters of Cotton Mather*, ed. Kenneth Silverman (Baton Rouge, La., 1971), p. 108ff. Parts of seven of them were published originally in *Phil. Trans.*, 29 (1714), pp. 62-65. For the correspondence in general, see George L. Kittridge, "Cotton Mather's Election into the Royal Society," *Proc. Mass. Hist. Soc.* (1916 reprint), pp. 1-42, and Raymond P. Stearns, *Science in the British Colonies in America* (Urbana, Ill., 1970), pp. 405-26. See also Theodore Hornenberger, "The Date, the Source, and the Significance of Cotton Mather's Interest in Science," *American Literature*, 6 (1934-35), pp. 413-20; Thomas J. Holmes, *Cotton Mather: A Bibliography of his Works* (3 vol., Cambridge, Mass., 1940.

35. Mather to Woodward, Nov. 17, 1712, Royal Society MS. Letter Book M 2, f. 21; Mather to Winthrop, Aug. 15, 1716, *Collections of the Mass. Hist. Soc.*, 4th ser., 8 (1868), pp. 419-20; Winthrop to Mather, Nov. 5, 1716, *ibid.*, 6th ser., 5 (1892), pp. 332-33; Woodward to Winthrop, April 3, 1721, and to Mather on the

same day, *Proc. Mass. Hist. Soc.*, 1873-75, pp. 110,110-11.

36. See "An Account or History of the Procuring the Small Pox by Incision or Innoculation as . . . practised at Constantinople," a paper sent to Dr. Woodward by Emanuel Timonius and communicated to the Society; *Phil. Trans.*, 29 (1714), pp. 72-82. The third series of *Curiosa* sent to Woodward in 1716 contained a report by Mather on the Smallpox. Mather, incidentally, also encouraged Woodward in his quarrel with Camerarius; *Selected Letters*, pp. 135-36.

37. Woodward to Mather, April 3, 1721, *Proc. Mass. Hist. Soc.*, 1873-75, pp. 110-11; and Mather to Woodward, quoted by Kittridge (n. 34 above), p. 107n.

38. Woodward to Scheuchzer, in Scheuchzer Corr., July 20, 1703; March 14, 1704; Jan. 15, 1707.

39. "Fossils" in Harris, *Lexicon Technicum* (London, 1705); Lhwyd to Richardson, July 16, 1704, B.M. MS. Sloane 4064, f. 24.

40. Harris, "Fossils"; Eyles, "Woodward" (ch. 2 above, n. 2), p. 409; Woodward to Newton, n.d., in Woodward, *Fossils*, app., pp. 1-5.

41. Woodward to Hoskins, n.d., in Woodward, *Fossils*, app., p. 22.

42. In his fossil catalog, Woodward typically ridicules Plot and Lhwyd for their fanciful names: "Those two Writers, Dr. Plot of meer Simplicity and Mr. Lhwyd of Design, *darken Council by Words*, Job xxxviii,2." Woodward, *An Attempt*, p. 101. Lhwyd's editor, William Huddesford, made much the same complaint in 1758; *Lit.Ill.*, IV, p. 472. Woodward enlarged upon the special difficulties in classifying minerals in the *Essay*, p. 171ff., and again in the letter to Newton (n. 40 above), pp. 3-4. For his achievement here, see North (n. 14 above), pp. 54, 74.

43. See especially "A Methodological Distribution of Fossils of all Kinds into their proper Classes," Woodward, *Fossils*, pp. 1-56; also the letters to Newton and Hoskins cited above (nn. 40-42), pp. 4, 72.

44. Woodward to Hoskins, in *Fossils*, app., pp. 13-14. Woodward credited Merret with at least one piece in his collection;

An Attempt, p. 168. Merret was a fellow member of the Royal Society; he died in 1695.

45. Nicolson to Lhwyd, Jan. 31, 1698, in Nicolson, *Letters*, pp. 103-05. Woodward's views about Buttner and the corals are set out at length in the Scheuchzer correspondence. They were criticized by James Parkinson, *Organic Remains* (London, 1808), II, p. 7.

46. *Fossils*, app., pp. 58-75.

47. He says he was fifty that year; Sir George Wheler, *Autobiography*, ed. E. G. Wheler (Birmingham, England, 1911), p. 1. From early in life, he wrote, however, that "nothing pleased me more than to make dayly Discoveries of the wonderfull operations of God in the creation of the visible world," p. 14.

48. I quote from the correspondence in the Durham Cathedral Library, MS. no. 13, Dec.-May 1700-01, nos. 41-42.

49. *Fossils*, app., preface. Their exchanges extended to antiquities also; see Woodward to Hearne, Sept. 27, 1711, Bodl. MS. Rawl. Lett. 12, f. 452.

50. Southwell to Sloane, Sept. 12, 1692, B.M. MS. Sloane 4036, f. 135, and Bodl. MS. Tanner 25, f. 380 – from which it appears, however, that Southwell did not yet know Woodward personally. The *Essay* is dedicated to Southwell, whom Woodward described in that year (1695) as "a person of very extraordinary Knowledge, Goodness and Worth, as well as my particular good friend." Woodward to Lhwyd, n.d., Bodl. MS. Eng. Hist. C 11, f. 115.

51. *The Petty-Southwell Correspondence*, ed. Marquis of Lansdowne (London, 1928), pp. 20-22.

52. Southwell to Cole, July 2, 1694, B.M. MS. Add. 18599, f. 5.

53. Cole to Southwell, Aug. 27, 1695, *ibid.*, f. 33. For Cole's own views, see Cole to Lhwyd, June 5, 1695, Bodl. MS. Ashmole 1830, f. 10; another with the date torn off in Bodl. MS. Ashmole 1829, f. 153; and again, March 9, 1698, Bodl. MS. Ashmole 1814, ff. 311-12. Lhwyd reported them fully to Lister, Nov. 10, 1695, *Lhwyd Letters*, pp. 288-89, and to Morton, B.M. MS. Sloane 4062, ff. 261-62.

Cole's collections are described in Bodl. MS. Ashmole 1820, ff. 64-69, 71. There is a brief mention of Cole in V. A. Eyles, "Scientific Activity in the Bristol Region in the Past," *Bristol and Its Adjoining Counties*, ed. C. M. MacInnes and W. F. Whitehead (Bristol, 1955), pp. 128-29.

54. Cole to Woodward, March 25, 1695, Camb. MS. Add. 7647, no. 90.

55. Woodward, *An Attempt*, p. 21; Cole to Southwell, May 10, 1696, Camb. MS. Add. 7647, n.p.; Southwell to Cole, April 25, 1696, B.M. MS. Add. 18599, f. 51.

56. *An Attempt*, Catalogue of English Fossils, p. 9. For brief sketches of each, see Richard Pulteney, *Historical and Biographical Sketches of the Progress of Botany in England* (2 vol., London, 1790).

57. T. Bobart to Lhwyd, n.d., Bodl. MS. Eng. Hist. C 11, f. 3. Ray to Sloane, April 13, 1700, B.M. MS. Sloane 4038. f. 4.

58. Nicolson to Thoresby, March 21, 1702, in Thoresby, *Letters* (1912), p. 120; Woodward to Newton, n.d., in *Fossils*, app., p. 1. Among Stonestreet's many contributions to Woodward's collections were some fossil teeth of the *Lupus Marinus* still in the jaw, "a very great and valuable Curiosity." *An Attempt*, p. 84. For a return from Woodward to Stonestreet, see Stonestreet to Lhwyd, Sept. 2, 1705, Bodl. MS. Ashmole 1817a, ff. 488-89. The Rev. William Stonestreet was rector of St. Stephen's Walbrook; he died in 1716.

59. Morton to Richardson, Nov. 9, 1704, *Lit. Ill.*, I, p. 325; Richardson, *Corr.*, pp. 84-86.

60. See especially Lhwyd's reply to Morton, April 26, 1696, B.M. MS. Sloane 4062, f. 264, which should be compared with Woodward to Morton, March 19, 1695, Bodl. MS. Eng. Hist. C 11, f. 117. For the attempt at mediation, see Morton to Lhwyd, March 25, 1696, Bodl. MS. Ashmole 1816, ff. 406-07.

61. Thus he spent a week traveling with Lhwyd to the Humber, and Lhwyd continued to think him "a very ingenious as well as good natured gentleman." Lhwyd to Richardson, March 27, 1703,

B.M. MS. Sloane 4063, f. 196. See also Morton to Sloane, June 22, 1706, B.M. MS. Sloane 4040, ff. 183-84. For Woodward's nomination of Morton, April 21, 1703, B.M. MS. Sloane 3342, f. 11.

62. Nicolson, *English Historical Library*, 2nd. ed. (London, 1736), p. 20; John Morton, *Natural History of Northamptonshire* (London, 1712), p. ii, and for the quotations, pp. 48, 49, 132, 265.

63. See, for example, Morton to Nicolson, April 24, 1706, in Nicolson, *Letters*, pp. 296-302; and Morton to Sloane, June 22, 1706, B.M. MS. Sloane 4040, ff. 183-84.

64. See Woodward to Locke, Jan. 6, 1695, and Feb. 14, 1697, Bodl. MS. Locke C 23, ff. 105, 107; Woodward to Evelyn, Oct. 5, 1697, Christ Church, Oxford MS. Upcott Coll., I, no. 100. At least once they dined together at their host's, the Earl of Pembroke, along with Southwell; see *The Diary of Ralph Thoresby*, ed. Joseph Hunter (2 vol., London, 1830), I, pp. 336-37.

65. See Newton's letters to Burnet, 1680-81, in the *Newton Correspondence* (ch. 3 above, n. 23), II, pp. 319, 321-22, 329-35; William Whiston, *Memoirs* (ch. 2, above, no. 76), I, p. 43. In the letter to Newton in Woodward's *Fossils*, Woodward writes of his work on method as "if not your own, wholly owing to you; it being begun, carried on, and finished at your request." (The work referred to here was published first in Latin as an appendix to Woodward's *Naturalis Historia Telluris*.)

66. Woodward, *An Attempt*, I, pt. 1, pp. 83-85; II, pt. 1, pp. 19-20; II, pt. 2, p. 8. The *lusus* was supposed to have medicinal qualities as a cure for the "stone". Dr. C. L. Forbes writes that a modern scientist would describe it as a fragment of a *septarian nodule*, this specimen consisting of grey mudstone encrusted with brown calcite crystals. Newton's copy of Woodward's *Essay* is in the possession of V. A. Eyles.

67. Newton called Sloane "a tricking fellow; nay a villain and rascal for his deceitful and ill usage of you." See the anonymous King's College letter cited in ch. 5 above, n. 43.

68. See Camb. MS. Add. 4007, f. 695^V; in general, Frank E. Manuel, *Isaac Newton, Historian* (Cambridge, Mass., 1963).

69. Joseph Spence, *Observations, Anecdotes, and Characters of Books and Men*, ed. James M. Osborn (2 vol., Oxford, 1966), p. 350.

70. These are the only verses known to have been penned by Bentley, and they derive from a defense of learning against the wits. They were published first in *The Grove or a Collection of Original Poems* (London, 1721), pp. 244-46. One subscriber for a copy in Royal paper was Dr. Woodward. Thomas Hearne possessed a manuscript copy which varied a little from the *Grove*, preferring "searching" to "delving." Hearne, *Remarks*, VII, p. 322. Bentley had expounded Newton's views publicly in the first Boyle Lectures and corresponded with him about the *Principia*. For Woodward's attitude to Bentley, see ch. 9 below, nn. 33 and 52.

71. Woodward to Scheuchzer, in Scheuchzer Corr., Jan. 25, 1708.

72. William Stukeley, *Memoirs of Sir Isaac Newton's Life*, ed. A. H. White (London, 1936), p. 67; Woodward to Scheuchzer, in Scheuchzer Corr., May 22, 1711, and Sept. 12, 1712.

73. Woodward to Scheuchzer, in Scheuchzer Corr., May 17, 1715.

74. Frank E. Manuel, *A Portrait of Sir Isaac Newton* (Cambridge, Mass., 1968), ch. 13, "The Autocrat of Science," pp. 264-91; Stukeley, pp. 63-64, 79-80.

75. Woodward to Scheuchzer, in Scheuchzer Corr., Nov. 29, 1715; Woodward, *Fossils*, app., pp. 1-5. There is a note by Newton in 1726, "Dr. Woodward desires to see the second edition of my optiques viz that in octavo. And also the chronological tables printed in France. Also the third edition of the Principles." Manuel, *Newton*, p. 259n.

76. Woodward to Scheuchzer, in Scheuchzer Corr., April 17, 1716; Nov. 26, 1717; June 24, 1718.

77. *Ibid.*, July 22, 1709.

78. Leibniz' work on the subject, the *Protogaea*, was written in 1691 but not published until 1749. A précis appeared,

however, in the *Acta Eruditorum* (Leipzig), Jan. 1693. Woodward seems to have discovered Leibniz' views in an article in the *Histoire de l'Academie des Sciences* (1706); see Woodward to Scheuchzer, in Scheuchzer Corr., July 22, 1709. Scheuchzer also corresponded with Leibniz; see Bodemann, *Briefwechsel des Leibniz* (ch. 5 above, n. 59), no. 809. In general, see Jacques Roger, "Leibniz et la theorie de la terre," *Leibniz: Aspects de l'homme et de l'oeuvre* (Paris, 1968), pp. 137-44.

79. The correspondence between Woodward and Leibniz is in the Niedersachischen Landesbibliothek, Leibniz Briefe 1015, ff. 1-14. Ward summarizes two other letters, 1712-14, which belonged then to Matthew Postlethwayte but which seem to have disappeared; *Lives*, p. 288. Finally, there is a letter from Woodward to Leibniz, June 25, 1711, summarized in *Goethes Autographensammlung Katalog*, ed. H. J. Schreckenbach (Weimar, 1961), p. 252.

80. For an account of Dr. Woodward's unpublished manuscripts, most of which have disappeared, see the letters of A. Taylor to Col. Richard King (Woodward's executor), May 15 and June 14, 1728, B.M. MS. Add. 6209, ff. 341-42. Apparently in Feb. 1726, Dr. Woodward "found himself declining" and gave instructions that the manuscripts be destroyed, although some were quite ready for publication. There is a catalog of them (drawn up by Taylor) in a letter of Col. King to another of Woodward's executors, May 15, 1728, B.M. MS. Add. 5860, ff. 154-55^V. It was used by Ward in *Lives*, p. 300. Some other works and copies of some of Woodward's foreign correspondence found their way into the hands of Matthew Postlethwayte and were also used by Ward; see B.M. MS. Add. 6209, ff. 345-46. For other lost manuscripts, see ch. 2 above, n. 115.

Chapter 7: The Virtuoso Satirized

1. The schoolmaster, Charles Hoole, writing of "Subsidiary books," suggests: "These should be laid up in the Schoole Library for every Form to make use on, as they shall have occasion. Some of these serve chiefly to the explication of Grammar, and are applyed to it; some are needful for the better understanding of Classical Authours." *A New Discovery of the Old Art of Teaching Schoole* (London, 1660), pp. 205-06.

2. Thomas Godwin's *Romanae Historiae Anthologia*, which I have used in the second edition (London, 1616), went through sixteen editions by 1696! Both Basil Kennett's work, *Romae Antiquae Notitia or the Antiquities of Rome* (London, 1696), and John Potter's work, *Archaeologia Graeca* (2 vol., London, 1696-97), went through many subsequent editions. See also Joannes Rosinus, *Antiquitatum Romanorum* (Paris, 1613), and Francis Rous, *Archaeologiae Atticae Libri Septem* (Oxford, 1637), both again with many later editions.

3. See my article, "Ancients, Moderns, and History" (ch. 2 above, n. 5), pp. 48-50.

4. Peacham's *Complete Gentleman*; I use the edition by G. S. Gordon (Oxford, 1906), p. 105. The work appeared, first without this chapter, in 1622.

5. *Ibid.*, pp. 123-24.

6. There is a useful but not entirely satisfactory article on the subject by Walter E. Houghton, "The English Virtuoso in the Seventeenth Century," *Journal of the History of Ideas*, 3 (1942), pp. 51-73, 190-219.

7. John Earle, *Micro-Cosmographie or a Piece of the World Discovered in Essays and Characters*, ed. I.G. (London, 1899), pp. 14-15. The work went through many subsequent editions.

8. Thus a writer considering the propriety of antiquities writes, " 'Tis true the last mention'd Subject is not, properly speaking, the business of the Royal Society; but as many of its Members are Antiquaries, and as several things in that venerable Study are no less entertaining, than useful and satisfactory; it is not to be questioned, but that all ancient Manuscripts, Paintings, Medals, Coins, Urns, Monuments, Inscriptions, or ought else, tending either to explain the Ceremonies, Customs, Arts and Sciences of our Ancestors; or to set any controverted Points of History in a clear Light, must be received with Gratitude." *A List of the Royal*

Society of London (London, 1718), pp. 9-10. Edward Chamberlayne (a member) described the regular business of the Society as "to View and Discourse upon the Productions and Rarities of Nature and Art." *Angliae Notitia or the Present State of England* (London 1694), pp. 636-37.

9. See, for example, Halley's tract on the date of Caesar's first landing in England. *Phil. Trans.*, 17 (1691), p. 495; Eugene F. MacPike, *Correspondence and Papers of Edmund Halley* (1932; reprinted London, 1937), p. 217.

10. Rouquet's *Etat des Arts en Angleterre* (1755), quoted by Austin Dobson, "Dr. Mead's Library," *Eighteenth Century Vignettes*, 3rd ser. (London, 1923), pp. 29-30.

11. Nicolson to Thoresby, Feb. 8, 1692, in Nicolson, *Letters*, pp. 21-25; Thoresby, *Diary*, I, p. 275. For Thoresby in general, see D. H. Atkinson, *Ralph Thoresby the Topographer: His Town and Times* (2 vol., Leeds, 1885); for Nicolson, see Francis G. James, *North Country Bishop* (New Haven, 1956).

12. Thoresby, *Diary* I, pp. 335-39 (1701). Compare Humfrey Wanley's similar account in his letter to Charlett, Dec. 19, 1700, Bodl. MS. Ballard 13, f. 91.

13. Pierre Gassendi, *The Mirrour of True Nobility and Gentility*, trans. W. Rand (London, 1657), sig. A4.

14. Thoresby, *Diary*, I, p. 299 (1695). According to Wanley, "Mr. Charleton has the most notable collection of coins in England for a private Gentleman," Wanley to Charlett, Aug. 22, 1697, Bodl. MS. Ballard 13, ff. 57-58.

15. Thoresby, *Diary*, I, p. 340 (1701). Woodward's gifts to Thoresby are catalogued in Yorks. Arch. Soc. MS. 27. They included over a hundred fossils and stones and many autographed letters from famous men.

16. Dr. Woodward began a manuscript catalog of his antiquities in 1725, but was unable to complete it; Camb. MS. Add. 7570. It became the basis for the published account drawn up for auction at Woodward's death by Robert Ainsworth; see ch. 14 below, Sect. 1.

17. Nicolson to Thoresby, March 21, 1702, in Thoresby, *Letters* (1912), pp.

119-21; Nicolson, *Diary*, Jan. 2, 1702. I have used a transcript of the latter through the courtesy of Mr. Clyve Jones of the Institute of Historical Research, who is editing it for the Clarendon Press; portions have already been printed, imperfectly, in *Trans. of the Cumberland and Westmorland Antiq. and Arch. Soc.*, new ser., 1-5 (1901-02), 35 (1935), 46 (1946), 50 (1950).

18. Nicolson to Thoresby, March 21, 1702, in Thoresby, *Letters* (1912), pp. 119-21. Nicolson found Stonestreet, also, "Nicely skill'd in Antiquities of all kinds and especially in Medals and Coins of which he has a good stock." Nicolson to Lhwyd, Feb. 25, 1702, Bodl. MS. Ashmole 1816, f. 523. For Woodward's relationship with Stonestreet, see ch. 6 above, n. 58, and Woodward to Scheuchzer, in Scheuchzer Corr., March 14, 1704.

19. Nicolson, Diary, Nov. 21, 1704.

20. *Ibid.*, Jan. 5, 1705.

21. This is the manuscript described above in ch. 5, n. 16. Woodward notes there that "These Discourses have been wrote now 20 years and yet I have not found Leisure to compare the Copy with the Original. But Mr. Taylor and Mr. Chace have done that carefully this Day [Aug. 13, 1724]." The copy is B.M. MS. Add. 25096.

22. Nicolson, Diary, Jan. 20, 1706.

23. *Ibid.*, Jan. 31, 1706.

24. The Earl of Carbury was talking to Nicolson one day about these matters. "This brought us into a Discourse about the several Systemes of Dr. Burnet, Dr. Woodward, etc., and his Lordship ingeniously enough observe'd that, since Des Cartes led the way Every New Philosopher thought himself wise enough to make a World." *Ibid.*, Jan. 14, 1706.

25. The work appeared under the authorship of "Samuel Sorbieres," the writer of an earlier English description that had been much criticized. Lister's work was entitled *A Journey to Paris in the Year 1698* and may be read in an annotated version with a helpful introduction by Raymond P. Stearns (Urbana, Ill., 1967).

26. *De Itinere suo Anglicano et Batavo*, 2nd ed. (Amsterdam, 1711), pp.

41-42; trans. London, 1711, p. 37. Erndtel or Erndl was a German physician who became doctor to the King of Poland and climbed the Alps with Scheuchzer. The English translator insisted that Woodward's *Essay* had been irrefutably answered by Arbuthnot and Wotton.

27. The relevant portion of Uffenbach's *Travels* was translated by J. E. B. Mayor in *Cambridge Under Queen Anne* (Cambridge, 1911), pp. 404-05; see also W. H. Quarrell and Margaret Mare, *London in 1710 from the Travels of Z. C. von Uffenbach* (London, 1934). There is a biography of Uffenbach describing his great collections in Chauffepié, IV, pp. 567-70.

28. This seems to have been a fossil fake.

29. Samuel Butler in *Character Writings of the Seventeenth Century*, ed. Henry Morley (London, 1891), pp. 324-25.

30. See Gimcrack's widow's letter to Isaac Bickerstaff in *The Tatler*, no. 221, Sept. 7, 1710. *The Virtuoso* was published in 1691; see Albert Bergman, *Thomas Shadwell* (New York, 1928), p. 160ff.; Claude Lloyd, "Shadwell and the Virtuosi," *PMLA*, 44(1929), pp. 472-94.

31. Dale to Lhwyd, Oct. 1, 1708, Bodl. MS. Ashmole 1814, f. 360; he is recalling a meeting of 1702.

32. Thoresby to Richardson, March 1, 1709, Bodl. MS. Radcliffe Trust C 2, f. 54; *Memoirs of the Life of Sir John Clerk* (Edinburgh, 1892), p. 127.

33. *Tatler*, no. 216, Aug. 26, 1710; Shaftesbury, *Characteristics* (London, 1732), III, pp. 156-60.

34. *An Essay in Defense of the Female Sex in which are inserted the Character of a Pedant, A Squire, A Beau, A Virtuoso . . .* (London, 1696). The original appeared in French by Madam de Pringy as *Les differens caracteres des femmes du siècle avec la description de l'amour propre contenant six caracteres et six perfections* (Paris, 1695). It did not, however, include the character of the virtuoso which appears here, pp. 92-108, and which seems to have been the addition of the translator. The English work has been variously ascribed to Mary Astell and Judith Drake, though the weight of opinion inclines

against the former, and I do not know what evidence there is for the latter. See A. P. Upham, "English *Femmes Savants* at the End of the Seventeenth Century," *JEGP*, 12 (1913), p. 262n.; Florence M. Smith, *Mary Astell* (New York, 1916), app., pp. 173-82. There were three editions by 1697. I find it quoted at length but without identification in a letter to Robert Wodrow by William Brodie; Wodrow, *Early Letters*, ed. L. W. Sharp, Scot. Hist. Soc., 3rd ser., 34 (1937), pp. xxiv-xxvii.

35. *Amusements Serious and Comical*, ed. Arthur L. Hayward (London, 1927), "Physic," pp. 78-82; "Virtuoso Country," pp. 76-77. There is an appreciation by Benjamin Brown, *Tom Brown of Facetious Memory*, Harvard Studies in English, 21, 1939.

36. Wotton, *Reflections*, "Of Ancient and Modern Natural Histories," pp. 256ff.

37. The text has been edited by Richard Morton and William Peterson, Lake Erie College Studies (Painesville, Ohio, 1961), along with *A Key to the New Comedy call'd Three Hours after Marriage* pp. 101-11, and *A Complete Key*, pp. 70-77. My account is derived largely from George Sherburn, "The Fortunes and Misfortunes of *Three Hours after Marriage*," *MP*, 24 (1926-27), pp. 91-109.

38. Thus Dr. Johnson, in his biography of Gay, saw that Dr. Woodward, the "Fossilist," was "a man not really or justly contemptible," Samuel Johnson, *Lives*, II, pp. 271-72.

39. Woodward to Wanley, March 9, 1712, B.M. MS. Harl. 3782, f. 214.

40. Woodward was introduced to Joshua Barnes through Thomas Hearne; among other encouragements, he secured him a patron in his friend, the Earl of Pembroke, and proposed him as a fellow of the Royal Society; see Woodward to Hearne, Aug. 19, 1710, Bodl. MS. Rawl. Lett. 12, f. 9, also ff. 94 and 96; Woodward to Barnes, April 5, 1710, Bodl. MS. Rawl. Lett. 40, f. 70. Woodward presented Wasse's work to the Royal Society, April 5, 1710, R.S. MS. Jour. Bk., X, p. 236. Among other things he tried to get a copy of the famous Sigean inscription (one of the oldest Greek inscriptions yet discovered) through its eventual edi-

tor, Edmund Chishull; see Chishull to Woodward, March 3, 1721, the Osborn Collection, Yale University.

41. Woodward to Hearne, Nov. 27, 1716, Bodl. MS. Rawl. Lett. 12, f. 118. A few months later he sent Hearne a list of nineteen scarce tracts for possible use in his edition of Camden; *ibid.,* ff. 119-20, 122. He told Wanley, however, that old manuscripts were not his special interest, and hinted that he might sell them to Harley; *The Diary of Humfrey Wanley*, ed. C. E. Wright and Ruth C. Wright (2 vol., London, 1966), May 17, 1723, II, p. 223.

42. *Phil. Trans.,* 25 (1706), pp. 291-92; Woodward to Hudson, Jan. 10, 1705, Bodl. MS. Rawl D. 316, ff. 8-9.

43. Hearne to James West, Oct. 30, 1728, B.M. MS. Lansdowne 778, f. 111. For a report on the auction itself, see below ch. 14, sect. 1. Pembroke, Fountaine, Coleraine, and Strype all enter our story later. For Chandos and Woodward, see letter of Nov. 20, 1718, in the Huntington Library, MS. St. 57, V. 16, p. 52, and the entries for June 30, 1697, and Nov. 30, 1698, in the Chandos Diary, MS. St. 26. See also C. H. Collins Baker, *The Life and Circumstances of James Brydges, First Duke of Chandos* (Oxford, 1949). Another example of Woodward's connoisseurship may be found in James Thompson to Cox Macro, June 20, 1717, B.M. MS. Add. 32556, f. 118.

44. Hearne, *Remarks*, X, p. 148. Miller was Woodward's neighbor for thirty years and at his death helped to pack the fossils for Cambridge University.

45. Coleraine to Woodward, Sept. 7 and Nov. 13, 1705, Camb. MS. Add. 7647, nos. 93-94. Coleraine was tutored in numismatics by William Charleton; see their correspondence in B.M. MS. Sloane 3962. There are two other letters of Coleraine to Woodward about coins in *Lit. Anec.,* IX, p. 762. For Coleraine's attainments, see John Nichols, *Biographical and Literary Anecdotes of William Bowyer* (London, 1782), pp. 106-07; *A Catalogue of the Genuine and Curious Collection . . . of the Lord Viscount Coleraine* (London, 1754).

46. *Catalogue of the Library, Antiquities, etc. of the late learned Dr. Woodward . . . to be sold at Auction . . . 11 November 1728* (Christopher Bateman), p. 277.

Chapter 8: Roman London

1. See, for example, the newsletter, Sept. 28, 1687, in *Wren. Soc. Pub.,* 13 (1936), p. xviii. In general, see Jane Lang, *Rebuilding St. Paul's After the Great Fire of London* (London, 1956), p. 65ff.

2. See the characteristic tributes to his universality by Evelyn and Hearne: Evelyn to Wren, Feb. 21, 1697, *Wren Soc. Pub.,* 13 (1936), pp. 36-37; Hearne to West, March 30, 1727, B.M. MS. Lansdowne 778, f. 34[v].

3. See his inaugural speech as Savilian Professor of Astronomy, *Parentalia,* ed. Stephen Wren (London 1750), pp. 200-06.

4. "Of London in Ancient Times," *Parentalia,* p. 265.

5. "Romano-British London," *VCH London,* I (1909), p. 113. Another large stone with inscription that went to Oxford is described by Charles Roach Smith, *Illustrations of Roman London* (London, 1859), pp. 22-23.

6. *Parentalia,* pp. 266-67.

7. *VCH London,* I, pp. 24-25.

8. There is a much fuller enumeration in John Strype, *A Survey of the Cities of London and Westminster* (London, 1720), II, app., p. 23.

9. *Parentalia,* p. 266; see also p. 296.

10. "Of the ancient cathedral Churches of St. Paul, from the first Age of Christianity to . . . 1666," *ibid.,* p. 271ff. See also *VCH London,* I, p. 33.

11. *VCH London,* I, p. 5.

12. William Maitland, *The History and Survey of London,* 3rd ed. (2 vol., London, 1760), II, p. 991.

13. "The Life and Character of Dr. Edward Stillingfleet," *Works* (6 vol., London, 1710), VI, pp. 1-46. According to Hearne it was by Dr. Timothy Godwin, the Bishop's chaplain; *Remarks,* X, p. 439. It was not, as it is sometimes supposed, by Richard Bentley; see James Nankiwell, *Edward Stillingfleet* (Worcester, 1946), p. 1n.

14. Thomas Burnet to the Electress Sophia, July 29, 1697, *State Papers and Correspondence,* ed. John Kemble (London, 1857), pp. 191-96.

15. "Life of Stillingfleet," p. 17. The text may be found in his *Works*, III, pp. 896-936.

16. One of Stillingfleet's first acts as Dean of St. Paul's was to launch an appeal for the rebuilding of the Cathedral; Nanki-well, pp. 5-6.

17. Stillingfleet, *Works*, III, p. 899.

18. An example of this weakness was pointed out by Maurice Johnson to Roger Gale, March 17, 1734, *Lit. Anec.*, IV, p. 366.

19. Stillingfleet, *Works*, III, p. 904. Stillingfleet refers here to William Somner's *A Treatise on the Roman Ports and Forts in Kent* (Oxford, 1693).

20. *Britannia* (1695), col. 315, 330-31; Stillingfleet, *Works*, III, p. 921ff.

21. See the intro. to Charles Leth-bridge Kingsford, ed., *A Survey of London* (2 vol., London, 1908).

22. Richard Chiswell (the publisher) to Strype, Aug. 25, 1694, Camb. MS. Baumgartner Papers, III, pt. 1, no. 23.

23. Tancred Robinson to Lhwyd, Sept. 22, 1702, Bodl. MS. Ashmole 1817a, f. 346. My student, John Morrison, will be publishing a full description of the composition of Strype's *Survey* in a forthcoming issue of the *London Journal*.

24. Strype, *Survey*, app., ch. 5, pp. 21-24.

25. On July 1, 1707, Strype wrote to Thoresby about "the ingenious Dr. Woodward, to whom I am much beholden for communicating to me an account of divers Roman antiquities found in and about London, which shall have a place in a new edition of Stow's Survey of London, which I am preparing." Thoresby, *Letters* (1832), II, pp. 56-58.

26. See the sketch of Bagford's life in Hearne's edition of *Hemingi Cartularium* (Oxford, 1723), p. 656; also Hearne to Browne Willis, May 8, 1716, *Remarks* V, p. 218.

27. Wanley to Covel, Aug. 30, 1701, B.M. MS. Add. 22911, ff. 1-2. There is additional correspondence between the two in B.M. MS. Harl. 4966.

28. See Wanley's description of Bagford's manuscripts in a letter to Sloane, May 6, 1707, *Lit. Anec.*, I, pp. 532-36. In general, see W. Y. Fletcher, "John Bagford and his Collections," *Transactions of the*

Bibliographical Society, 4 (1896-98), pp. 185-201; Cyril Davenport, "Bagford's Notes on Bookbindings," *ibid.*, 7 (1904), pp. 123-59.

29. "A Letter to the Publisher written by the ingenious Mr. John Bagford in which are many curious Remarks relating to the City of London and some things about Leland," Hearne, Leland, *Coll.*, I, pp. lviii-lxxxvi. A second letter to Hearne on the same subject is in Bodl. MS. Rawl. D 400, f. 128ff. Since he "wrote a bad hand and spelt very ill," I will not pretend to have deciphered it; see *Lit. Anec.*, II, pp. 462-65.

30. Strype, *Survey*, II, app., p. 23; see also Bagford's "Letter," pp. lxi, lxiii.

31. Bagford, "Letter," pp. lix, lxxiv-lxxv. For more about Kemp, see below.

32. *Ibid.*, p. lxvii.

33. *Ibid.*, p. lxiv; Strype, *Survey*, II, app., p. 24. Some of Bagford's notes on London are in B.M. MS. Harl. 5953, including "Mr. Conyer's Observations," ff. 112-13.

34. "Letter," pp. lxxvii-lxxviii. The judgment of John Lewis (in 1740) on Bagford is probably not unfair; he "had not learning and knowledge enough for what he undertook." *Lit. Ill.*, IV, pp. 172-73.

35. *The Athenian Mercury*, Nov. 21, 1691, quoted by David Murray, *Museums* (3 vol., Glasgow, 1904), I, pp. 134-35. Conyers frequently contributed to the activities of the Royal Society and was the inventor, among other things, of an improved hydroscope, pump, and hearing trumpet. See also Richard Gough, *British Topography* (London, 1780), I, pp. 718-20.

36. B.M. MS. Sloane 958, f. 105ff. On March 2, 1685, Lister reported Conyer's views on urns to the Royal Society; see Thomas Birch, *The History of the Royal Society* (4 vol., London, 1756-57), IV, p. 371.

37. John Woodward, *An Account of Some Roman Urns* (London, 1713), p. 6. See also Woodward, *An Attempt*, p. 86. It was noticed in the new edition of Camden that much of Conyer's collection had already passed into Woodward's possession; *Britannia* (London, 1695), p. 334. Strype lists many of the Bagford antiq-

uities in Woodward's hands; *Survey*, II, app., p. 23.

38. Strype, *Survey*, II, app., p. 21.

39. *Portland MSS.*, VI, p. 81; Woodward to Hearne, Nov. 22, 1716, Bodl. MS. Rawl. Lett. 12, f. 118.

40. Woodward sent Strype "a Mass of Materials for you to work upon, mould, and bring into any Form you please." Nov. 1, 1705, Camb. MS. Baumgartner Papers, IV, no. 83. It was followed by more; Nov. 12, 1705, and Jan. 16, 1706, *ibid.*, nos. 84-85.

41. Woodward to Strype, Feb. 3, 1707, *ibid.*, III, pt. 3, no. 295. See also, on the same subject, III, pt. 3, nos. 307, 312; IV, pt. 1, no. 25.

42. The letter is dated Nov. 30, 1711, and is prefixed to the published tract, pp. i-xii.

43. There was a pirated edition in 1723 as well; see Hearne, *Remarks*, VIII, p. 67. A quarrel soon arose between Woodward and Hearne over the number of copies printed, and involving the unscrupulous printer Curll, but it was finally resolved; see Edward Steele to Hearne, Nov. 13, 1712, Bodl. MS. Rawl. Lett. 16, f. 430.

44. Christopher Erndtel, *Relation of a Journey* (London, 1712), pp. 41-42.

45. In 1707 an apothecary named Joseph Miller informed him of the new discovery; see Strype, *Survey*, II, app., p. 23; Woodward, *Urns*, p. 7.

46. John Edward Price, *On a Bastion of London Wall: or Excavations in Camomile St. Bishopsgate* (Westminster, 1880), p. 18; see also *VCH London*, I, pp. 45, 94-59.

47. *Miscellaneous Works*, ed. Thomas Birch (2 vol., London, 1737), I, pp. 165-356. Greaves had also been a Gresham professor; see Birch's life prefixed to the *Works*, and the biography in Ward's, *Lives*, I, pp. 135-53.

48. Woodward, *Urns*, p. x.

49. *Ibid.*, p. 25.

50. *VCH London*, I, p. 43ff.; Jocelyn M. C. Toynbee, *Death and Burial in the Roman World* (Ithaca, N.Y., 1971), p. 40; A. D. Nock, "Cremation and Burial in the Roman Empire," *Harvard Theological Review*, 25 (1932), p. 321ff. Lord Coleraine was one contemporary who admired

Dr. Woodward's method; *Lit. Anec.*, IX, p. 762.

51. Woodward, *Urns*, p. 19.

52. *Ibid.*, p. 23.

53. *Ibid.*, p. 31. Strype was impressed by Woodward's argument and was sure that Stillingfleet would have been persuaded; *Survey*, II, p. 141.

54. "What I shall collect relating to the Antiquities in Lombard Street you shall know . . . but my Business will not give me leave to attend the Digging there as often as I wish." Woodward to Strype, Jan. 19, 1717, Camb. MS. Baumgartner Papers, pt. 2, no. 180.

55. The manuscript is quoted at length by James P. Malcolm, *Londinium Redivivum* (3 vol., London, 1807), III, pp. 509-12. It was in the possession of Alexander Chalmers in 1817 but has since disappeared; see "Woodward" in *The General Biographical Dictionary* (London, 1817), XXXII, p. 277.

56. The statue was purchased, at Woodward's death, by James West; see West to Hearne, April 15, 1729, Bodl. MS. Rawl. Lett. 11, f. 145. Where it went thereafter I cannot say.

57. Conyers Middleton, according to the Earl of Oxford, April 10, 1725, said about the Horace that it was "doubtless antique but [he] much questions whether the present inscription upon it be as old as the urn itself." Woodward's evidence, besides the inscription, was "the several emblems of a lyric poet that are carved about." *Portland MSS.*, VI, p. 81. Woodward identified the Terence by a drawing found in a Vatican manuscript of the comedies and published by Fulvius Ursinus in 1570; see Woodward to Hearne, Nov. 27, 1716, Bodl. MS. Rawl. Lett. 12, f. 118.

Chapter 9: Dr. Woodward's Shield

1. Woodward to Hearne, Dec. 11, 1712, Bodl. MS. Rawl. Lett. 18, ff. 155-56. "The Roman Shield was bought by Mr. Conyers of a Smith in Rosemary Lane, who bought all the Waste-Things in the Tower at the New-Fitting up of the Armoury, at the latter end of the reign of K. Charles 2d. The Shield probably came thence. . . ." See the note by Woodward in

B.M. MS. Add. 6127, f. 81. Thomas Smith reported that Woodward had paid seven or eight pounds for it; Hearne, *Remarks* I, p. 318.

2. All this is now in B.M. MS. Add. 6127, but some of the original file seems to have been separated and copied in the 18th century; it is now in Bodl. MS. Gough Misc. Antiq. 10. The former followed the various owners of the shield; the latter passed through the hands of the Woodward Professor, Charles Mason, to Michael Lort (the editor of Woodward's essay on the Egyptians), and at his death to Richard Gough. Ward used it for his biography.

3. Aug. 16, 1699, R.S. MS. Jour. Bk., IX, p. 170; Woodward to Sperling, Sperling Corr., Feb. 8, 1700. The Woodward-Sperling correspondence is in the Royal Library, Copenhagen, as follows: MS. Gl. kgl. Saml. 3092 4°, vol V, 2 (eight draft letters of Sperling to Woodward); vol. VI, 2 (nine letters of Woodward to Sperling); MS. Kabl. 539 4° (ten letters of Woodward to Sperling). I am grateful to the librarian, Mr. Tue Gad, for making a photocopy available to me. Only one of Sperling's letters survives in the British Museum's Woodward correspondence. In 1788 the Museum acquired a large collection of Sperling's manuscripts, B.M. MS. Add. 5812-5206; Add. 5813 has some notes on arms.

4. Woodward to Sperling, Sperling Corr., Feb. 8 and June 6, 1700; Sperling to Woodward, *ibid.*, Sept. 2, 1700. Sperling's work on coins appeared that year as *Dissertatio de Nummis non Cusis* (Amsterdam, 1700); it was presented to the Royal Society Jan. 15, 1701, by Woodward.

5. Woodward to Sperling, B.M. MS. Add. 6127, ff. 4-6; the original is in Copenhagen.

6. Crixus, we learn, who claimed descent from Brennus, possessed a shield which showed the Gauls weighing the gold on the Tarpeian Hill; Silius Italicus, *Punica*, IV, 11, 150-53.

7. Sperling to Woodward, Sperling Corr., Oct. 20, 1703.

8. Plutarch, "Life of Camillus," *Lives,*

trans. by "several hands" (5 vol., London, 1700), I, pp. 423-98; Livy, *The Roman History*, trans. Edmund Bohun (London, 1686), p. 149 (V, xxxviii-xlix).

9. *The Roman History*, preface, sig. A.

10. See, for example, Robert M. Ogilvie, *A Commentary on Livy, Books 1-5* (Oxford, 1965), pp. 719-29, 736-38. 630-31; Jean Gagé, "La Balance de Kairos et l'Epée de Brennus à propos de la Rancon de l'*Aurum Gallicum* et de sa Pesée," *Revue Archaeologique*, 6th ser., 43 (1954), pp. 141-76; J. Briscoe, "The First Decade," *Livy*, ed. T. A. Dorey (London, 1971), pp. 1-20.

11. On this point see Arnaldo Momigliano, "Ancient History and the Antiquarian," *Studies in Historiography* (New York, 1966), pp. 1-39.

12. *The Roman History*, preface, sig. A.

13. George Cornwall Lewis, *An Inquiry into the Credibility of the Early Roman History* (2 vol., London, 1855), I, pp. 1-2. See also Thomas Arnold in *The History of Roman Literature*, ed. Henry Thompson, 3rd ed. (London, n.d.), pp. 351-52.

14. Barthold Niebuhr, *Lectures on the History of Rome,* trans. Leonhard Schmitz (London, 1853), p. 2.

15. *Arrian's History of Alexander's Expedition*, trans. Rooke (2 vol., London, 1729), I, ded., sig. A4ᵛ.

16. Besides Momigliano (n. 11 above), see H. J. Erasmus, *The Origins of Rome in Historiography from Petrarch to Perizonius* (Assen, 1962).

17. For Hardouin, see the summary of his views in *Gent. Mag.,* Jan.-Feb. 1734, pp. 8, 83, 240-41; also "Hardouin" in Chauffepié; and more recently G. Martini, "La Stravaganza Critiche di Padre Jean Hardouin," *Scritti di Palaeographia et Diplomatica in Onore di V. Frederici* (Florence, 1944), pp. 351-64. The fruits of this skepticism may be seen in the *Spectator,* July 2, 1712, no. 420, where Addison shows himself suspicious of Livy's veracity, though all admiration for his style.

18. *Gent. Mag.,* Jan. 1734, p. 8;

Hearne, *Remarks*, I, p. 179 and XI, p. 280; Spanheim to Nicaise, July 5, 1698, *Lettres de Divers Savants à l'Abbé Claude Nicaise*, ed. E. Caillemer (Lyons, 1885), pp. 120-21.

19. Einar Gjerstad, "Legends and Facts of Early Roman History," *Scripta Minora* (Lund, 1962), p. 3.

20. He was anticipated by the slighter work of Lawrence Echard, *The Roman History* (London, 1698-99) and by the more limited work of William Wotton, *The History of Rome from the Death of Antoninus Pius to the Death of Severus Alexander* (London, 1701). Echard complained of an earlier writer that "his often mixing of Critical Learning, makes him far less pleasant than otherwise he might be." *Roman History*, sig. A2ᵛ. Compare Hearne, "Eachard hath a good Pen, but he does not look into, much less follow, Original Authors." *Remarks*, VI, p. 170; VII, pp. 247-48. For Wotton, see my article, "Ancients, Moderns and History" (ch. 2 above, n. 5).

21. Nathaniel Hooke, *The Roman History*, 2nd ed. (London, 1751), I, p. iv. The controversy over the authority of Livy appears in the *Memoires de Litterature de l'Academie Royale des Inscriptions*, 6 (1729); much of it appeared later in *Philological Miscellany* 1 (London, 1761), pp. 87-157.

22. *Roman History*, I, pp. 395ff., 407n.

23. Woodward to Sperling, Sperling Corr., July 29, 1703.

24. B.M. MS. Add. 6127, ff. 63-68 (in French and in English).

25. The engraving was made by P. van Gunst from a drawing by K. Howard (Amsterdam, 1705). The inscription reads: *Clypeus antiquus exhibens Romam a Gallis, Duce Brenno, captam et incensam. Auri pro Capitolio redimendo pacti pensationem, adventum Camilli, metum fugamque Gallorum. Aedificia varia publica, Equites, Galeas, Saga, Caligas, Ephippia, Clypeos, Gladios, Pila et Vexilla, omnia mira Opificis Arte elaborata.*

26. B.M. MS. Add. 6127, f. 111. As late as 1726 Woodward was still trying to

place his plaster casts in the Italian museums. See Woodward to Conyers Middleton, Nov. 14, 1726, and Middleton to Giusto Fontanini, n.d., B.M. MS. Add. 32, 457, ff. 49, 27-30.

27. "Brennus" in *Harper's Dictionary of Classical Literature and Antiquities*, ed. Harry T. Peck (New York, 1896, 1923, 1965, etc.?) Clarke's Caesar was almost universally praised for its beauty; see the *Spectator*, May 1, 1712, no. 367; and *Lit. Ill.*, I, p. 295. In 1712 Woodward was able to add another illustration: M. Nilant's edition of Baldwin de Calceo; see Woodward to Hearne, Feb. 14, 1712, in Hearne, *Remarks*, III, p. 299.

28. Valckenaer (here spelled Vaulknier) to Woodward, July 8, 1707, Bodl. MS. Gough Misc. antiq. 10, ff. 5-6. Valckenaer writes also of his "figurated stones" and looks forward to a box of fossils promised by Woodward. There are also letters recounting Valckenaer's efforts on Woodward's behalf in B.M. MS. Add. 6127, ff. 16, 23. In Woodward's fossil catalog there are many specimens from his Dutch friend; e.g., *An Attempt*, pt. 2, pp. 23-24.

29. *Biographie Universelle* (Paris, 1823), vol. 38, pp. 378-80. See also D. Th. J. Meijer, *Kritick als Herwaardering het Levenswerk van Jacob Perizonius (1651-1715)* (Leyden, 1971); Arnaldo Momigliano, "Perizonius, Niebuhr and the Character of the Early Roman Tradition," *Journal of Roman Studies*, 47 (1957), p. 104ff.

30. Valckenaer to Woodward, July 8, 1708, B.M. MS. Add. 6127, f. 16.

31. Valckenaer to Woodward, Oct. 14, 1717, *ibid.*, f. 23. Van Dale had recently published nine dissertations, "serving to illustrate the Roman and chiefly the Greek Antiquities and Marble Inscriptions." Reviewed in the *Works of the Learned*, Oct. 1702, pp. 614-21. See also *Bibliothèque Choisie*, 3 (1704), p. 106ff., and the *éloge* in *ibid.*, 17 (1709), pp. 309-12.

32. Woodward note on Valckenaer letter, B.M. MS. Add. 6127, f. 23.

33. Hearne to Smith, July 15, 1705, in Hearne, *Remarks*, I, p. 7; also II,

pp. 185-186, and VI, p. 82. Dr. Woodward couldn't decide whether Gronovius or Bentley was the more "insolent and haughty critic." April 2, 1709, *ibid.*, II, p. 179.

34. *Mémoires pour servir à l'Histoire des Hommes savantes Illustre dans la Republique des Lettres*, 10 (1729-30), pp. 87-88.

35. B.M. MS. Add. 6127, ff. 31-35^V; Bodl. MS. Gough Misc. antiq. 10, ff. 6^V-10^V (dated Oct. 4, 1707).

36. Le Clerc to Valckenaer, June 28, 1707, B.M. MS. Add. 6127, ff. 10-11; Le Clerc to Woodward, June 28, 1707, Bodl. MS. Gough Misc. antiq. 10, ff. 18-21. For Le Clerc, see Annie Barnes, *Jean Le Clerc et la Republique des Lettres* (Paris, 1938); *An Account of the Life and Writings of Mr. John Le Clerc* (London, 1712).

37. July 7, 1707, B.M. MS. Add. 6127, ff. 14-15; July 21, 1707, Bodl. MS. Gough Misc. antiq. 10, ff. 3-3^V. For Reland, see Niceron, I, pp. 332-41; Chauffepié, art. "Reland," where he is called, "un des plus illustres Scavans du XVIII siècle."

38. Scheuchzer to Woodward, May 30, 1708 and April 22, 1710, B.M. MS. Add. 6127, ff. 37, 42. For Spanheim and Cuper, see below. Another prominent scholar, Jacques Basnage de Beauval, wrote in favor of the monument; Basnage to Valckenaer, July 6, 1707, *ibid.*, ff. 12-13 (in French); Gough Misc. antiq. 10, ff. 5-7 (in English). Basnage's specialty was ecclesiastical history and he wrote a celebrated history of the Jews; see *Bibliothèque Choisie*, 13 (1707), pp. 410-12; Niceron, pp. 206-13; Chauffepié, I, pp. 108-11; and the slight but not very helpful biography by E-André Mailhet (Geneva, 1880).

39. See, for example, Thomas Smith – who, however, thought that the shield needed further explication than Spon had given it. He wrote Hearne that he would send him an engraving of Dr. Woodward's shield as soon as it was ready; Smith to Hearne, Dec. 1, 1705, Bodl. MS. Smith 127, f. 103.

40. Momigliano complains that there is no adequate study of Spon; "Ancient History" (n. 11 above), p. 35n. There is a brief notice in the *Nouvelles de la Re-publique des Lettres*, 1 (1684), pp. 496-505.

41. Jacob Spon, *Recherches Curieuses d'Antiquité* (Lyons, 1683); it received a Latin translation in 1685.

42. Gilbert Burnet, *Travels, or Letters Containing an Account of What Seemed Most Remarkable in Switzerland, France, Italy, Germany, etc.* (Amsterdam, 1687), I, pp. 2-3.

43. The discussion occupies the first dissertation of the *Recherches*, pp. 1-26.

44. Livy, XXVI, i, 1-14.

45. "When you read historical facts, think of them within yourself, and compare them with your own notions. For example, when you read of the first Scipio . . . are you not struck with the virtue and generosity of that action?" Chesterfield, *Letters*, ed. Charles Strachey, 3rd ed. (London, 1932), no. lx (1740), p. 75.

46. I use the translation of Spon that is given in *Reflexions on Dr. Gilbert Burnet's Travels* (London, 1688), p. 21. The author of the original Latin work, Varillas, tried to discredit Burnet's antiquarian knowledge.

47. R.S. MS. Jour. Bk., X, loose sheets inserted after p. 146, from a volume of rough minutes.

48. Many years later, Woodward tried unsuccessfully to obtain some of Covel's collection, Woodward to Covel, June 25, 1706, Camb. MS. Mm 6 50, ff. 265-66 and B.M. MS. Add. 22911, ff. 57^V-58. See also *Cambridge Under Queen Anne* (ch. 7, n. 27 above), pp. 147-52, 470-77; Woodward to Strype, Feb. 3, 1707, Camb. MS. Baumgartner Papers, III, pt. 3, no. 292; Strype to Woodward, Feb. 6, 1707, B.M. MS. Add. 4276, f. 136.

49. Woodward to Hudson, May 27, 1707, Bodl. MS. Rawl. D 316, f. 91; April 1, 1707, *ibid.*, f. 134. Several years later Woodward sent five copies to Oxford, one to be framed and hung in the Ashmolean Museum; Woodward to Whiteside, Nov. 12, 1718, Bodl. MS. Ashmole 1821, f. 3. Apparently it hung there until 1845 anyway.

50. Woodward note in B.M. MS. Add. 6127, ff. 82-88. Most of these characters will appear later. For Benedetti, see Uffenbach's unflattering description in

London in 1710, trans. W. H. Quarrell and Marjory Mare (London, 1934), pp. 70-72: "He does a lively trade in antiques, and manages to swindle the English shockingly, palming off on them for prodigious sums articles which he gets from France for nothing. He is not only well known to all who have collections but is also their prime counsellor in all matters concerning these medals and antiquities. He has a most excellent collection of medals, cameos, and of copper engravings."

51. There is a draft dated Prid. Kal., Sept. 1707, in B.M. MS. Lansdowne 1041, ff. 55-56, and a fair copy at ff. 45-46. Woodward sent another copy to his friend Cuper (see n. 74 below). Newton received profuse compliments for it from Rome, *ibid.;* 77-78. Some years later Woodward wrote to Hearne, "The Antiquaryes in Italy are much taken with the Clypeus. I have sent several Icons of it thither, and others to France." May 7, 1712, Bodl. MS. Rawl. Lett. 18, f. 144.

52. Woodward to Thoresby, Jan. 13, 1710, Yorks. Arch. Soc. MS. 8, n.p. As far as I can determine, one of the few scholars Dr. Woodward did *not* send a print to was Richard Bentley, who however was not really an antiquary. "Dr. Bentley is my old acquaintance," he wrote to Joshua Barnes, "and there are many things in the Ideas I retain of his Disposition and Demeanour that I wish I could forget too." April 5, 1710, Bodl. MS. Rawl. Lett. 40, f. 241. It was not likely that these two difficult men would get along. Typically, Woodward thought to disparage Bentley's Horace when that controversial work appeared; see Woodward to Hearne, Feb. 14, 1712, Bodl. MS. Rawl. Lett. 18, ff. 137-38.

53. Once Thoresby recorded a nightmare in his diary about an old manuscript delicately printed and gilded; not that he found that sinful in itself, "but that it plainly argues that my mind is too much set upon these things, also why might not it have been the Bible." Thoresby, *Letters* (1912), p. xn.

54. *Phil. Trans.*, 19 (1696), pp. 319, 663, 738. See D. H. Atkinson, *Ralph Thoresby the Topographer* (2 vol., Leeds, 1885), ch. 10, pp. 392-443.

55. The catalog is appended to his *Ducatus Leodiensis* (London, 1715). The

British Museum copy seems to contain Thoresby's own notes (shelfmark c21 e8).

56. *Phil. Trans.*, 20 (1698), pp. 206-08; *Ducatus*, pp. 564-65.

57. June 29, 1698, R.S. MS. Jour. Bk., IX, p. 100. This accords with the account of a similar shield by S.R. Meyrick, "Description of two Ancient British Shields," *Archaeologia*, 23 (1831) pp. 92-97. See Albert Way, *Catalogue of Antiquities in the Possession of the Society of Antiquaries* (London, 1847), p. 16. The shield seems to have come into possession of the Society in 1759, *Lit. Ill.*, III, p. 386.

58. Richardson to Thoresby, Feb. 28 and April 1, 1709, and Woodward to Thoresby, April 22, 1707; Thoresby, *Letters* (1912), II, pp. 194-95, 198-99, 142-43.

59. Woodward to Hudson, Nov. 30, 1705, Hearne, *Remarks*, I, p. 109.

60. Woodward to Batteley, Aug. 2, 1707, *Lit. Ill.*, IV, pp. 102-03.

61. See the first part of the Museum Woodwardianum in the *Catalogue of the Library, Antiquities, etc. of the late Dr. Woodward* (London, 1728).

62. So too Benedetti; see Woodward note in B.M. MS. Add. 6127, f. 83^V. Woodward thought Benedetti "the best judge in England of these Things." Woodward to Hearne, April 6, 1708, Bodl. MS. Rawl. Lett. 12, f. 421.

63. Robert Hooke, July 16, 1684, in Birch, IV, p. 316. The modern authority is I.A. Richmond, "Trajan's Army on Trajan's Column," *Papers of the British School at Rome* (London, 1935), p. 4. See also Lino Rossi, *Trajan's Column and the Dacian Wars* (Ithaca, 1971), pp. 14-18.

64. Niceron, IV, pp. 374-75; Henry Hallam, *Introduction to the Literature of Europe* (4 vol., London, 1882), IV, p. 13.

65. J. Vignoli, *De Columna Antonini Pii Dissertatio* (Rome, 1705); Woodward, *Catalogue*, no. 2881. See Woodward to Batteley, Feb. 25, 1706, *Lit. Ill.*, IV, pp. 100-01; Woodward to Hudson, Jan. 10, 1705, Bodl. MS. Rawl. D 316, ff. 8-9; Hearne, *Remarks*, I, p. 109.

66. B.M. MS. Add. 6127, ff. 82-83.

67. John Batteley, *Antiquities of Richborough and Reculver* (London, 1774), pp. 134-36; *Lit. Ill.*, IV, p. 87n.

The work was praised by the distinguished 19th-century scholar Charles Roach Smith in his own *Antiquities of Richborough, Reculver and Lymne* (London, 1850), p. 201ff.

68. Batteley to Woodward, Jan. 28, 1706, Camb. MS. Add. 7647, no. 45; Woodward note, B.M. MS. Add. 6127, f. 82. There is another dealing with some inscriptions, April 10, 1707, *ibid.*, no. 46, and three letters, Woodward to Batteley, 1706-07, in *Lit. Ill.* IV, pp. 101-02.

69. Woodward to Hearne, April 6, 1708, Bodl. MS. Rawl. Lett 12, f. 421; Woodward to Batteley, Aug. 2, 1707, *Lit. Ill.*, IV, pp. 101-02.

70. Woodward thought Seller "one of the best scholars of the age," and noticed that he had promised a dissertation on the shield; Woodward to Batteley, Aug. 2, 1707, *Lit. Ill.* IV, pp. 101-02. For Seller's fine collection of books and coins, see Hearne, *Remarks,* I, pp. 53-54. Hearne's opinion of Seller rose and fell, especially after Seller quarreled with his close friend Thomas Smith; see Hearne, *Remarks,* II, p. 388; III, p. 15. In after years Woodward tried to retrieve some papers about the shield that Seller had written, apparently without success. Woodward to Hearne, May 1, 1712, Bodl. MS. Rawl. Lett. 18, f. 8; Hearne to Woodward, Aug. 8, 1712, Rawl. Lett. 39, f. 38; Woodward to Hearne, Aug. 19, 1712, Rawl. Lett. 18, f. 9. For a qualified opinion of Seller's abilities, see Smalridge to Gough, Nov. 10, 1696, *Lit. Ill.*, III, p. 253.

71. When the first volume of the new edition of *De Praestantia* appeared, Woodward wrote to Hudson, "As he is an excellent Scholar, so he has a mighty Mastery in that Particular Science and I wish we had the 2nd Volume...." June 29, 1706, Bodl. MS. Rawl, D 316, f. 96. It was not to be; Spanheim died in 1710 and was buried in Westminster Abbey. The second volume had to be brought out posthumously in 1717 (London); it was prefixed with a life of the author. There is a copy of a letter from Woodward to Spanheim (misdated 1712) about the shield in B.M. MS. Add. 6194, ff. 3^V-4, and there is a letter from Spanheim to Woodward indicating his intention of making some notes on the subject; Sept. 1,

1709, B.M. Add. 6127, f. 39. Spanheim was then in his eightieth year. Hearne spoke for everyone when he wrote at Spanheim's death, "I know of no one hardly that was better qualified ... than this Great Man." Hearne to Fothergill, Feb. 5, 1711, Bodl. MS. Rawl. D 1166, f. 71. Spanheim's reputation remained high long afterwards; see Hallam, IV, p. 14; Monk, *Bentley,* I, pp. 189-90. See also Victor Loewe, *Ein Diplomat und Gelehrter Ezechiel Spanheim (1629-1710)* (Berlin, 1924); R. Doebner, "Spanheim's Account of the English Court," *Eng. Hist. Rev.,* 2 (1887), pp. 757-73.

72. See Cuper's diary for 1706, *Het Dagbook van Gisbert Cuper Gedeputeerde Te Velde Gehouden in de Zuidelijke Nederlanden in 1706,* ed. with a useful intro. by A. J. Veenendahl (The Hague, 1950). See also *Lettres Inédités de Gisbert Cuypert à Daniel Huet et divers correspondents,* ed. with intro. by Leon-G. Palissier (Caen, 1903); and A. J. van der Aa, *Biographisch Woordenboek der Nederlanden* (Haarlem, 1852-78), III, pp. 923-29. There are helpful contemporary descriptions in the *Histoire de l'Academie des Inscriptions,* 2 (1717), pp. 95-110, and Niceron, VI, pp. 88-96.

73. In a letter to Graevius, Cuper complains that his public duties interfere with his scholarship, but goes on to defend the idea that the loftiest purpose of study is to serve the public interest and administer justice; Aug. 29, 1696, Bodl. MS. D'Orville 478, ff. 126-27.

74. Cuper Corr., Oct 1, 1712. This correspondence is in the Koninklijke Bibliotheek, The Hague; I am grateful to Mrs. G. Dekker-Piket of the Dept. of MSS. for making copies available to me. MS. 68 B 10 contains six letters from Woodward to Cuper (1707-12) and nine draft returns (1708-16), as well as a copy of the Latin poem of Henry Newton discussed in n. 51 above.

75. Only a fraction of Cuper's correspondence has appeared in print. The most useful collection remains Cuper, *Lettres,* published in Amsterdam in 1742. I have not seen P. Bosscha, *Opgave en beschrijving van de Handschriften, magele-ten door Gisb. Cuperus* (Deventer, 1842).

76. Sperling to Cuper, Aug. 29 and

Oct. 17, 1705, Joannes Polenus, *Utriusque Thesauri Antiquitatem Romanorum Graecorumque Nova Supplementa* (Venice, 1737), IV, col. 183, 189; Cuper to Witsen, June 21, 1707, B.M. MS. Add. 6127, ff. 8-9. The latter was sent by Valckenaer to Woodward. For Witsen's fine collections, see the auction catalog, *Catalogus van de uitmuntende en zeer vermaarde konsten natur-kabinetten* (Amsterdam, 1728).

77. Cuper to Bignon, Aug. 1, 1708, *Lettres de Critique*, pp. 192-97. See also June 6, 1708, *ibid.*, pp. 189-91. Cuper first heard from Woodward about the shield in a letter, Cuper Corr., Feb. 17, 1707. There is a fragment of a letter between them copied by Ward and dated 1706 in B.M. MS. Add. 6194, f. 7. Cuper was soon corresponding with Basnage and Van Dale about it; *Lettres de Critique*, pp. 399, 563-64.

78. Cuper to Witsen, June 21, 1707, B.M. MS. Add. 6127, ff. 8-9. There are five volumes of Cuper-Witsen correspondence in the Bibliothek de Universiteit, Amsterdam.

79. Cuper to Valckenaer, July 23, 1707, B.M. MS. Add. 6127, ff. 17-20.

80. Polenus, col. 86ff. (1700-04).

81. B.M. MS. Add. 6127, ff. 25-30; Bodl. MS. Gough Misc. antiq. 10, ff. 21-29.

82. The letter was from Goswin Vilenbroek and is translated here by M. Con from the Dutch, Bodl. MS. Gough Misc. antiq. 10, f. 17.

83. Cuper to Bignon, June 6 and Aug. 1, 1708, *Lettres de Critique*, pp. 189-91, 192-97. According to an extract of a letter from Bignon to Scheuchzer which Scheuchzer sent on to Woodward, the Abbé only came to learn about the Doctor's scientific views in about 1710; B.M. MS. Add 6127, f. 42.

84. *Lettres de Critique*, n.d., p. 121. Cuper reported his doubts continuously to Dr. Woodward; see Cuper Corr., April 14, 1708; Dec. 24, 1709; Sept. 1, 1711.

85. Woodward to Scheuchzer, Scheuchzer Corr., July 3, 1710, and May 22, 1711. Woodward answered Cuper, Cuper Corr., Nov. 23, 1708; Sept. 9, 1709; Oct. 30, 1711. In the last he tried to call it off. On Dec. 19, 1711, Cuper complained

of their flagging correspondence and pointedly noted that Bourguet continued to write despite a similar disagreement.

86. See the correspondence, with many references to the shield, in *Sylloge Epistolarum mutuarum D. Jo. Jacobi Scheuchzeri et Gisberti Cuperi* in J. G. Schelhorn, *Amoenitatis Historiae* (Frankfort, 1738), II.

87. René Kerviler, "Les Bignons," *La Bibliophile Francais*, 6 (1872), pp. 300-12. See the *éloge* in *Histoire de l'Académie des Inscriptions*, 2 (1726), pp. 376-93; and the tribute in *Histoire Critique de la Republique des Lettres*, 3 (1712-18), p. 302n.

88. Kuster to Bentley, Oct. 20, 1714, *Bentley Corr.*, I, pp. 490-95.

89. My copy lacks the date, but it is either 1712 or 1713. The original is in Camb. MS. Add. 7647, with several others in Latin and French, 1712-27, nos. 50-63. After exchanging greetings in Latin, they agreed to write to each other in their respective languages. I have not discovered Woodward's English replies, but they are very likely among the great caches of Bignon correspondence in the Bibliothèque Nationale.

90. Cuper Corr., June 26, Sept. 9, and Dec. 24, 1709.

91. Cunningham to the Earl of Oxford, Oct. 23, 1711, *Portland MSS.*, V, pp. 99-100. There were two Alexander Cunninghams in this period, often confused. Ours is 1654-1737; see the *DNB*. His interests as an antiquary may be traced in the *Portland MSS.*, IV-V.

92. May 14, 1707, Hearne, *Remarks*, II, p. 13; see also Hearne to Smith, July 5, 1707, *ibid.*, p. 24. For Gregory, see Agnes G. Stewart, *The Academic Gregories* (Edinburgh, 1901), pp. 52-76; P.D. Lawrence, "David Gregory's Inaugural Lecture at Oxford," *N.R.*, 25 (1970), pp. 143-78; W. G. Hiscock, ed., *David Gregory, Isaac Newton and their Circle* (Oxford, 1937). For the connection with Arbuthnot, see the letters to Charlett at Gregory's death in 1708, *Letters Written by Eminent Persons* (3 vol., London, 1813), I, pp. 176-78, 178-80.

93. Woodward to Bignon, April 23, 1713, in Ward, *Lives*, pp. 139-42, from a manuscript then in the possession of Matthew Postlethwayte.

94. Bignon to Woodward, Jan. 1714, Camb. MS. Add. 7647; Cuper to Sperling and reply, June 5, 1713, and March 16, 1715, Polenus, col. 292, 300; Sperling to Woodward, in Sperling Corr., Sept. 12, 1711, and Jan. 9, 1712. See also Woodward to Hearne, Oct. 28, 1712, describing Sperling as "one of the best Antiquaryes now living" and as an advocate of the shield; Bodl. MS. Rawl. Lett. 18, f. 149.

Chapter 10: Enter Thomas Hearne

1. The only life of Hearne remains Huddesford, I, pp. 1-124, based largely upon Hearne's autobiography. For the composition of this work, see *Lit. Ill.,* pp. 683-84, and the interleaved copy in the British Museum, C 45 e 17. There is a brief hostile biography of little value entitled *Impartial Memorials of the Life and Writings of Thomas Hearne by Several Hands* (London, 1736). Hearne's autobiography appears as an appendix to his *Remarks,* XI, pp. 467-79.
2. Hearne, *Remarks,* IV, p. 280; VI, p. 325; IX, pp. 350, 365; Huddesford, p. 1.
3. Hearne, *Remarks,* X, p. 236.
4. See Eliza Berkeley, ed., *Poems of the Late George M. Berkeley* (London, 1797), pp. ccccxviii-ccccxlix (esp. p. ccccxxxi); "Cherry" in *DNB.*
5. Hearne, *Remarks,* VI, p. 220.
6. Huddesford, p. 3; Hearne, *Remarks,* IX, pp. 132-33; and X, p. 261. See J. M. Bulloch, "Hearne's First Master, the Rev. Patrick Gordon," *NQ,* 169 (1935), pp. 344-46.
7. Hearne, *Remarks,* X, p. 259.
8. Huddesford, p. 4.
9. *Ibid.,* pp. 5-6; see also Hearne, *Remarks,* I, pp. 1-2; IV, p. 121; V, p. 196; IX, p. 349; X, pp. 121, 196, 221, 236-39.
10. Huddesford, p. 10.
11. Hearne, preface, *A Collection of Curious Discourses* (Oxford, 1720), p. lxv; W. D. Macray, *Annals of the Bodleian Library* (Oxford, 1890), pp. 173, 197-98. For an appreciation, see J. Grafton Milne, "Oxford Coin Collectors of the Seventeenth and Eighteenth Centuries," *Oxoniensia,* 14 (1949), p. 58.
12. Hearne, *Remarks,* VII, p. 370. See

also, "An Extract and particular Account of the rarities in the Anatomy School transcribed from the Original in Thomas Hearne's hands," Bodl. MS. Rawl. C 865, printed in R. T. Gunther, *Early Science in Oxford* (Oxford, 1925), pp. 264-74.
13. W. D. Macray, "Prayers of Thomas Hearne," *NQ,* 2nd ser., 12 (1861), pp. 165-66.
14. Cherry to Hearne, Dec. 12, 1700, *Letters from the Bodleian,* ed. J. Walker (London, 1813), pp. 117-19.
15. Huddesford, pp. 14-16.
16. Hearne, *Remarks,* VI, p. 262.
17. Hearne, ed., *Joannis Lelandi Antiquarii De Rebus Britannicis Collectanea* (6 vol., Oxford, 1715), II, p. 55.
18. So at least he speaks of his correspondence with Thomas Smith, a letter every other Saturday recapitulating much of the diary: "Perhaps when You and I are in our graves, they will be of admirable use in the History of these Times." Hearne, *Remarks,* II, pp. 311-12, 345.
19. Hearne's account is in *Remarks,* XI, pp. 16-17; see also Hearne to Rawlinson, Dec. 15, 1730, *ibid.,* X, p. 363. In Baker's transcript of Hearne's autobiography, there is a piece bearing on this episode which does not appear in the Huddesford version; see Camb. MS. Mm 47, ff. 294-96. The work was entitled *A Vindication of those who take the Oath of Allegiance* (London, 1731).
20. Hearne, *Remarks,* X, pp. 328, 310, 395.
21. *Vindication,* p. 2.
22. *Ibid.,* p. 43.
23. Vallemont's work was entitled *Les Élémens de l'Histoire* (Paris, 1696) and passed through six editions by 1745.
24. See "A List of the Books written and published by me," Bodl. MS. Rawl. Lett. 39, ff. 40-43, and the letters of Childe to Hearne in I. G. Philip, "The Genesis of Thomas Hearne's *Ductor Historicus,*" *Bodleian Library Record,* 7 (1966), pp. 251-64. The *Ductor* was republished 1714, 1723, and 1724, without Hearne's permission. "Some Collections in order to a IIId vol of Ductor Hist" appears in Bold. MS. Rawl, D 370. Hearne also

proposed an English history, "in such stile as to please as well as instruct the vulgar." Bodl. MS. Rawl. Lett. 14, ff. 381, 383, 385. There is a specimen in Bodl. MS. Rawl. D 1171, ff. 10-39. And there are some "Historical Collections relating to the English Antiquities made by an Undergraduate," Bodl. MS. Rawl. B 246, perhaps related to this exercise.

25. Bodl. MS. Rawl. Lett. 39, ff. 40-43. Pufendorf's work was entitled *Introductio ad Historiam Europaeam;* it received an English translation (London, 1699) and was immediately popular, with five editions by 1702.

26. Hearne, *Remarks,* III, p. 219.

27. *Ductor,* II, pp. 112-19.

28. Even so, these two chapters were both dropped for the third edition (London, 1714) as too advanced for young students.

29. *Ductor,* I, p. 134.

30. Bodl. MS. Rawl. D 370, ff. 56ff., 77^{V}ff.

31. *Ductor,* I, p. 16.

32. *Ductor* (1698), pp. 203-05; *Ductor* (1704), I, pp. 168-72.

33. *Ductor* (1704), p. 421ff.

34. Hearne to Rawlinson, Dec. 20, 1713, Hearne, *Remarks,* IV, pp. 276-77. See also Hearne to Murray, Sept. 6, 1726, *ibid.,* IX, p. 190; "Specimen . . . towards an Epitome of English History," Bodl. MS. Rawl. D 1171, f. 13.

35. In 1714, however, there is a brief notice taken from Stow's *Survey; Ductor* (1714), I, pp. 447-50.

36. In fact, it was the earlier of the two volumes to be published and so is really Hearne's first general statement about history.

37. Hearne, "To the Reader," *Reliquiae Bodleianae* (London, 1703).

38. Tancred Robinson was one admirer; Robinson to Lhwyd, April 3, 1705, Bodl. MS. Ashmole 1817a, f. 354. They were still being quoted and the editions approved in the 19th-century; see Joseph W. Moss, *A Manual of Classical Philology* (2 vol., London, 1837), II, pp. 132-33, 197, 436; Thomas Frognell Dibdin, *An Introduction to the Knowledge of Rare and Valuable Editions of the Greek and*

Latin Classics, 4th ed., (2 vol., London, 1827), II, pp. 3, 331-32. See also the note on Hearne's Eutropius in *The Works of the Learned,* 6 (1704), pp. 97-99.

39. *Oratio de Fide Historiarum contra Pyrrhonismum Historicum* (Lugduni, 1702); Hearne, *Remarks,* I, p. 52. In 1706 Elisha Smith sent Hearne "a compleat dissertation against the general Error of the Learned that Livy was a superstitious Writer believing all the wonders and stories he relates." Unfortunately it was by the freethinker John Toland, and Hearne (who might otherwise have sympathized) rejected it; see Bodl. MS. Rawl. C 146, ff. 47-48, and Rawl. D 401, no. 8; also F. H. Heinemann, "Prolegomena to a Toland Bibliography," *NQ,* 185 (1943), pp. 182-86. The work was later published separately, enlarged, as *Adeisidaimon sive Titus Livius a superstitione vindicatus* (The Hague, 1709).

40. Hearne, *Remarks,* I, pp. 199-200.

41. Hearne, *T. Livii Patavini Historiarum* (Oxford, 1708), I, "To the Reader" (May 1708). The chronology was supplied by Dodwell and Cherry. Hearne's basic text was by Gronovius, 3rd ed. (Amsterdam, 1679). The principles of the edition are set out in a letter to his old teacher Patrick Gordon, June 14, 1710, who had accused Hearne (Hearne thought unfairly) of being overly fastidious; Bodl. MS. Rawl. D 1166, f. 39.

42. Hearne to Cherry, Dec. 14, 1708, *Remarks,* II, pp. 157-58.

43. *Ibid,; NQ,* 185, (1943), pp. 165-66.

44. Woodward to Hudson, Jan. 10, 1705, and June 29, 1706, Bodl. MS. Rawl. D 316, ff. 8-9, 96; Hearne, *Remarks,* I, p. 109.

45. Woodward to Hudson, Aug. 16 and Oct. 29, 1706; and April 1 and May 27, 1707; Bodl. MS. Rawl. D 316, ff. 138, 68-69, 134, 91. Benedetti thought the gem contemporary with Scipio, and Hearne eventually printed it; see Woodward to Hearne, April 16, 1708, Bodl. MS. Rawl. Lett. 12, f. 421.

46. Hearne, *Remarks,* II, pp. 35-36.

47. *Ibid.,* Hearne to King, Feb. 21, 1708, Bodl. MS. Gough Misc. antiq. 10, f. 19.

48. "What I value it most upon, is the several Habits of that Age which are to be known from it." Hearne, *Remarks,* I, pp. 6-7.

49. Smith to Hearne, Sept. 6, 1707, Hearne, *Remarks,* II, p. 41.

50. Woodward to Hearne, Sept. 18, 1707, Bodl. MS. Rawl. Lett. 12, ff. 411-12.

51. Hearne to Woodward, Sept. 26, 1707, Hearne, *Remarks,* II, pp. 52-53.

52. "I am afraid the objections rais'd rather upon some personal account, than any real Shew of its being Spurious in the Shield it self." Hearne to King, Feb. 21, 1708, Bodl. MS. Gough Misc. antiq. 10, f. 19. See also Hearne to Thoresby, July 16, 1708, in Thoresby, *Letters* (1832), II, pp. 107-09.

53. Woodward to Hearne, Nov. 12, 1707, Bodl. MS. Rawl. Lett. 12, f. 413, and Woodward's note on the back of Hearne's letter, B.M. MS. Add. 6127, f. 22.

54. Hearne, *Remarks,* II, pp. 151-52.

55. *Ibid.,* X, p. 444. It was praised in the *Bibliothèque Choisie,* 20 (1707), pp. 196-98, and by Perizonius in a letter to Hudson, Bodl. MS. Rawl. Lett. 45, f. 26. The print appears in Hearne's Livy, VI, p. 195, with a note at I, p. 326. The Woodward gem was reproduced alongside Spon's shield, the only other illustration in the work; *ibid.,* III, p. 246.

Chapter 11: Mr. Dodwell Discovers the Shield

1. Dodwell to Hearne, Aug. 8, 1707, Bodl. MS. Rawl. Lett. 25, f. 454.

2. Bodl. MS. Rawl. B 246, ff. 492-93. A contemporary life of Dodwell was written by his friend Francis Brokesby, *The Life of Mr. Henry Dodwell* (London, 1715). There is no modern biography except the *DNB.* Some of his correspondence with foreign savants is in Bodl. MS. Cherry 23. There are other notices in Niceron, I, pp. 138-54; Chauffepié, II, pp. 31-35.

3. Edmund Calamy, *An Historical Account of My Own Life,* ed. John T. Rutt (2 vol., London, 1829), I, pp. 281-83.

4. The Nicolson letter was copied by Hearne, who does not say to whom it was

addressed; Sept. 16, 1703, *Remarks,* I, p. 209.

5. Thomas Macaulay, *The History of England* (London, 1858), III, p. 461. Gibbon is quoted in "Dodwell" in the *DNB.*

6. W. Gill to ?, Sept. 20, 1701, B.M. MS. Stowe 747, f. 147.

7. Francis Brokesby, *Dodwell,* p. 47.

8. "He has left us a most noble Example behind him of Piety, Humility, Constancy, Learning, and Industry." Hearne to Cherry, June 15, 1711, Bodl. MS. Rawl. Lett., 36, f. 351. See also Bodl. MS. Rawl. B 246, f. 493; Hearne, *Remarks,* III, pp. 176-77; Hearne, Leland, *Itin.,* V (1711), pp. 109-13.

9. Bodl. MS. Rawl. B 246, f. 493; see also Hearne, *Remarks,* X, pp. 236-38.

10. Brokesby, *Dodwell,* p. 523; Hearne, Leland, *Itin.,* V, pp. 111-12.

11. Hearne, *Remarks,* VII, p. 13; Brokesby, *Dodwell,* pp. 524, 535.

12. Lloyd to Dodwell, March 22 and April 18, 1688, Bodl. MS. Eng. Lett. 29, ff. 100, 102; Mill to Dodwell, April 2, 1688, "The Missing Dodwell Papers," *St. Edmund Hall Magazine* (1934), p. 88.

13. Hearne, *Remarks,* X, p. 154.

14. Dodwell to Charlett, March 11, 1690, Bodl. MS. Ballard 34, ff. 2-10.

15. Anthony Wood, *Athenae Oxoniensis,* ed. Philip Bliss (London, 1820), IV, "Fasti," cols. 404-06. For Dodwell's defense of the Church, see his letter to the editor of *The Works of the Learned,* Bodl. MS. Eng. Lett. C 28, ff. 106-07; Brokesby, *Dodwell,* pp. 43, 177.

16. Henry Dodwell, *Praelectiones Academicae in Scola Historices Camdeniana* (Oxford, 1692).

17. Cuper to Graevius, Dec. 18, 1691, Bodl. MS. D'Orville 478, ff. 65-67.

18. Wood, n. 15 above.

19. "Joseph Scaliger," Pattison, *Essays* (ch. 3 above, n. 26), I, p. 131; see also Anthony Grafton, "Joseph Scaliger and Historical Chronology: The Rise and Fall of a Discipline," *History and Theory,* 14 (1975), pp. 156-85.

20. *The Works of the Learned,* 6 (1704), p. 211.

21. *Ibid.,* pp. 211-12; Pattison, "Scaliger," p. 137.

22. Dodwell to Smith, May 23, 1667,

Bodl. MS. Smith 49, ff. 43-45. The modern spelling is usually Sanchuniathon but I have kept the 18th-century usage.

23. Richard Cumberland, *Sanchoniathon's Phoenician History,* ed. S. Payne (London, 1720), p. xvii. The editor assigns the work to the time of the Glorious Revolution.

24. Francis A. Yates, *Giordano Bruno and the Hermetic Tradition* (1964, reprinted New York, 1969), p. 401. "Isaac Casaubon was one of the greatest Men that ever lived." Hearne, *Remarks,* VI, p. 250.

25. "The Discourse concerning Sanchoniathon's Phoenician History" was printed as an appendix to Dodwell's *Two Letters of Advice,* 3rd ed. (London, 1691). My quotation is from the conclusion, pp. 114-15.

26. Sa. Parker to Dodwell, Nov. 13, 1680, Bodl. MS. Cherry 23, ff. 325-27.

27. In his work on the Egyptians, Woodward insisted that the authority of Sanchoniathon was not very great; Woodward, "Egyptians," p. 290.

28. It should be noted that the recent discoveries of Ugaritic and Hittite texts seem to give some weight to Philo's claims; see W. F. Albright, *Archaeology and the Religion of Israel,* 3rd ed. (Baltimore, 1953), p. 70.

29. For bibliography, see my article, "Ancients, Moderns and History," (ch. 2 above, n. 5), pp. 17-18; A. T. Bartholomew, *Richard Bentley: A Bibliography* (Cambridge, 1908), pp. 26-41; Alexander Dyce, ed. *The Works of Richard Bentley* (3 vol., London, 1836), I, pp. xi-xiv.

30. Bentley, *Dissertations upon the Epistles of Phalaris,* ed. Wilhelm Wagner (London, 1883).

31. Dodwell to Bentley, May 1698, Bodl. MS. Cherry 22, f. 58.

32. Perizonius to Dodwell, April 13, 1706, Bodl. MS. St. Edmund Hall 13, ff. 33-35.

33. Dodwell agreed that Bentley's method of "hypothesis of Synchronisms of Persons" was "the most accurate way of judging them," but he said nothing about Phalaris. (This was, of course, only one of Bentley's arguments.) On the other hand, he thought that Bentley was still mistaken about the dates of Pythagoras, and told him he still disliked his tone; see

Dodwell to Bentley, March 22, 1699, Bodl. MS. Cherry 22, ff. 61^V-62^V. To Smith he acknowledged that Bentley's treatment of Phalaris raised "good usefull subjects for a review," but no more; March 13, 1699, Bodl. MS. Smith 49, f. 163.

34. *Bentley Corr.,* intro.; Nicéron, I, pp. 142-43. See also Bentley to Perizonius (1695) in *Some Letters of Richard Bentley,* ed. Elfriede H. Pol (Leyden, 1959), p. 27.

35. Hearne, *Remarks,* I, pp. 193-94.

36. Henry Dodwell, *De Veteribus Graecorum Romanorumque Cyclis* (Oxford, 1701), pp. 675-80. On the early Greek history, see also Dodwell to Bentley, May 14, 1698, Bodl. MS. Cherry 22, f. 56.

37. The full title of *De Cyclis* in contemporary English translation runs as follows: "Ten Dissertations concerning the Ancient Cycles of the Greeks and Romans together with some short Account of the Jewish Cycle in the time of Christ: To which are added some necessary Tables, wherein are inserted such Fragments of the Ancients as relate to Chronology never before Publish'd. Being a Treatise very requisite for the due understanding of Ancient History both the Greek, Roman, and Sacred." *The Works of the Learned* (1701), pp. 565-68. In a letter to Lloyd, July 1700, Dodwell explained the kind of thing he was up to. "My Book de Cyclis encreases beyond expectation. It is already a 100 sheets and will take up a good many more before it is finished. I have endeavored to settle all the Chronological passages of Thucydides according to his own Principles. . . . I have fixt the Archonship of Themistocles otherwise than Lydiat, and of Aristides not yet known and have settled the times of Themistocles, Cimon, Aristides, Pericles, and Alcibiades nearer than hath hitherto been that I know attempted." The "modern" boast is unmistakable. The letter is one of several to Lloyd at Hardwick Court, this one in Box 74, f. 27. I am grateful to the present owner, Miss O.K. Lloyd-Baker, and to Col. A.B. Lloyd-Baker for permission to see and to use them.

38. Hearne recalled copying over Dodwell's *Paraenesis* at Shottesbrook twice before he was old enough to understand it;

Hearne to Smith, Oct. 8, 1709, Bodl. MS. Rawl. D 1169, ff. 38-39. *De Cyclis* is dedicated to Cherry for his help.

39. Dodwell to Hearne, July 21, 1701, Bodl. MS. Rawl. Lett. 25, f. 2.

40. Hearne, Leland, *Itin.*, V, p. 111.

41. Hearne, *Remarks*, I, pp. 247-48 and p. 211. Apollonius' *Life* is now thought to be genuine.

42. Hearne, *Remarks*, VI, pp. 200-01 (1718).

43. For example, on a reading of Pliny's *Epistles*, Dodwell to Hearne, Nov. 12, 1702, Bodl. MS. Rawl. Lett. 25, f. 421.

44. Hearne, *Remarks*, VII, p. 38, and X, p. 371. For Dodwell's small personal collection of inscriptions, see Dodwell to Hearne, July 11, 1706, Bodl. MS. Rawl. Lett. 25, f. 431.

45. Dodwell to Hearne, Sept. 11, 1705, Bodl. MS. Rawl. Lett. 25, f. 433; Thomas Cherry to Hearne, April 17, 1706, Bodl. MS. Rawl Lett. 4, f. 473; Dodwell to Hearne, Oct. 22, 1706, Bodl. MS. Rawl. Lett. 25, f. 439; Hearne to Dodwell, Nov. 26, 1705, Bodl. MS. Eng. Lett. C 28, ff. 76-77.

46. Dodwell to Lloyd, July 1700, MS. Hardwick Court Box 74 (see above n. 46), f. 27. "My chronology of Pythagoras," he wrote with obvious relief, "is very little concerned in what has been opposed by the Dr."

47. *Exercitationes Duae: Prima De Aetate Phalarides; Secunda De Aetate Pythagorae* (London, 1704). There is a long review by Le Clerc in the *Bibliothèque Choisie*, X (1706), p. 130ff.

48. Perizonius was impressed by Dodwell's erudition, but still wanted to know whether he thought the letters of Phalaris genuine and what their date was; see Perizonius to Dodwell, July 17, 1705, Bodl. MS. St. Edmund Hall 13, ff. 29-30.

49. See his letter to Jeremy Mills (1699) advising a course of reading in ecclesiastical history for Balliol College; Bodl. MS. Cherry 23, f. 137.

50. "Invitation to Gentlemen to Acquaint themselves with Antient History," appended to Degory Wheare's *Method and Order of Reading Histories*, trans. Edmund Bohun (London, 1694). (Wheare had

been the first Camden Professor, and Dodwell's work was probably prepared while he held that post.) The other work was entitled "An Apology for the Philosophical Writings of Cicero" and appeared in Jeremy Collier's translation of *Tully's Five Books De Finibus* (London, 1702).

51. Hearne, *Remarks*, I, p. 210. Brokesby remains the best account of the published works.

52. Richard King to Dodwell, Nov. 13, 1705, Bodl. MS. Eng. Lett. C 28, ff. 74-78; Hearne, *Remarks*, VI, p. 198.

53. Another example is the Greek voyager and geographer Hanno. Like Sanchoniathon, there was supposed to be a Punic original for him extant only in Greek, and Dodwell rejected it, although Hanno is generally accepted today; see Thomas Falconer, *The Voyage of Hanno* (London, 1797), p. 40.

54. Nicéron, I, pp. 142-43. Dodwell's opinions on chronology still carried great weight in the 19th-century; see Henry Fines Clinton, *Fasti Hellenici* (Oxford, 1834), II, pp. xxv, 327-36, *et passim*.

55. Dodwell to Hearne, Feb. 21, 1710, Bodl. MS. Rawl. Lett. 25, f. 499.

56. Hearne to Dodwell, March 3, 1710, Bodl. MS. Rawl. D 1166, f. 22; Bagford to Hearne, Jan. 31, 1710, in Hearne, *Remarks*, II, p. 340. See also Hearne to Smith, Feb. 4, 1710, Bodl. MS. Rawl. D 1166, f. 9^V; Hearne to Thoresby, *ibid.*, f. 10^V; and Thoresby, *Letters* (1832), II, pp.227-29.

57. Nicolson's account, which looks as though it once belonged to his diary, is dated Feb. 5, 1712; Bodl. MS. Top. Gen. C 27/1. The catalog was drawn up by Robert Ainsworth, *Monumenta Vetustatis Kempiana* (London, 1720). There is another helpful description of the collection by John Ward in a letter to John Pointer, Aug. 17, 1719, Bodl. MS. Eng. Lett. D 77, f. 28. Ward thought it unequaled as a private collection.

58. Mr. John Cherry of the British Museum first called the present existence of the helmet to my attention; he writes that it was bequeathed to the Museum at the same time as Dr. Woodward's shield (1818) by the owner, John Wilkinson, for whom see ch. 14, below, esp. n. 63.

59. Dodwell to Hearne, March 22, 1710, Bodl. MS. Rawl. Lett. 25, f. 101.

Chapter 12: De Parma Equestri

1. Smith to Hearne, Jan. 18, 1707, in Hearne, *Remarks*, I, p. 318. The following year Smith wrote, "Dr. Hudson whom you deservedly call your *Friend and Patron* his judgment must be submitted to," this despite the fact that Hearne must be taken from English antiquities; *ibid.*, II, p. 186.

2. Hearne to Richard Rawlinson, Feb. 20, 1715, *Remarks*, V, p. 27. By 1708 Smith too was speaking of Hearne's "genius and inclination toward English History and Antiquity." *Ibid.*, II, pp. 158-59.

3. Hearne resolved upon an edition in Sept. 1708, encouraged by the success of Livy; at the end of 1712 it had been ready for the press "for some time"; in 1715 he was still looking for a patron. He never found sufficient interest; see *Remarks*, II, pp. 128-29; V, pp. 42-43; XI, p. 317. Proposals for printing it are given in Huddesford, app., pp. 119-20; see also *Lit. Anec.*, III, p. 684.

4. Hearne to Smith, June 4, 1709, *Remarks*, II, p. 205.

5. Hearne to Charlett, Sept. 19, 1715, *ibid.*, V, pp. 113-14. Hearne had already transcribed Leland in 1707; Hearne to Smith, March 29, 1707, *ibid.*, II, p. 2.

6. Hearne to Cherry, Sept. 18, 1710, *ibid.*, III, p. 49; Stratford to Harley, Feb. 28, 1719, *Portland MSS.*, VII, p. 249.

7. See Hearne's account in *Remarks*, II, pp. 179-83. The offending passage appears in a long note by Hearne in his text of *The Life of Alfred the Great* by John Spelman (Oxford, 1709), pp. 177-80. See also Smith to Charlett, March 31, 1709, in *Remarks*, II, p. 438; Smith, *Annals of University College* (Newcastle, 1728), p. 207ff.; James Parker, *The Early History of Oxford* (Oxford Hist. Soc., 1885), p. 52ff. Hearne never conceded; when Smith's book appeared he complained to his friend West in despair that "such a studied Rhapsody of Lyes should ever come from a Member of this University." July 17, 1728, B.M. MS. Lansdowne 778, f. 95.

8. *Life of Alfred*, pp. 226-28.

9. *Athenian Mercury*, suppl., 2 (1691), pp. 25-27. Exactly Hearne's sentiments! See *Remarks*, IV, pp. 51, 95; V, pp. 32-33.

10. J. G. Graevius, *Inscriptiones Antiquae Totius Orbis Romani* (4 vol., Amsterdam, 1707). Hearne admired Gruter enormously for his precision, "so exact in what he transcribed himself that he even gives the false lections and the Position of the letters." He was less happy with the new edition of Graevius. *Remarks*, II, pp. 39, 89.

11. On one occasion Pembroke purchased a number of them incidental to some more valuable antiquities, "But these (tho' they are on Marble) having nothing material but the unknown names he gave to Dr. Woodward." B.M. MS. Stowe 1018, f. 11.

12. *Remarks*, VII, p. 131.

13. *Life of Alfred* (n. 7 above), p. 226. See also Hearne, Leland, *Itin.*, ix, p. 154; *Remarks*, II, p. 70.

14. John Urry (the Chaucer editor) sent a drawing, transcription, and some remarks to Lhwyd, Nov. 2, 1708, Bodl. MS. Ashmole 1817b, f. 182. For the rest, see Gale, *Antonini Iter Britannicarum* (London, 1709), p. 134; Musgrave and Dodwell, *Julii Vitalis Epitaphium* (Exeter, 1711); Hearne, Leland, *Itin.*, VIII, p. xxxii; Horsley, *Britannia Romana* (London, 1732), p. 323. Later Hearne made some marginal comments, defending his accuracy, on the letter to him from Samuel Gale, Aug. 1, 1714, Bodl. MS. Rawl. Lett. 15, ff. 13-14.

15. R. G. Collingwood and R. P. Wright, *The Roman Inscriptions of Britain* (Oxford, 1965), I, no. 156; F. J. Haverfield in *VCH Somerset*, I, p. 275.

16. "My confinement to the Library, and the several new Curiosities I continually light upon, together with the trouble of Republican times, make me keep close to Oxford." Hearne to Cherry, May 28, 1710, Bodl. MS. Rawl. Lett. 36, f. 336. Hearne's description of the Bath inscription is in his *Life of Alfred*, pp. 226-38.

17. Dodwell to Hearne, July 13, 1706, Bodl. MS. Rawl. Lett. 25, f. 431.

18. Musgrave to Sloane, July 7, 1711, B.M. MS. Sloane 4042, ff. 309-10.

19. Brokesby to Hearne, n.d., Bodl. MS. Rawl. Lett. 3, f. 123.

20. Dodwell to Hearne, Jan. 24 and April 13, 1709, Bodl. MS. Rawl. Lett. 25, ff. 30, 38. It was ready for publication by Sept. 1710; *ibid.*, f. 52.

21. Dodwell to Hearne, May 16 and June 4, 1709, and Jan. 4, 1710; *ibid.*, ff. 40, 41, 485.

22. The work appeared as *Julii Vitalis Epitaphium* (Exeter, 1711). "The Doctor was a better Physician than Antiquary" was Hearne's final comment; *Remarks,* VII, p. 314. There is a long review in the *Bibliothèque Choisie,* 25 (1712), pp. 220-28, and a summary in *Phil. Trans.,* 28 (1713), pp. 283-85.

23. Cuper to Le Clerc, Sept. 27, 1711, in Cuper, *Lettres,* p. 371; "An Account of the Roman Eagles," *Phil. Trans.,* 28 (1713), pp. 145-50. See also Musgrave to Sloane, July 28, 1713, B.M. MS. Sloane 4043, ff. 166-67; Musgrave to Charlett, July 17, 1713, Bodl. MS. Ballard 24, ff. 129-30. For Musgrave's subsequent career, see the *DNB.*

24. Dodwell to Hearne, May 20 and Nov. 21, 1710, Bodl. MS. Rawl. Lett. 25, ff. 51, 53.

25. Hearne to Dodwell, June 15, 1710, Bodl. MS. Rawl. D 1166, f. 40; *Remarks,* III, p. 136.

26. Hearne to Smith, Aug. 7 and 13, 1709; Hearne to Brokesby, Aug. 18, 1709, Bodl. MS. Rawl. D 1166, ff. 29, 29^V-30, 30.

27. Dodwell to Hearne, May 20, 1710, Bodl. MS. Rawl. Lett 25, f. 505, and March 29, 1710, Bodl. MS. Eng. Lett. C 28, f. 50; Hearne to Dodwell, May 7, 1710, *ibid.,* ff. 86-87; Hearne to Cherry, May 28 and Oct. 21, 1710, Bodl. MS. Rawl. D 1166, ff. 35-36, 56^V-57.

28. Dodwell to Woodward, Jan. 3, 1711, B.M. MS. Add. 6127, f. 40.

29. *Remarks,* IX, p. 175. Elsewhere we learn from Hearne that Downes was a traveling tutor to Lord Scudamore and wrote some religious polemic; *ibid.,* VI, p. 23 and II, pp. 103, 430. There are some letters from Downes to Hilkiah Bedford, 1702-04, describing his Italian journey, in Bodl. MS. Rawl. Lett. 42, ff. 36-38.

30. Downes to Dodwell, n.d., Bodl. MS. St. Edmund Hall 14, ff. 56-59.

31. *Ibid.,* f. 59; Dodwell to Woodward, April 14, 1711, B.M. MS. Add. 6127, f. 43.

32. Theophilus Downes, *De Clypeo Woodwardiano Stricturae Breves.* As we shall see (ch. 14 below), it was printed only many years afterward (1729) by Richard Rawlinson. There is a ms. copy in B.M. MS. Add. 6127, f. 59, that belonged to Dr. Woodward. The work appears to have been written in 1708.

33. Downes to Dodwell, n.d., Bodl. MS. St. Edmund Hall 14, ff. 56-59.

34. Dodwell to Downes, *ibid.,* ff. 60-61.

35. "If my aylings will permit me to come up to London next month, I shall then endeavor to make use of the liberty you have been pleased to grant me, of examining your pillars, especially that of Titus, in order to the Explication of Dr. Woodward's shield. The Doctor has other Pillars but wants it. But it I take to come nearest the time of the shield." Dodwell to Nelson, Feb. 14, 1711, *ibid.,* ff. 44-45.

36. B.M. MS. Add. 6127, ff. 43-45.

37. *Remarks,* III, p. 176; see also Hearne, Leland, *Itin.,* V, pp. 111-12.

38. Brokesby to Hearne, June 27, 1711, Bodl. MS. Rawl. Lett. 3, f. 122.

39. Woodward to Hearne, Sept. 27, 1711, Bodl. MS. Rawl. Lett. 12, ff. 452-53.

40. Hearne to Woodward, July 6, 1711, Bodl. MS. Rawl. D 1170, f. 41; Woodward to Hearne, July 17, 1711, Bodl. MS. Rawl. Lett. 12, f. 449; Hearne to Woodward, July 24, 1711, Bodl. MS. Rawl. D 1170, f. 42^V. Hearne's one problem was that Dodwell had left specific instructions not to publish anything unfinished; on this he had to be reassured by Brokesby. See Hearne to Woodward, n.d., Bodl. MS. Rawl. Lett. 39, f. 38; Woodward to Hearne (conveying the reassurance), n.d., Bodl. MS. Rawl. Lett. 18, f. 139.

41. Hearne to Woodward, Dec. 8, 1710, Bodl. MS. Rawl. D 1188, ff. 45^V-46.

42. Hearne to Woodward, Oct. 23,

1711 and Jan. 24, 1712, in *Remarks,* III, pp. 24, 295. Woodward induced Roger Gale to lend Hearne his manuscript of Leland; Woodward to Hearne, March 3, 1711, Bodl. MS. Rawl. Lett. 12, f. 441.

43. Thoresby's essay and Hearne's letter were read together and printed in *Phil. Trans.,* 26 (1709), pp. 510-18. See also Hearne to Thoresby, Dec. 5 and 23, 1709, in Thoresby, *Letters* (1832), II, pp. 207-08, 210-12; Thoresby to Sloane, Oct. 13, 1711, B.M. MS. Sloane 4042, f. 356.

44. Hearne, Leland, *Itin.,* I, pp. 99-114; Woodward to Hearne, Jan. 5, 1710, Bodl. MS. Rawl. Lett. 12, ff. 426-27. In a later volume of the Leland, Hearne reviewed the discussion and printed a letter from Richard Richardson disputing with him and suggesting that they were "Perhaps . . . Axes used in sacrificing some of the smaller Quadrupeds by the ancient Britons." *Itin.,* IX, pp. 138-44. Eventually he wrote to Richardson that his views were meant only as a "conjecture," though he held to them still; Feb. 18, 1712, *Lit. Ill.,* I, p. 301.

45. Woodward to Hearne, Jan. 21, 1710, Bodl. MS. Rawl. Lett. 12, ff. 430-31; Hearne to Woodward, Jan. 27, 1710, Bodl. MS. Rawl. D 1166, f. 49^V; Hearne to Sloane, Jan. 25, 1710, R.S. MS. Letter Book H 3, f. 90; Hearne to Woodward, Aug. 2 and Dec. 11, 1710, Bodl. MS. Rawl. D 1166, ff. 45, 62^V.

46. Hearne to Cherry, Sept. 1, 1711, in *Remarks,* III, pp. 218-19.

47. *Ibid.,* III, p. 232.

48. Hearne had disavowed any "advantage" from the project, Woodward insisted that he take whatever profit might be in it; Woodward to Hearne, Sept. 18, 1712, Bodl. MS. Rawl. Lett. 18, f. 147.

49. Woodward to Hearne, Jan. 6, 1713, Bodl. MS. Rawl. D 1029, f. 48.

50. *Ibid.,* ff. 49-50. Compare it with the letter reproduced in Ward, *Lives,* pp. 139-42, and discussed above in ch. 9, sect. 8.

51. Woodward to Hearne, Oct. 28, 1712, Bodl. MS. Rawl. Lett. 18, f. 149.

52. Woodward to Hearne, Bodl. MS. Rawl. B 206, f. 2.

53. Woodward to Hearne, Dec. 15, 1711, Bodl. MS. Rawl. Lett. 18, f. 142; Sept. 27 and Dec. 25, 1711, Bodl. MS. Rawl. Lett. 12, ff. 452-53, 455.

54. Hearne to Woodward, Dec. 8, 1712, in *Remarks,* III, p. 497.

55. *Ibid.,* III, pp. 109-10; Thoresby to Hearne, May 16, 1711, *ibid.,* III, p. 161.

56. *Remarks,* III, pp. 395, 408. A full description was read to the Royal Society, which took a consuming interest in the new discovery, judging from a letter addressed to Halley, Feb. 20, 1712, R.S. MS. Classified Papers 1660-1710, XVI, no. 41. See also R.S. MS. Jour. Bk, X, pp. 362, 366, 369, 396, and 501 (where Hearne's draft was presented).

57. Hearne, Leland, *Itin.,* VIII, pp. ix-xxxix with a plate by Burghers. Hearne's chief opponent was John Pointer of Merton College who published a separate tract on the monument, *An Account of a Roman Pavement Lately found at Stunsfield* (Oxford, 1713). Pointer was also a virtuoso with a good collection of "Rarities both Natural and Artificial" – see "Museum Pointerianum" in R. T. Gunther, *Early Science in Oxford* (Oxford, 1925), III, p. 454ff. I intend to publish a separate account of the pavement shortly; meanwhile see M. V. Taylor, "The Roman Tesselated Pavement at Stonesfield, Oxon.," *Oxoniensia,* 6(1941), pp. 1-8; *VCH Osfordshire,* I, pp. 315-16.

58. Woodward to Hearne, Aug. 19, 1712, Bodl. MS. Rawl. Lett. 18, f. 146; Thoresby to Hearne, Aug. 20, 1712, Bodl. MS. Rawl. Lett. 17, f. 7.

59. Hearne, Leland, *Itin.,* VIII, p. iv.

60. Gale to Hearne, Nov. 15, 1712, Bodl. MS. Rawl. Lett. 15, f. 20.

61. Hearne, Leland, *Coll.,* II (1715), p. 61ff.

62. *Itin.,* VIII, p. xi.

63. *Remarks,* II, p. 372.

64. Hearne to Dodwell, April 7, 1710, Bodl. MS. Rawl. D 1166, f. 28; see also the letter of March 3, and Hearne to Bagford, May 1, 1710, *ibid.,* ff. 22, 25^V-26^V.

65. *Remarks,* X, pp. 118, 130.

66. *Ibid.,* III, pp. 432-33, 441-42.

67. Woodward to Hearne, Oct. 22, 1712, in *Remarks*, III, p. 475.
68. *Ibid.*, III, p. 488, and IV, p. 42.
69. Hearne to Woodward, Dec. 15, 1712, *Remarks*, IV, pp. 31-32. In fact, Hearne seems to have printed only 120 copies of Leland.
70. Woodward to Hearne, Nov. 13 and Dec. 13, 1712, Bodl. MS. Rawl. Lett. 18, ff. 151-52, 157; Hearne to Woodward, Feb. 16, 1713, *Remarks*, IV, pp. 74-75.
71. Hearne to Woodward, March 3, 1713, *ibid.*, IV, p. 92.
72. Woodward to Hearne, April 11, 1713, *ibid.*, p. 154. Dodwell's opponent seems to have been John Mill, Principle of St. Edmund Hall, with whom Hearne also had a grudge.
73. "About Suppressing my Book and the Occasion of it," *Remarks*, IV, pp. 108-31. I confess I do not know whether this Mollineux is the Doctor, Thomas, whom we met earlier; his nephew Samuel; or someone altogether unrelated.
74. *Ibid.*, p. 115.
75. Hearne's elaborate account of this affair, again the result of a publishing indiscretion, is given in *Remarks*, VI, pp. 341-413.
76. *Remarks*, IV, p. 117. Hearne had called Dodwell's book on the oaths *aureus tractatus*.
77. *Ibid.*, p. 126.
78. Hearne, "Nota de Vexillis", in Dodwell, *De Parma Equestri Woodwardiana Dissertatio* (Oxford, 1713), p. xxviii.

Chapter 13: The Shield of Martinus Scriblerus

1. I am especially indebted in this chapter to the excellent edition with introduction and commentary by Charles Kirby-Miller of *The Memoirs of Martinus Scriblerus* (1950; reprinted New York, 1966). See the review by Herbert Davis, *PQ*, 30 (1951), pp. 254-56. See also Robert J. Allen, *The Clubs of Augustan London* (Harvard Studies in English, 7, 1933), pp. 260-83.
2. For Swift's friendship and praise, see his *Corr.*, II, p. 82, and *The Journal to*

Stella, ed. George A. Aitkin (London, 1901), *passim*. For Arbuthnot, see Kippis, *Biog. Brit.*, 1, pp. 238-43; Aitkin, *Arbuthnot;* Beattie, *Arbuthnot*; and most recently, Claude Bruneteau, *John Arbuthnot et les idées au debut du dix-huitième siècle* (2 vol., Service de Reproduction des Thèses, Université de Lisle, III, 1974). For other references to the founding of the Scriblerians, see George Sherburn, *The Early Career of Alexander Pope* (1934; reprinted New York, 1963), pp. 73-76.
3. Kippis, *Biog. Brit.*, I, pp. 241n, 242n. N. See also Chesterfield, *Letters*, ed. Viscount Mahon (London, 1845), II, p. 446.
4. Johnson, *Lives*, III, p. 177 and app. L, pp. 373-74.
5. See, for example, *NQ*, 6th ser., 7 (1883), pp. 406, 451-52, 469, 498.
6. "Memoirs of the Life of Dr. Arbuthnot," *Miscellaneous Works* (2 vol., London, 1770), p. xv.
7. William Warburton, ed., *The Works of Alexander Pope* (9 vol., London, 1751), VI, p. 96.
8. Spence, *Anecdotes*, I, p. 56.
9. Thus the ideal in the *Memoirs of Scriblerus:* "a discreet man, sober in his opinions, clear of Pedantry, and knowing enough both in Books and in the World, to preserve a due regard for whatever was useful or excellent." *Memoirs*, p. 113.
10. Warburton, Pope, *Works* (n. 7 above), VI, p. 97.
11. Swift was not unaware of the irony; see his letter to Pope and Bolingbroke, April 5, 1729, Pope, *Corr.*, III, pp. 27-30.
12. Spence, *Anecdotes*, I, p. 235.
13. April 16, 1713, *Berkeley and Percival: the Correspondence* (Cambridge, 1914), pp. 112-15. Arbuthnot admired Berkeley's philosophy and Berkeley claimed him as his "first prosolyte." *Ibid.*, pp. 114, 119-21, 123. For a similar characterization, see Lord Orrery, *Remarks on the Life and Works of Dr. Jonathan Swift*, letter no. 20, quoted by Aitkin, *Arbuthnot*, pp. 164-65; for the modern view, Sherburn, *Pope* (n. 2 above), p. 72.
14. See Beattie, *Arbuthnot*, pp. 334-38.

15. Arbuthnot, *Works* (Glasgow, 1751), pp. 1-39; see Beattie, *Arbuthnot*, pp. 339-46. The "Argument" appeared in *Phil. Trans.*, 27 (1710), pp. 186-90.

16. Arbuthnot, *Works*, pp. 17-18.

17. *Ibid.*, pp. 18-19. (See ch. 7 above, n. 9.)

18. *Tables of Coins Weights and Measures Explain'd* (London, 1727), preface, n.p. Proposals for the new work were read to the Society of Antiquaries, Dec. 22, 1725; MS. Minute Book, I, p. 177.

19. *Tables*, preface.

20. In a dissertation dealing with the price of Roman food, *ibid.*, p. 134ff.

21. See, for example, Arbuthnot to Charlett, Jan. 25, 1698, in Aitkin, *Arbuthnot*, pp. 22-24.

22. *Tables*, p. 110.

23. *Ibid.*, p. 143; see also p. 112.

24. It received a new edition in 1754 and a Latin translation in 1756 (reprinted 1764), and it was used in the schools until the 19th-century; see Robert Hussey, *An Essay on the Ancient Weights and Money and the Roman and Greek Liquid Measures* (Oxford, 1836), pp. 6-7.

25. Hussey, *Essay*, p. 6. It should be compared with the work of William Smith, *Litterae de Re Nummaria* (Newcastle, 1729). This work, which was composed in the form of letters — several to Ralph Thoresby — was written out in old age, but despite a garrulous style it showed effectively how the coins themselves might be used along with the literary evidence. Greaves, whose work on the Roman foot Arbuthnot knew well, had done even better.

26. Smith, p. 243.

27. E.g., Benjamin Langwith, *Observations on Dr. Arbuthnot's Dissertations* (London, 1747), p. 39. See also Beattie, *Arbuthnot*, pp. 349-50.

28. Aitkin, *Arbuthnot*, pp. 65-67.

29. *Memoirs of Scriblerus*, p. 97.

30. Arbuthnot, *Tables*, pp. 134-35.

31. *Memoirs*, pp. 103-04.

32. Hearne, *Remarks*, IV, p. 45. See also "Some Account of the Author and his Writings," *Remains of the Late Learned and Ingenious Dr. William King* (London, 1732); "King," in *Biog. Brit.;* John

Nichols, "Memoirs of Dr. King," *The Original Works of Dr. William King* (London, 1776); Samuel Johnson, "King," in *Lives*, II, pp. 26-31.

33. Monk, *Bentley*, I, pp. 99-100, 125-30, 264-65. See also Colin J. Horne, "The Phalaris Controversy: King vs. Bentley," *RES*, 22(1946), pp. 289-303; "Early Parody of Scientific Jargon," *NQ*, 184(1943), pp. 66-69; "Dr. King's *Miscellanies*," *The Library*, 4th ser., 25(1944-45), pp. 37-45.

34. Johnson, *Lives*, II, p. 26.

35. See his "Adversaria" in King, *Works*, I, pp. 223-24. His biographer says in the *Remains* that King read through 22,000 books and manuscripts to compile the *adversaria*, to which Dr. Johnson replied that "the books were certainly not very difficult, nor the remarks very large." Later King attempted his own scholarly publication, like the rest of the Scriblerians, *An Historical Account of the Heathen Gods and Heroes* (1711), which was very popular but again not very scholarly.

36. Bentley replied in his *Dissertations upon the Epistles of Phalaris*, ed. Wilhelm Wagner (London, 1883), pp. 18-22; see Monk, *Bentley*, I, pp. 125-26.

37. William King, *Dialogues of the Dead relating to the present Controversy concerning the Epistles of Phalaris* (London, 1699); *Works*, p. 133ff. King may also have written *A Short Account of Dr. Bentley's Humanity* (London, 1699) which in any case does include a second letter by King describing Bentley's insolence; see Nichols, "Memoir," I, p. 140; Monk, *Bentley*, I, p. 130.

38. "Modern Learning," *Dialogues*, pp. 52-67.

39. *Some Remarks on The Tale of a Tub* (1704) in *Works*, I, p. 209ff.

40. John Gay *The Present State of Wit*, ed. Donald Bond (Augustan Reprint Soc., 1947), p. 1.

41. "A Voyage to Cajamai in America," *Useful Transactions*, May-Sept. 1709.

42. "Meursius's Book of the Plays of the Grecian Boys," *ibid.*, pp. 40-47; *Memoirs*, pp. 72-73, 221-22. Kirby-Miller (n. 1 above) thinks that both pieces might

possibly have been composed by Dr.
Arbuthnot; in general he seems to me to
underestimate King's influence.
   43. King, *Works,* I, pp. 41-102.
   44. King, *Journey to London,* p. 26.
(See ch. 7 above, n. 25.) For a typical
spoof on the antiquaries' love of rust, see
Thomas Durfey, *Madame Fickle* (London,
1677), p. 26.
   45. See, for example, Swift to Ambrose Philips, March 8, 1709, in Swift,
Corr., I, p. 129.
   46. *The Censor,* no. 5, April 20, 1715
(London, 1717), pp. 29-32.
   47. Pope to Arbuthnot, July 11, 1714,
in Pope, *Corr.,* I, pp. 233-35; Aitkin,
Arbuthnot, p. 69.
   48. Parnell and Pope, joint letter to
Arbuthnot, Sept. 2, 1714, in Aitkin,
Arbuthnot, pp. 79-80; Pope, *Corr.,* I,
249-51.
   49. Arbuthnot to Pope, Sept. 7, 1714,
in Pope, *Corr.,* I, p. 251.
   50. He was fortunate also to be spared
some later shafts by Pope who could not
quite forget his memory. See, for example,
The Fourth Satire of Dr. John Donne
(written about 1713 but published only in
1733), Pope, *Works,* III, p. 29. In the
Dunciad of 1742 there is an incident that
centers on one Mummius who retrieves a
collection of priceless coins that had been
swallowed under duress, after a debate
between physicians as to whether purga-
tions or vomits would be more effective.
Mummius sounds very like Dr. Woodward,
though the generic resemblance among
antiquaries makes positive identification
impossible and perhaps undesirable. See
Marjorie Nicolson and George Rousseau,
who prefer Dr. Mead; *This Long Disease
My Life: Alexander Pope and the Sciences*
(Princeton, N.J., 1968), pp. 123-29. The
editor of the Twickenham *Works* prefers
the Fifth Earl of Sandwich; Pope, *Works,*
V, pp. 449-50.
   51. Warburton note quoted in Pope,
Works, VI, p. 106. The *Memoirs* appeared
first in the *Works of Mr. Alexander Pope
in Prose* (London, 1741), II.

Chapter 14: The Decline and
Disappearance of Dr. Woodward's Shield

   1. Hearne, *Remarks,* X, p. 11.
   2. At a cost of 120 guineas; B.M. MS.

6193, f. 27. For Woodward's failure to
make the French Academy, see Woolhouse
to Jurin, July 26, 1727, R.S. MS. Letter
Book W 3, no. 119. Bignon was active in
the translation of Woodward's *Natural
History;* see Hugh Bethel to Woodward,
March 22, 1725, Camb. MS. Add. 7647,
no. 49. The theological passages created a
difficulty. See also Bignon to Woodward,
ibid., nos. 61-63.
   3. Brome to Richard Rawlinson,
March 3, 1729, Bodl. MS. Rawl. Lett. 31,
f. 271. According to the Worcester news-
paper, June 8, 1745, Brome died that year
at 82, "a Gent of excellent Learning, a
skilful Antiquary . . . though he was much
inclin'd to a retired Life." He left "a
curious collection of Coins." Bodl. MS.
Ballard 41, f. 283. He had known Dr.
Woodward; Brome to Hearne, c. 1715,
Bodl. MS. Rawl. Lett. 13, f. 230.
   4. Hearne, *Remarks,* VIII, p. 62.
   5. Minutes of the Society of Antiquaries, B.M. MS. Egerton 1041, *passim;*
Hearne, *Remarks,* XI, p. 372. Ainsworth
was elected to the Society on Jan. 11,
1721.
   6. *Monumenta Vetustatis Kempiana*
(London, 1720); see Ainsworth to Strype,
Jan. 26, 1719, Camb. MS. Mm VI, f. 49.
Ainsworth was assisted on the *Monumenta*
by Woodward's later biographer, John
Ward, who wrote a long essay for it, *De
asse.* There is an elaborate presentation
copy to Lord Coleraine in the British
Museum, with two notes by Ainsworth.
   7. Hearne, *Remarks,* IX, p. 4. Perhaps
Dodwell's shortsightedness did not help;
see *Lit. Anec.,* V, p. 252, and IX, p. 798.
   8. *London Journal,* no. 507, April 19,
1729, signed "Atticus." Hearne approved;
Hearne to West, May 14, 1729, B.M. MS.
Lansdowne 778, f. 127. Ainsworth's subsequent work was entitled *De Clypeo
Camilli Antiqui* (London, 1734) and was
dedicated to Col. King. There is a synopsis
in B.M. MS. Add. 6127, ff. 56-58.
   9. *London Journal,* B.M. MS. 6127,
f. 189; see also *ibid.,* f. 72^V, for the
manuscript.
   10. Pope's lines (11. 41-42) appeared
first in his own *Works* (1720) and then in
Addison's *Works* (1721) prefixed to the
Dialogues. Norman Ault believes that the
first half of the poem, 11. 1-44 (less 11.
5-10 added in 1726), was originally a

separate poem on antiquities written 1713-15 and adapted afterward to the specific occasion of Addison's *Dialogues;* see his "Pope and Addison," *RES,* 17 (1941), pp. 428-51, and *New Light on Pope* (London, 1949), pp. 121-23. See also Peter Smithers, *The Life of Joseph Addison,* 2nd ed. (Oxford, 1968), p. 237; Howard Erskine-Hill, "The Medal Against Time: A Study of Pope's Epistle to Mr. Addison," *JWCI,* 28 (1965), pp. 274-98. For text and commentary, see Pope, *Works,* VI, p. 202ff. Warburton annotated "Vadius" thus: "See his history and that of his Shield, in the Memoirs of Scriblerus." Warburton, Pope, *Works,* III, p. 327.

11. Hearne, *Remarks,* X, p. 123.

12. For West, see Soc. Antiq. MS. Min. Bk., I, April 17, 1729, p. 234. For the difficulty in procuring a copy of Dodwell, see Brome to Rawlinson, Bodl. MS. Rawl. Lett. 31, ff. 282, 287, 326, 331, 332, 334, 335, 336; Brome to Rawlins, Bodl. MS. Ballard 19, ff. 19, 23-24, 27, 29-30, 31-32; Rawlins to Ballard, Bodl. MS. Ballard 41, ff. 52-53, 102.

13. Hearne, *Remarks,* X, pp. 118, 124; Baker to Rawlinson, Bodl. MS. Rawl. Lett. 30, ff. 124, 126. So too William Brome to Thomas Rawlins: "Those two Gent. had been to Rome and studyed Trajans Pillar and all Antiquities there." April 5, 1735, Bodl. MS. Ballard 19, ff. 23-24.

14. Hearne, *Remarks,* IX, p. 112.

15. Moyle to Reynolds, Aug. 18, 1718, St. John's College, Cambridge MS. K 27, ff. 74-75.

16. See R. T. Gunther, *Early Science in Oxford* (Oxford, 1937), XI, app. D, pp. 341-58; F. A. Turk, "Natural History Studies in Cornwall," *Journal of the Royal Institution of Cornwall,* new ser., III, 1959, pp. 234-37. For Moyle in general, see Caroline Robbins, *Two English Republican Tracts* (Cambridge, 1969), pp. 21-39: *Bibliotheca Cornubiensis* (London, 1874), I, pp. 375-77.

17. Moyle, *Works* (2 vol., London, 1726), I, p. 148.

18. Caroline Robbins, *The Eighteenth-Century Commonwealthman* (1959; reprinted New York, 1968), pp. 105-08.

19. Moyle, *Works,* I, p. 414.

20. "Some Account of Mr. Moyle and

his Writings," *The Whole Works* (London, 1721), pp. 37-49.

21. St. John's Col. Camb. MS. K 27, pp. 41-46 and elsewhere; Moyle to King, *Gent. Mag.,* IX (1938), pp. 248, 607-09; XII (1941), pp. 254, 577-81.

22. "I intend to inscribe it to Dr. Bentley, who shall peruse it in Manuscript." Moyle, *Works,* I, p. 364.

23. Hearne, *Remarks,* XI, p. 96. For the *Philopatris,* see Moyle, *Works,* I, pp. 285-364; for Marcus Aurelius, "Mr. Moyle's Discourse to prove Marcus Antoninus a Persecutor," *The Theological Repository,* 1 (London, 1773), pp. 147-73; for the "thundering legion," *Works,* II, pp. 79-390.

24. Moyle to Musgrave, July 14, 1709, and June 22, 1713, in Moyle, *Works,* I, pp. 174-75, 210. Moyle's essay is in Bodl. MS. Locke C 38, ff. 14-15.

25. It is dated Feb. 21, 1719; Moyle, *Works,* I, pp. 255-64, esp. 260-64.

26. Hearne to Woodward, June 30, 1726, in Hearne, *Remarks,* IX, p. 156; Bodl. MS. Rawl. Lett. 12, f. 112.

27. Hearne, *Johannis Glastoniensis Chronica* (2 vol., Oxford, 1726), app., pp. 649-52; Hearne, *Vindication* (ch. 10 above, n. 19), pp. 38-39.

28. Curll, *An Apology for the Writings of Walter Moyle* (London, 1727), p. 15.

29. Alexander Gordon, *Itinerarium Septentrionale* (London, 1726), preface.

30. John Clerk, *Memoirs,* ed. John M. Gray, *Scot. Hist. Soc.,* XIII (1892), pp. 15-16, 27, 73, 74, 237.

31. "Journey to England in April 1724," Penicuik MS. GD 18, no. 2106.

32. See the letters of Gordon to Clerk, 1723-24, *ibid.,* no. 5023.

33. Lord Hertford to Clerk, April 3, 1705, *ibid.,* no. 5028. According to Gordon, Hertford "stood up and harangued much in your laud as the Grand Maecenas of the North." Gordon to Clerk, March 4, 1725, *ibid.,* no. 5023.

34. Clerk to Stukeley, March 22, 1725, *Lit. Ill.,* II, pp. 797-98.

35. Horsley to Clerk, Sept. 16, 1729, and Dec 7, 1731, Penicuik MS. GD 18, no. 5038.

36. *Ibid.,* no. 2107. For Gale generally, see "Reliquiae Galeana," *Bib. Top. Brit.* (London, 1790), III; *Lit. Anec.,* IV, pp. 543-50.

37. Penicuik MS. GD 18, no. 5030/5.
38. May-Nov. 1729, *ibid.*, no. 5030/13-15.
39. *Ibid.*, no. 5030/13.
40. *Ibid.*, 5030/18. It appeared abridged by Roger Gale in *Phil. Trans.*, 38 (1731), pp. 157-63, and complete in Graevius' *Thesaurus;* see the bibliographical note in the intro. to Clerk's *Memoirs* (n. 30 above), p. xxvi.
41. Penicuik MS. GD 18, no. 2107. There is no pagination. All the quotes that follow are from this small volume.
42. May 14, 1727, *ibid.*
43. Lethieullier to Clerk, Aug. 8, 1727, Penicuik MS. GD 18, no. 5032/1.
44. Clerk to Lethieullier, "Reliquiae Galeana" (n. 36 above), pp. 253-55.
45. Gale to Clerk, Nov. 1, 1729, Penicuik MS. GD 18, 5030/15; Clerk to Gale, Dec. 22, 1729, "Reliquiae Galeana," pp. 255-57; Gale to Clerk Jan. 17, 1730, Penicuik MS. GD 18, 5030/11.
46. William Brome remembered hearing Freind, though he "knew not whether in earnest or jest, or owt of Pique . . . his notion was applauded by the laughter of the company." Brome to Rawlinson, April 21, 1729, Bodl. MS. Rawl. Lett. 31, f. 272.
47. "Dissertation sur la Prise de Rome les Gaulois," *Mémoirs de Litterature,* 15 (1738), pp. 1-21. For Melot, see the *éloge* in *Histoire de l'Académic des Inscriptions,* 29 (1764), pp. 360-71.
48. *Catalogue des Libres du Cabinet de M. de Boze* (Paris, 1753), no. 2284. De Boze's letter to Melot is quoted on pp. 16-18 of the "Dissertation."
49. The sculptor, Joseph Nollekens, was one of the few to remember it; *Nollekens and his Times,* ed. Wilfred Whitten (2 vol., London, 1920), I, p. 37.
50. Hearne, *Remarks,* VIII, p. 363. See *Lit. Anec.,* V, pp. 517-27; Thomas Birch, *An Account of the Life of John Ward* (London, 1766).
51. In 1768, according to a note by Richard Gough in his copy of Ward's *Lives* (Bodl. Gough London 141), p. 292. For the forty pounds, see *Nollekens,* I, p. 37n.
52. I have been able to find only a draft of the letter from Wilkinson to

Hunter offering the shield; it is dated Aug. 17, 1779. Wilkinson sent Hunter an engraving and offered the correspondence as well; B.M. MS. Add. 6127, ff. 238-38^V. For Hunter and his collections, see Samuel F. Simmons, *An Account of the Life and Writings of the Late William Hunter* (London, 1783), pp. 59-61; R. Hingeston Fox, *William Hunter* (London, 1901), app. by John Young, p. 37ff.; Anne S. Robertson, *The Hunterian Museum* (Glasgow, 1954) and *Roman Imperial Coins in the Hunter Coin Cabinet* (Oxford, 1962), p. vii.
53. B.M. MS. Add. 6127, f. 210ff.
54. *Ibid.,* ff. 82, 201.
55. *Ibid.,* ff. 221, 223.
56. For the following, see *Lit. Anec.,* VI, pp. 262-343, and Gough's autobiographical memoir, *ibid.,* IV, pp. 613-26.
57. See the sales catalogs drawn up at his death, for his books (1810), and for his prints, drawings, coins, and medals (1812). Much of this went to the Bodleian Library.
58. British Topography, 2nd ed. (London, 1780), I, pp. 718-20.
59. Bodl. MS. Gough Misc. antiq. 10; see ch. 9 above, n. 2.
60. See also his note in Ward's *Lives* (n. 51 above).
61. *An Historical Description of the Tower of London and its Curiosities* (London, 1754), p. 40.
62. Wilkinson's note, B.M. MS. Add. 6127, f. 239; for the meeting of the Society, see the Minute Book, (n. 5 above), X, pp. 326-27.
63. The note in the register is brief and enigmatic: "Woodwardian Shield and a helmet beq. by Dr. John Wilkinson." It is dated Nov. 14, 1818. The note on the title page of B.M. MS. Add. 6127 says that it was "bought" at the sale of Wilkinson's library, lot 665, Oct. 29, 1819, but Wilkinson's will indicates that it too was bequeathed. I owe this information to Mr. John Cherry of the British Museum. Writing in 1829, John Thomas Smith impugned the artistry of the shield and assigned it confidently to the time of Henry VIII. Smith was appointed keeper of prints and drawings at the Museum in 1816, Nollekens (n. 49 above), p. 37n.

Chapter 15: Fakes and the Progress of Modern Scholarship

1. *The Works of the Learned,* Nov. 1701, summarizing *De Veteris Numismata* (Paris, 1701), ch. VI. For modern accounts, see Louis Courajod, "Le Imitation et la Contrefaçon des Objets d'Art Antiques au xve et au xvie siècle," *Gazette des Beaux-Arts,* 34 (1886), pp. 188-201, 312-30; John Evans, "The Forging of Antiques," *Longman's Magazine,* 23 (1893-94), p. 149.
2. Hearne, *Remarks,* XI, p. 246.
3. *The Works of the Learned,* May 1701, reviewing Father Philip Bonani's *Numismata Pontificum Romanorum* (Rome, 1700), pp. 260-61.
4. *Ibid.*
5. Scipio Maffei, *Complete History,* p. 136.
6. I must here remind the reader of one fake in Dr. Woodward's collection, the Muscovy sheep described above in ch. 5.
7. For the following, see Melvin E. Jahn and Daniel J. Woolf, *The Lying Stones of Dr. Johann Bartholemew Adam Beringer* (Berkeley and Los Angeles, 1963).
8. "An Account of the Beringer Hoax and of the Subsequent Trials," *ibid.,* pp. 125-41.
9. James Parkinson, *Organic Remains of a Former World* (3 vol., London, 1804-11), I, p. 26.
10. So B. Ehrhart, *De Belemnitis* (Lugduni, 1727), pp. 24-25; see Jahn and Woolf, *Lying Stones,* p. 126.
11. For the following, see Melvin E. Jahn, "Some Notes on Dr. Scheuchzer and on Homo *diluvii testis,*" *Toward a History of Geology,* ed. Cecil J. Schneer (Cambridge, Mass., 1969), pp. 192-93.
12. The passage is translated in Jahn and Woolf, *Lying Stones,* pp. 173-76.
13. *Phil. Trans.,* 34 (1726), pp. 38-39; *Journal des Scavans,* 81 (1726), pp. 24-28; *New Memoirs of Literature,* 3 (1726), pp. 440-42; *Sammlung von Naturund Medicin-Geschichte,* 32 (1726), pp. 406-08. Jahn and Woolf reproduce the first and the last in *Lying Stones.*
14. *Homo diluvii testis* (Tiguri, 1726).
15. The relevant passage is from the

3rd ed. (1825), V, pp. 431-40, and is translated by Jahn and Woolf, *Lying Stones,* pp. 210-13. For Cuvier, see Mrs. R. Lee, *Memoirs of Baron Cuvier* (London, 1833); "Éloge de Cuvier," M. Flourens (1834), in G. Cuvier, *Éloges Historiques* (Paris, n.d.), pp. i-lviii; John Vienot, *La Napoleon de l'Intelligence* (Paris, 1932).
16. Quoted in *Lying Stones,* p. 210.
17. See William R. Coleman, *Georges Cuvier, Zoologist* (Cambridge, Mass., 1964), p. 45ff.
18. Two of the most famous of the fossil elephants had been recently discovered: the Siberian mammoth and the mastodon of Ohio. See the "Mémoirs sur les espèces d'elephans vivantes et fossiles," *Mémoirs de l'Institut, Classe math. et phys.,* 2 (1799), pp. 1-22, summarized in an appendix to Cuvier's *Essay on the Theory of the Earth,* trans. Robert Kerr (Cambridge, 1813), pp. 231-39. Patrick Blair, who had contributed a report of his dissection of an elephant in Dr. Woodward's day, had almost no comparative evidence to employ; *Phil. Trans.,* 27 (1710), p. 53ff.
19. Cuvier, *Essay,* pp. vii, 5.
20. *Ibid.,* pp. 124-25 *et passim;* see Frank Bourdier, "Saint-Hilaire versus Cuvier," *Toward a History of Geology,* ed. Cecil J. Schneer (Cambridge, Mass., 1969), pp. 36-61.
21. Cuvier's *Essay* was composed originally as a *discours préliminaire* to the first volume of the *Recherches sur les ossements fossiles de quadrupèdes* (Paris, 1812). It was rewritten in 1821 and issued separately in 1825 as *Discours sur les revolutions de la surface du globe.*
22. Cuvier, *Essay,* pp. 1-3, 54-55.
23. Cuvier knew and cited most of his predecessors including Dr. Woodward, *ibid.,* p. 39.
24. "A nice and scrupulous comparison of their [fossil] form, of their contexture, and frequently even of their composition, cannot detect the slightest difference between these shells and the shells which still inhabit the sea." Cuvier, *Essay,* p. 9. Exactly Dr. Woodward's argument!
25. For the "Fossil Elk of Ireland,"

see the appendix to Cuvier's *Essay,*
p. 245ff.

26. From Scheuchzer's article in the
Sammlung von Naturund, trans. Melvin E.
Jahn (see n. 13 above), pp. 406-08.

27. Cuvier, *Essay,* pp. 51-52.

28. I discover an analogy to my
observation in a remark about Niebuhr.
"In the way in which he sometimes
brought some important relation to light
from a few mutilated lines, he resembled
such a naturalist as Cuvier, who from the
fragments of a bone . . ." Prof. Loebell in
Chevalier Bunsen et al., *The Life and
Letters of Barthold Niebuhr* (New York,
1854), p. 541.

29. See, for example, E. M. Butler,
The Tyranny of Greece over Germany
(Cambridge, 1935); and Henry Hatfield,
*Winckelmann and his German Critics,
1755-1781* (New York, 1943) and
*Aesthetic Paganism in German Literature
from Winckelmann to the Death of
Goethe* (Cambridge, Mass., 1964).

30. Stosch to Hollis, Jan. 22, 1762,
Soc. Antiq. MS. Min. Bk., VIII, p. 412.
Winckelmann was elected a member April
9, 1761, *ibid.,* p. 323. See Lesley Lewis,
*Connoisseurs and Secret Agents in
Eighteenth Century Rome* (London,
1961), p. 196.

31. See, for example, Wolfgang Lepp-
mann, *Winckelmann* (London, 1971),
and Nikolaus Himmelmann, *Winckelmanns
Hermeneutik* (Wiesbaden, 1971), reviewed
by Leonard P. Wessell, Jr., *Lessing Yearbook,* 5 (1973), pp. 287-89. The Society
of Antiquaries received a report from him
Feb. 19, 1761; MS. Minute Book, VIII,
pp. 300-02.

32. *Description des Pierres Gravées du
feu Baron de Stosch* (Florence, 1760).
Stosch himself presented the Society with
a copy; MS. Minute Book, VIII, p. 317.
For Stosch, see Lewis, *Connoisseurs and
Agents,* p. 194ff.

33. Winckelmann, *Lettres Familières*
(Amsterdam, 1781), pp. xi-xxxiv. "Perhaps there is no one today who knows the
ancients better than Winckelmann, who
has better studied the admirable monuments which remain to us and is more
sensible to their beauty." *Bibliothèque des*

Sciences et des Beaux Arts (1762), quoted
by David Irwin, *Winckelmann: Writings on
Art* (London, 1972). p. 9. The standard
biography of Winckelmann is by Carl Justi
(3 vol., Leipzig, 1866-72; reprinted 1923).

34. Soc. Antiq. MS. Min. Bk., X,
pp. 394, 554.

35. Winckelmann, *The History of
Ancient Art,* trans. G. Henry Lodge
(2 vol., London, 1881), I, pp. 109-10. The
original appeared in 1764 as *Geschichte
der Kunst des Althertums* and may be
consulted in the *Samtliche Werke,* ed.
Joseph Eiselein (12 vol., Donauoschingen,
1825-29), bd. 3-6. See earlier, Winckelmann to Stosch [July 1760], *Briefe,* ed.
Walther Rehm (4 vol., Berlin, 1952-57), II,
pp. 91-92. A contemporary catalog notes
that the four statues were a present from
the Pope to Richelieu. "Three of them
were judg'd at Rome to be of the same
Work as the Venus of Medici by Cleomenes." B.M. MS. Stowe 1018, ff. 18^V,
41^V-42.

36. Winckelmann, History, pp.
110-13.

37. *Ibid.,* p. 113. "We live in an iron
country as antiquaries; it is in Italy alone
that we must make researches; never can
we surpass the Romans but in Rome. I
blush a thousand times a day at those
infinitely little relicks which are preserved
in our infinitely little cabinets of antiquities. . . ." Abbé Barthélemy to the
Comte de Caylus, Nov. 5, 1755, in Barthélemy's *Travels in Italy* (London, 1802),
p. 29.

38. Winckelmann, *History,* p. 121.

39. "It was Winklemann who first
urged on us the need of distinguishing
between various epochs and tracing the
history of styles in their gradual growth
and decadence." Goethe, *Italian Journey,*
trans. W. H. Auden (London, 1962),
pp. 155-56.

40. Winckelmann, *History,* p. 107.

41. André Tibal, *Inventaire des Manu-
scrits de Winckelmann deposés a la
Bibliothèque Nationale* (Paris, 1911),
p. 120.

42. Winckelmann, *Monumenti Antichi
Inediti* (2 vol., Rome, 1767), reprinted in
Kunsttheoretische Schriften, Bd. 4

(Studien zur Deutsches Kunstgeschichte, Baden-Baden, 1967, Bd. 345), preface, p. xxiii; *Versuch einer Allegorie* (Dresden, 1766), in *Kunsttheoretische Schriften*, bd. 339 (1964), p. 12n., where De Boze is cited.

43. James Kennedy, *A Description of the Antiquities and Curiosities in Wilton House* (Salisbury, 1769), pp. iv, ix.

44. Wanley to Charlett, Dec. 19, 1700, Bodl. MS. Ballard 13, f. 91.

45. Finch to Batteley, Oct. 12, 1700, *Lit. Ill.*, IV, pp. 95-96; Wasse in *Bibliotheca Literaria*, 3 (1723), pp. 27-28, noted by Hearne, *Remarks*, VIII, p. 52; Gordon, *Itinerarium Septentrionale* (London, 1726), preface; Clerk, *Memoirs* (see ch. 14 above, n. 30), pp. 127-28; Gale to Clerk, Sept. 6, 1726, "Reliquiae Galeana," *Bib. Top. Brit.*, III, pp. 25-52.

46. Chindonax Britannicus to Galgacus (William Stukeley to Alexander Gordon), Sept. 25, 1723, Penicuik MS. GD 18, 5023, no. 1.

47. Hearne, *Remarks*, II, p. 93; Hearne to West, B.M. MS. Lansdowne 778, ff. 259, 261, 263, 265. The inscription is described in the manuscript catalog, B.M. MS. Stowe 1018, f. 50.

48. Penicuik MS. GD 18, 5030, nos. 26 and 2107 (Clerk's "Journey to London, 1727"). See also Gale to Clerk, Sept. 6, 1726, "Reliquiae Galeana," *Bib. Top. Brit.*, III, p. 251.

49. The work was entitled *The Marble Statues of the Right Hon. Earl of Pembroke*, n.d. See Gale to Clerk, Sept. 8, 1730, Penicuik MS. GD 18, 5030, no. 16; Hearne, *Remarks*, VIII, p. 190.

50. Clerk, "Journey to London," Penicuik MS. GD 18, 2107, April 23, 1727.

51. Kennedy, *Description* (n. 43 above), p. xii.

52. B.M. MS. Stowe 1018, f. 11^V.

53. Clerk to Gale, Sept. 22, 1732, "Reliquiae Galeana," *Bib. Top. Brit.*, III, pp. 300-01.

54. For this and a thorough modern assessment, see Adolf Michaelis, *Ancient Marbles in Great Britain*, trans. C.A.M. Fennell (Cambridge, 1882), p. 43ff. Michaelis compares, for example, the

sculpture as it was in the Mazarin collection (where many of Pembroke's pieces derived) with its reappearance at Wilton, to show the additions.

55. Pembroke's coins were also engraved by N.F. Haym with descriptions, but Gale was equally dissatisfied, "the same as in his book of statues incorrect as to spelling and frequently not intelligible." Haym's work was not published. See Gale to Clerk, Dec. 9, 1732, Penicuik MS. GD 18, 5030, no. 28; *Aedes Pembrokianae* (11th ed., Salisbury, 1788), pp. 128-29.

56. Clerk, "Journey" (n. 50 above); Creed (no title) no. 15; Kennedy, *Description*, p. xxx; B.M. MS. Stowe 1018, ff. 41^V-42; Richard Cowdry, *A Description of the Pictures, Statues, Bustoes, Basso-Relievos, and other Curiosities at Wilton* (2nd ed., London, 1752), p. 101.

57. For the story in Livy, VIII, vi; for the modern judgment, Michaelis, *Ancient Marbles*, p. 689. The first modern reassessment of these marbles was Charles Newton, *Notes on the Sculptures at Wilton House* (London, 1849).

58. A characteristic treatment is the Abbé Massieu, "Dissertation sur les Boucliers Votifs," *Memoirs de Litterature de l'Academie des Inscriptions et Belles Lettres*, 3 (1719), pp. 227-43.

59. In the preface to the *Monumenti* and the essay on allegory, where Winckelmann treats also of Dr. Woodward's shield; see n. 42 above.

60. "Dissertation sur un Disque d'argent du Cabinet des antiques connu sous le nom de Bouclier de Scipion," *Monumens Antiques inédits* (2 vol., Paris, 1802), I, pp. 69-96.

61. "Some believe it to have been only a large silver dish, and indeed the embossing being all inward does much favor this opinion." Richard Richardson, to Thoresby, April 1, 1709, Thoresby, *Letters* (1912), II, pp. 198-99. In the same letter, Richardson praised Dr. Woodward's shield as "a very fine and valuable piece of antiquity."

62. Ernest Babelon, *Guide Illustré au Cabinet du Medailles et Antiques de la Bibliothèque Nationale* (Paris, 1900), no. 2875, pp. 17-18; *Notice Historique et*

Guide du Visiteur (Paris, 1924), pp. 207-08. For a similar shield unearthed in 1714, placed eventually in the Cabinet du Roi and also misunderstood, see the article "Sur un Bouclier Votif mis depuis peu au Cabinet du Roi," *Mémoirs de Litterature de l'Academie des Inscriptions et Belles Lettres,* 9 (1731-33), pp. 152-57; Babelon, pp. 19-20.

63. Kazimierz Bulas, *Les Illustrations Antiques de l'Iliade* (Eus Supplementa, III, Lwow, Poland 1929), pp. 82-84. Bulas does not believe in the traditional iconography either, preferring to think that it represents the enslavement of Briseis in which Agamemnon (i.e., Scipio) becomes a herald!

64. Bashford Dean, *Handbook of Arms and Armor in the Metropolitan Museum,* ed. Stephen V. Granscay (New York, 1930), p. 45. See also the article "Clipius" by Maurice Albert in the *Dictionnaire des Antiquités Grecques et Romaines,* ed. Charles Daremberg and Edmund Saglio (Paris, 1887), I, pt. 2, pp. 1248-60; and Paul Couissin, *Les Armes Romaines* (Paris, 1926). For shields discovered in England, see J.M.C. Toynbee, *Art in Roman Britain,* 2nd ed. (London, 1963), p. 299.

65. There is a fascinating discussion of some of these, quite analogous to our shields, in an article by Anthony Radcliffe, "Bronze Oil Lamps by Riccio," *Victoria and Albert Museum Yearbook,* 3 (1972), pp. 29-58.

66. See the accounts (not entirely consistent) in Harold A. Dillon, "Arms and Armour at Westminster, the Tower, and Greenwich, 1547," *Archaeologia,* 51 (1888), p. 222n.; and Charles Ffoulkes, *Inventory and Survey of the Armouries of the Tower of London* (2 vol., London, 1916), p. 69.

67. "Gothick Manner of Building," Harris, *Lexicon Technicum* (London, 1710); Addison, "Remarks on Italy," in *Works,* ed. H. G. Bohn (London, 1901), I, p. 485. See the President de Brosses on St. Mark's Venice: "C'est un vilain monsieur, s'il en fut jamais, massif, sombre et gothique, de plus méchant gout." "C'est une église a la grecque, basse, impénétrable a la lumière, d'un gout misérable." *Lettres*

Familières sur l'Italie, ed. Yvonne Bezard (2 vol., Paris, 1931), I, pp. 193, 194.

68. Sir James Mann, *European Arms and Armour in the Wallace Collection* (London, 1962), pp. 201-02.

69. Francis Grose, *A Treatise on Ancient Armour* (London, 1786), p. vii. The shield was then owned by the antiquary and fossilist, Gustavus Brander. Grose's view was modified by Smuel Pegge in *Gent. Mag.,* 72 (May 1802), pp. 406-07.

70. Guy F. Laking, "The European Arms And Armour of the Wallace Collection, IV," *The Art Journal* (1903), p. 44. There is, however, a problem in the literature on this shield. Sir James Mann gives its provenance as Dr. Mead to Samuel Tyssen (through an intermediate owner) in 1802, and refers to another similar shield in the Julienne sale at Paris in 1767. Cripps-Day and Laking believe the Julienne shield to be the one in the Wallace Collection and trace it back to a purchase by Dr. Ward in Italy. According to Mann, there are still other shields extant with the same subject, so that it may be that there were once two shields in England like the one that Grose portrayed. Where then did the other one go? See Sir Guy Francis Laking, *A Record of European Armour and Arms through Seven Centuries* (London, 1921), IV, p. 225; Francis Henry Cripps-Day, *A Record of Arms Sales 1881-1924* (London, 1925), p. xxxi, n.

71. Laking, *Record,* p. 233ff.

72. Ffoulkes, p. 193; Arthur Richard Dufty, *European Armour in the Tower of London* (London, 1968), pl. cxlii. The only armor collector of renown in Dr. Woodward's day seems to have been one Don Saltero, alias James Salter, once a servant of Hans Sloane's, who displayed his very mixed collections (including stuffed animals, manuscripts, etc.) in a coffeehouse in Cheyne Walk. See the *Tatler,* no. 34, and the Salter catalog which ran through 48 editions after 1729.

73. *The Family Memoirs of the Rev. William Stukeley* (Surtees Soc., 80, 1885), p. 411. The shield was described to Clerk by Jo. Johnstoune who continued, "If this had been the German Artificer you Wrytte off he ought to been a thousand

tymes hangd for being so good a counter-
fitter. Since no moderat knower but such
only of your thorough Knowledg and
Sagacity could fail being imposed on by
him." March 7, 1737, Penicuik MS.
GD 18, no. 5048.

74. Besides Dillon (n. 66 above), see
Samuel Rush Meyrick, "Description of the
Engravings on a Suit of Armour made for
King Henry VIII in the Tower of Lon-
don," *Archaeologia*, 22 (1829),
pp. 106-13; J. G. Mann, "Notes on the
Armour of the Maximillian Period and the
Italian Wars," *ibid.*, 79 (1929), p. 238ff.;
J. Starkie Gardner, *Foreign Armour in
England* (London, 1898), p. 12ff.; Mar-
garet Mitchell, "Works of Art from Rome
for Henry VIII," *JWCI*, 34 (1971),
pp. 178-201.

75. The history of the history of
modern armor may be said to have begun
with the works of Meyrick and John
Hewitt in the 19th-century. See Joseph
Skelton, *Engraved Illustrations of Ancient
Armour from the Collection of Llewelyn
Meyrick* (London, 1830), I, p. viii; Charles
J. Ffoulkes, *The Armourer and his Craft
from the XIth to the XVIth Century*
(1912; reprinted New York, 1967).

76. The only published description of
the shield in modern times appears to be
the note from the Assistant Keeper of the

Department of British and Medieval
Antiquites, Mr. A. B. Tonnochy, to Lester
Beattie, giving it a late 16th-century
origin; see also Kirby-Miller, *Memoirs of
Scriblerus*, p. 206. The present Assistant
Keeper, Mr. John Cherry, informs me that
he too believes it 16th-century, and with
Mr. A. V. B. Norman of the Wallace
Collection plans to submit a note to *The
Journal of Arms and Armour* to that
effect. The state of contemporary knowledge about 16th-century armor in general,
however, is still too primitive even now to
permit any exact notion of its place of
origin.

77. This will be hard to reconcile with
(among others) the views of Thomas Kuhn
in his recent and very influential book,
The Structure of Scientific Revolutions,
2nd ed. (Chicago, 1970). If my argument
sounds a bit old-fashioned, it is because I
think that the reaction to the old positivist
views about the progress of science and of
history has gone too far – not because I
am myself a positivist or because I think
that *all* knowledge advances in these fields
in this fashion.

78. The "philological circle" has been
traced back as far as Schliermacher; see
Leo Spitzer, *Linguistics and Literary History* (New York, 1962), pp. 19-20, 25-26.

Index

Abraham: 64
Abydenis: 205
Academy of Inscriptions: 173, 178, 255, 267-68, 283, 344
Academy of Sciences: 178
Account of a Strange and Wonderful Dream: 15
Account of the Sickness and Death of Dr. Woodward: 15
Achilles: 165, 283, 287
Acta Eruditorum: 194, 222, 238
Addison, Joseph: 56, 257, 289, 344-45
Aeneas: 197
Agamemnon: 287, 350
Agricola, Johannes: 83
Ainsworth, Robert: 256-58, 269, 323, 338, 344
Ajax: 162
Alban Kings: 259
Aldovrandi, Ulisse: 245
Alexander the Great: 53, 164, 214
Alfred, King: 65, 195, 216-18, 220
Allen, Dr. John: 202-03
All Souls College: 185
Allucius: 164
Alphonso of Castille: 191
Ammianus Marcellinus: 137, 176
Anatomy: 10, 15, 18, 22, 30, 84, 277-78, 280
Ancients and Moderns: 10, 14, 19-21, 33, 85-6, 115-16, 125-26, 134, 148, 157, 193, 240, 242, 247, 275, 291
Anne, Queen: 182, 239
Antidiluvian man: see *Homo Diluvii*
Antiquarianism: 114-24, 145, 171, 217, 242, 250-51
Antonine Column: 170, 195, 222, 224
Antonine Itineraries: 137

Antoninus Pius: 145-46, 236
Apicius, Caelius: 249
Apollo: 152, 230
Appeal to Common Sense: 15
Arbuthnot, Dr. John: 14-15, 40-42, 80, 127, 179, 238-44, 246, 249, 251-52, 302, 308-09, 333, 342-43
Archer, John: 35
Aristotle: 20, 52, 67, 214, 238
Arundel Marbles: 71, 78, 116, 134, 137, 140
Arundel, Thomas Howard, Earl of: 116, 119, 218, 284
Ashe, Dr. St. George: 307
Ashmolian Museum: 22, 27, 95, 118, 318, 330
Astell Mary: 324
Atheists: 56, 62, 65, 66, 72
Athenagoras: 184
Atossa: 212
Atterbury, Francis: 248
Aubrey, John: 73-74, 303-04
Augustine, St.: 66, 203, 260
Augustinus, Antonius (Antonio Agustin): 224, 227
Augustus: 224, 232, 250, 262
Avebury: 73
Avignon: 165

Bacchus: 230
Bacon, Francis: 12, 21, 23, 25, 33, 53, 54, 93, 110
Bactrian Kings: 260
Bagford, John: 118, 139-42, 144, 146, 170, 214, 229, 231, 249, 326
Baker, Thomas: 31, 58-62, 66, 72, 255, 258, 311
Baldwin de Calceo: 232, 329

353